TESTICULAR CELL DYNAMICS AND ENDOCRINE SIGNALING

ANNALS OF THE NEW YORK ACADEMY OF SCIENCES

Volume 1061

TESTICULAR CELL DYNAMICS AND ENDOCRINE SIGNALING

Edited by Matthew P. Hardy and Michael D. Griswold

The New York Academy of Sciences
New York, New York
2005

♾ *The paper used in this publication meets the minimum requirements of the American National Standard for Information Sciences—Permanence of Paper for Printed Library Materials, ANSI Z39.48-1984.*

Library of Congress Cataloging-in-Publication Data has been applied for.

GYAT / PCP

ISBN 1-57331-537-0 (cloth)
ISBN 1-57331-538-9 (paper)
ISSN 0077-8923

ANNALS OF THE NEW YORK ACADEMY OF SCIENCES

Volume 1061
December 2005

TESTICULAR CELL DYNAMICS AND ENDOCRINE SIGNALING

Editors

MATTHEW P. HARDY AND MICHAEL D. GRISWOLD

This volume is the result of a conference entitled the **XVIIIth North American Testis Workshop: Testicular Cell Dynamics and Endocrine Signaling** held on March 30 through April 2, 2005, in Seattle, Washington.

CONTENTS

Preface. *By* MATT HARDY AND MIKE GRISWOLD ix

Part I. Development

Hormonal Regulation of Male Reproductive Tract Development. *By* SUNITA VERMA-KURVARI, SERGE NEF, AND LUIS F. PARADA 1

Sox9 in Testis Determination. *By* AKIO KOBAYASHI, HAO CHANG, MARIE-CHRISTINE CHABOISSIER, ANDREAS SCHEDL, AND RICHARD R. BEHRINGER 9

Seminiferous Cord Formation and Germ-Cell Programming: Epigenetic Transgenerational Actions of Endocrine Disruptors. *By* MICHAEL K. SKINNER AND MATTHEW D. ANWAY 18

In Vitro Generation of Germ Cells: New Techniques to Solve Current Issues. *By* JOSHUA M. BAUGHMAN AND NIELS GEIJSEN 33

Part II. The Endocrine Axis

Structure-Function Relationships of the Luteinizing Hormone Receptor. *By* DAVID PUETT, YONGSHENG LI, KRASSIMIRA ANGELOVA, GENEVA DEMARS, THOMAS P. MEEHAN, FRANCESCA FANELLI, AND PREMA NARAYAN 41

Transgenic Rescue of SF-1-Null Mice. *By* TATIANA KARPOVA, R.R.M. MARAN, JEREMY PRESLEY, SERGE P. SCHERRER, LOVELLA TEJADA, AND LESLIE L. HECKERT 55

Estrogens in Testis Biology. *By* JAYNE E. SIERENS, SHARON F. SNEDDON, FRANCES COLLINS, MICHAEL R. MILLAR, AND PHILIPPA T.K. SAUNDERS 65

Foxa1 and Foxa2 Interact with the Androgen Receptor to Regulate Prostate and Epididymal Genes Differentially. *By* XIUPING YU, APARNA GUPTA, YONGQING WANG, KICHIYA SUZUKI, JANNI MIROSEVICH, MARIE-CLAIRE ORGEBIN-CRIST, AND ROBERT J. MATUSIK 77

Role of Glial Cell Line–Derived Neurotrophic Factor in Germ-Line Stem Cell Fate. *By* LAURA BRAYDICH-STOLLE, COURTNEY NOLAN, MARTIN DYM, AND MARIE-CLAUDE HOFMANN 94

Part III. Steroidogenesis

P450 Oxidoreductase Deficiency: A New Disorder of Steroidogenesis. *By* WALTER L. MILLER, NINGWU HUANG, AMIT V. PANDEY, CHRISTA E. FLÜCK, AND VISHAL AGRAWAL 100

Neuroendocrine Regulation of Leydig Cell Development. *By* P.J. O'SHAUGHNESSY, P.J. BAKER, AND H. JOHNSTON 109

Mitochondrial Function in Leydig Cell Steroidogenesis. *By* DALE B. HALES, JOHN A. ALLEN, TRISTAN SHANKARA, PAUL JANUS, STEVE BUCK, THORSTEN DIEMER, AND KAREN HELD HALES 120

Determining Structure and Function of Steroid Dehydrogenase Enzymes by Sequence Analysis, Homology Modeling, and Rational Mutational Analysis. *By* WILLIAM L. DUAX, JAMES THOMAS, VLADIMIR PLETNEV, ANTHONY ADDLAGATTA, ROBERT HUETHER, LUKAS HABEGGER, AND CHARLES M. WEEKS 135

Part IV. Spermatogenesis

Postnatal and Pubertal Development of the Rhesus Monkey (*Macaca mulatta*) Testis. *By* TONY M. PLANT, SURESH RAMASWAMY, DAVID SIMORANGKIR, AND GARY R. MARSHALL 149

The mRNA Encoding τCstF-64 Is Expressed Ubiquitously in Mouse Tissues. *By* ZANE HUBER, ROBERTO R. MONAREZ, BRINDA DASS, AND CLINTON C. MACDONALD 163

Drivers of Germ Cell Maturation. *By* K.L. LOVELAND, C. HOGARTH, S. MENDIS, A. EFTHYMIADIS, J. LY, C. ITMAN, S. MEACHEM, C.W. BROWN, AND D.A. JANS 173

Role of Nuclear Receptors in *INSL3* Gene Transcription in Leydig Cells. *By* JACQUES J. TREMBLAY AND NICHOLAS M. ROBERT 183

Analysis of the Human Sperm Proteome. *By* DANIEL S. JOHNSTON, JOE WOOTERS, GREGORY S. KOPF, YONGCHANG QIU, AND KENNETH P. ROBERTS 190

Part V. Clinical Correlates

Male Gamete Contributions to the Embryo. *By* RYUZO YANAGIMACHI 203

The Androgen Microenvironment of the Human Testis and Hormonal Control of Spermatogenesis. *By* JONATHAN P. JAROW AND BARRY R. ZIRKIN 208

Sperm Chemotaxis: A Primer. *By* E. HILDEBRAND AND U.B. KAUPP 221

Epigenetic Regulation of Testis-Specific Gene Expression. *By* JOHN R. MCCARREY, CHRISTOPHER B. GEYER, AND HIROTAKA YOSHIOKA 226

Index of Contributors 243

Financial assistance was received from:

- AMERICAN SOCIETY OF ANDROLOGY
- AMERICAN SOCIETY OF REPRODUCTIVE MEDICINE
- CONRAD
- NATIONAL INSTITUTES OF HEALTH (NICHD AND NIEHS)
- SERONO REPRODUCTIVE BIOLOGY INSTITUTE
- WYETH-AYERST

Preface

The XVIII biennial North American Testis Workshop took place in Seattle, Washington, on March 30th through April 2nd, 2005, and this volume assembles the platform presentations in printed form.

The theme of the meeting was "testicular cell dynamics and endocrine signaling," recognizing the importance of cell-cell interactions in steroid synthesis and spermatogenesis. Endocrine signaling in the testis was also a timely topic for Seattle 2005, as investigators equipped with the powerful technologies of mouse genetics and genomic analysis came within graspable distance of the goal of understanding hormonally regulated gene expression in the male reproductive system. Since the meeting occurs every other year and there have been 17 previous programs, the Workshop is both a spotlight on current knowledge and part of a 30-year wave of progress in testis biology. It was, therefore, especially fortuitous that Ryuzo Yanagimachi (Yana to everyone in the field) could deliver the keynote lecture, summarizing a career-long analysis of fertilization and the sperm's contributions to early postnatal development. Yana's presentation, with its insightful historical perspective, was forged amid hardship after a flash flood on the campus of the University of Hawaii nearly destroyed his lab. Yana's work and that of all of the contributing authors herein we trust will be of lasting value for reference and inspiration. As the editors, we owe a debt to several individuals without whose help these proceedings would not have been possible. The meeting itself was the product of a group effort led by us, the executive committee of the Testis Workshop, the scientific organizing committee, and the business office of the American Society of Andrology. Funds for the meeting came from the National Institutes of Health and several companies and organizations, and its magnitude points to the continued robust health of the Workshop. The names of the people involved in this collective effort are listed on the ensuing pages. Meghan Howard, on loan from the editorial office of the *Journal of Andrology*, stepped into a number of roles including general administrator, editorial coordinator, and all-around booster. Thanks also go to Linda Mehta, Acquisitions Editor for the New York Academy of Sciences, for assistance in bringing the Workshop proceedings back into print as a volume of their *Annals* series.

—MATT HARDY
—MIKE GRISWOLD

Ann. N.Y. Acad. Sci. 1061: ix (2005). © 2005 New York Academy of Sciences.
doi: 10.1196/annals.1336.026

Preface

system have allowed us to uncover new functions of insulin family signaling in reproduction.

INSL3 GENE FUNCTION AND CRYPTORCHIDISM

Cryptorchidism, a condition of impaired testicular descent, is a common congenital birth defect affecting 2–3% of newborn males. Complications associated with undescended testes include infertility and testicular cancer.[3] In mammals, testicular descent takes place in two phases: intraabdominal phase and inguinoscrotal phase.[4] In the intraabdominal phase (E15.5–E17.5 in mice), testes descend from their original location in the abdomen by the kidney to the bottom of the abdomen. During the inguinoscrotal phase (between the second and third postnatal weeks in mice) the testes become relocated from the base of the abdomen to the scrotum. Embryonic gonads are connected to the abdominal wall by two mesenchymal structures: gubernaculum and cranial suspensory ligament (CSL). Development of both gubernaculum and CSL is sexually dimorphic. Development and differentiation of the gubernaculum and regression of the CSL in the male during the intraabdominal phase lead to positioning of the testes to the base of the abdomen. In the female, gubernaculum does not differentiate and remains thin, whereas development of the CSL allows the ovaries to remain at their original position, just below the kidney.

Insl3, also known as relaxin-like factor, or Leydig insulin-like hormone is a member of the insulin/relaxin family of peptide hormones.[5,6] The *Insl3* gene shows a sexually dimorphic expression pattern during fetal development. Expression is observed only in Leydig cells during embryogenesis. After birth, expression is also observed in the theca cells of the ovary. To understand the function of the *Insl3* gene, we generated mice lacking a functional *Insl3* gene by homologous recombination.[7] Mice lacking the *Insl3* gene are viable, and homozygous null females are fertile. By contrast, homozygous null males are infertile. Anatomical inspection of mutant males revealed a poorly developed scrotum and free-moving intraabdominal testes.[7] Although there was no difference in the size of the testes at birth, testes in adult mutant mice were substantially smaller.[7] Histological examination of adult testes showed comparable number and morphology of Leydig cells in the mutant testes. Seminiferous tubules in mutant mice, however, were much smaller and showed little or no spermatogenesis. There was no reduction in the number of Sertoli cells and spermatogonia, but the number of primary spermatocytes was reduced and a large number contained picnotic nuclei. Because testosterone deficiency can also lead to impaired testicular descent, we checked testosterone levels in *Insl3* mutant mice. Testosterone levels and other androgen-dependent characteristics were found to be normal in *Insl3* mutant mice, suggesting that testosterone deficiency cannot be attributed to failure of testicular descent. To determine the anatomical basis of failure of testicular descent, we examined the structure of gubernaculum in newborn male pups. Wild-type pups had a well-developed gubernacular bulb and a thick gubernacular cord anchoring the testis firmly to the basis of the abdomen. By contrast, *Insl3* mutant males had a thin, flat, gubernacular bulb and a poorly developed gubernacular cord, similar to the gubernaculum in the female.[7] Together, these data support the hypothesis that Insl3, a hormone secreted by the testis, mediates the first phase of testicular descent via outgrowth and contraction of the gubernaculum.

effect on *Insl3* expression in Leydig cells, suggesting a difference in regulation of *Insl3* expression in fetal vs adult Leydig cells.[13]

In addition to estrogen, many related molecules, including xenoestrogens, have been considered responsible for the increasing incidence of cryptorchidism. To determine if estrogen-related molecules could mediate cryptorchidism via Insl3, we examined for the presence of *Insl3* transcripts from estriol, 17α-estradiol, and DES testes at postnatal day 0. Our preliminary results indicate that several xenoestrogens (DES) or compounds with reduced estrogenic activity (estriol and 17α-estradiol) can promote cryptorchidism via downregulation of *Insl3* transcripts.[13] We currently do not know which transcription factors and which pathways are affected by various estrogenic compounds to downregulate *Insl3* expression. Experiments to determine the potential importance of the Insl3 pathway in human cryptorchidism are currently underway in several laboratories.

INSULIN RECEPTOR FAMILY SIGNALING AND MALE SEX DETERMINATION

The most severe forms of reproductive disorders result from mutations in genes necessary for gonad formation. In mammals, the male sex-determining process is set in motion by the sex-determining region of the Y chromosome (*Sry*), which triggers differentiation of the Sertoli cell lineage.[14] In turn, Sertoli cells function as organizing centers and direct differentiation of the testis. In the absence of *Sry* gene expression, neither XX nor XY gonads develop into testes, and alterations in *Sry* expression are often associated with abnormal sexual differentiation.[15,16] In addition to *Sry*, several genes are required for testis and ovary formation, and mutations in many of these genes are responsible for dysgenetic gonads in newborn babies. Mutations in *Lim1, Wt1, Sf1, Emx2,* and *Lhx9* block gonad formation, whereas mutations in *Sry, Sox9, Wt1, Dmrt1,* and *Wnt4* lead to sex reversal in mice and/or humans (reviewed in Refs. 15 and 17). Mutations in genes for sex hormones and genes responsible for hormone synthesis, including *Mis, Insl3, Cyp17,* and *Cyp19,* result in gonadal dysfunction.[1] From our work with mice, we have identified a new pathway that is essential for testis formation in mice.[18]

The insulin family signaling pathway is essential for normal growth and development in animals.[19,20] The vital roles of insulin and insulin-like growth factors 1 and 2 (Igf1and Igf2) are mediated by receptors belonging to the tyrosine kinase family: insulin receptor (Ir), insulin-like growth factor 1 receptor (Igf1r), and insulin receptor-related receptor (Irr).[19] Insulin binds to Ir, Igf1 binds only to Igf1r, whereas Igf2 can signal through Igf1r and less efficiently through Ir. The ligand for Irr has currently not been identified. Upon ligand binding, the receptors and downstream effectors, including insulin receptor substrate (Irs) and Sh2-B proteins, become phosphorylated. The effector proteins, in turn, activate the downstream signaling cascade.[20] In humans, mutations in the *IR* gene lead to leprechaunism, hypoglycemia, and retarded growth.[21–23] There are no reported homozygous *IGF1R* mutations in humans, possibly because of embryonic lethality. Deletions of *INSULIN* and *IGF1* genes lead to metabolic abnormalities and growth retardation.[19,24] In mice, *Ir* and *Igf1r* are both necessary for pre- and postnatal growth and development. *Ir* mutant pups die within 4 days of birth because of ketoacidosis.[25,26] *Igf1r* mutant mice die at birth due to respiratory failure and exhibit multiple abnormalities.[27] By contrast, mice mutant for *Irr* are viable and do not exhibit overt deficiencies.[28] In addition to mediating growth and metabolism, insulin signaling pathways have important functions in reproduction.[1,29,30] Polycystic ovary syndrome in women is associated with

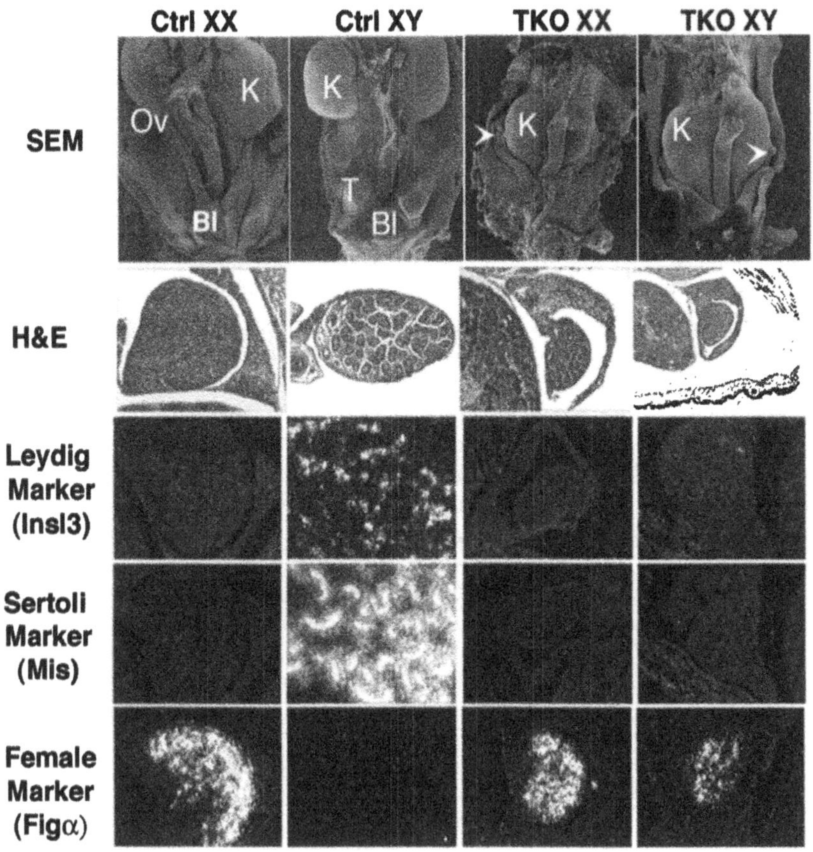

FIGURE 1. Male-to-female sex reversal in mice lacking *Ir/Igf1r/Irr*. Analysis of control (Ctrl) and triple mutant (TKO) XX and XY mice at E17.5. Scanning electron micrographs (SEM) show ovary situated below the kidney and testis situated by the bladder neck in ctrl XX and XY embryos, respectively. The size of TKO embryos is ~30% of the wild-type embryos. As expected, in TKO XX embryos, ovaries (*arrowhead*) are situated by the kidney. In TKO XY embryos, gonads are situated by the kidney as well. Histologically (ascertained by hematoxylin and eosin [H&E] staining), ovaries (Ctrl XX) show no overt differentiation before birth, whereas testis (Ctrl XY) show well-defined seminiferous tubules. Ovaries in TKO XX mice look like the controls, but the gonads in TKO XY mice look like ovary. At the molecular level, testes express male-specific markers (Insl3 [Leydig cells] and Mis [Sertoli cells]), whereas ovaries express only female-specific markers (Figα). TKO XX gonads express only female-specific markers, but TKO XY gonads do not express any male-specific markers and express only female-specific markers. Bl, bladder; K, kidney; Ov, ovary; T, testis

impaired insulin signaling and hyperandrogenemia.[30] Mice mutant for *Igf1, Irs2,* and *Irs4* and mice with neuron-specific deletion of *Ir* have reproductive abnormalities including ovarian and uterine maturation.[31–34] The role of insulin signaling in gonadogenesis in mice was recently discovered by us and is described below.[18]

Through the use of mice mutant at each of the known insulin receptor tyrosine kinase family genes (*Ir, Igf1r*, and *Irr*), we show that insulin receptor family function is a precondition for Sry-dependent processes to induce testicular differentiation. Male mice mutant for all three receptors develop ovaries and display a complete female phenotype (FIG. 1). Anatomically, the XY mutant gonads do not descend to reside by the bladder; instead, they stay close to the kidney. Histolgically, the XY triple mutant gonads are indistinguishable from the XX gonads and show no evidence of seminiferous tubules. At the molecular level, the triple mutant XY gonads do not express any testis-specific genes, but express all ovary-specific genes (FIG. 1).

Analysis of gonads at an early stage of development indicated that the male sex determination pathway is derailed early on. In mice, bipotential gonads arise as a paired structure from the urogenital ridge just before E10.5. Expression of *Sry* between E10.5 and E12.5 turns on the male pathway, including upregulation of *Sox9* expression. In triple mutant XY mice, there is a drastic reduction in *Sry* and *Sox9* expression and no expression of downstream male markers such as *Mis* and *P450scc*.[18] Therefore, in mice, insulin receptor family signaling is required for normal *Sry* expression and, hence, the initiation of the male determination pathway. In addition to the triple mutant, several compound mutants (null for two receptors and heterozygous or wild type for the third receptor) display defective testicular differentiation, including ovotestes formation.[18] The contribution of each receptor, however, is not equal, Igf1r being most important, Ir being less important, and Irr being least important.

In the future, we would like to delete insulin receptor family function only in the gonad using the Cre-LoxP conditional deletion system. Mice containing the floxed alleles of *Ir* and *Igf1r* have previously been generated,[35,36] and we have generated *Sf1-cre* (in collaboration with K. Parker at UTSW) and *Sry-cre* mice (in collaboration with H. Tojo at the University of Tokyo).

INSULIN 6

Insulin 6 (Insl6) is a member of the insulin/relaxin family of peptide hormones to which no function has thus far been assigned.[37–39] The *Insl6* gene is expressed at very high levels in the postpubertal testis, and expression is observed in pachytene spermatocytes and round spermatids. To understand its function, we have generated mice with a deletion in the *Insl6* gene by homologous recombination. The first coding exon containing signal peptide, the B- and part of the C-chain, were replaced by a neomycin gene cassette. This strategy ensured generation of null mice that do not express *Insl6*. Both male and female mice lacking *Insl6* are viable and fertile. Although *Insl6* is expressed at high levels in germ cells in the male, the lack of a phenotype suggests that other factors may compensate for Insl6 function. We are currently pursuing a more thorough analysis of the reproductive structures in *Insl6* mutant mice.

ACKNOWLEDGMENTS

This work was supported by an Excellence in Education Grant to L.F.P.

REFERENCES

1. ACHERMANN, J.C. *et al.* 2002. Genetic causes of human reproductive disease. J. Clin. Endocrinol. Metab. **87:** 2447–2454.
2. NEF, S. & L.F. PARADA. 2000. Hormones in male sexual development. Genes Dev. **14:** 3075–3086.
3. HUTSON, J.M. *et al.* 1994. Hormonal control of testicular descent and the cause of cryptorchidism [Review]. Reprod. Fertil. Dev. **6:** 151–156.
4. HUTSON, J.M., S. HASTHORPE & C.F. HEYNS. 1997. Anatomical and functional aspects of testicular descent and cryptorchidism [Review]. Endocr. Rev. **18:** 259–280.
5. ZIMMERMANN, S. *et al.* 1997. Mouse Leydig insulin-like (Ley I-L) gene: structure and expression during testis and ovary development. Mol. Reprod. Dev. **47:** 30–38.
6. ADHAM, I.M. *et al.* 1993. Cloning of a cDNA for a novel insulin-like peptide of the testicular Leydig cells. J. Biol. Chem. **268:** 26668–26672.
7. NEF, S. & L.F. PARADA. 1999. Cryptorchidism in mice mutant for Insl3. Nat. Genet. **22:** 295–299.
8. GILL, W.B. *et al.* 1979. Association of diethylstilbestrol exposure in utero with cryptorchidism, testicular hypoplasia and semen abnormalities. J. Urol. **122:** 36–39.
9. BERNSTEIN, L. *et al.* 1988. Maternal hormone levels in early gestation of cryptorchid males: a case- control study. Br. J. Cancer **58:** 379–381.
10. HADZISELIMOVIC, F., R. GENETO & L.R. EMMONS. 2000. Elevated placental estradiol: a possible etiological factor of human cryptorchidism. J. Urol. **164:** 1694–1695.
11. GROCOCK, C.A., H.M. CHARLTON & M.C. PIKE. 1988. Role of the fetal pituitary in cryptorchidism induced by exogenous maternal oestrogen during pregnancy in mice. J. Reprod. Fertil. **83:** 295–300.
12. PEREZ, M.C. *et al.* 1996. Effects of in-utero exposure to zeranol or diethylstilboestrol on morphological development of the fetal testis in mice. J. Comp. Pathol. **114:** 407–418.
13. NEF, S., T. SHIPMAN & L.F. PARADA. 2000. A molecular basis for estrogen-induced cryptorchidism. Dev. Biol. **224:** 354–361.
14. CAPEL, B. 2000. The battle of the sexes. Mech. Dev. **92:** 89–103.
15. BRENNAN, J. & B. CAPEL. 2004. One tissue, two fates: molecular genetic events that underlie testis versus ovary development. Nat. Rev. Genet. **5:** 509–521.
16. CAPEL, B. *et al.* 1993. Deletion of Y chromosome sequences located outside the testis determining region can cause XY female sex reversal. Nat. Genet. **5:** 301–307.
17. KOOPMAN, P. 2001. The genetics and biology of vertebrate sex determination. Cell **105:** 843–847.
18. NEF, S. *et al.* 2003. Testis determination requires insulin receptor family function in mice. Nature **426:** 291–295.
19. NAKAE, J., Y. KIDO & D. ACCILI. 2001. Distinct and overlapping functions of insulin and IGF-I receptors. Endocr. Rev. **22:** 818–835.
20. WHITE, M.F. 2003. Insulin signaling in health and disease. Science **302:** 1710–1711.
21. WERTHEIMER, E. *et al.* 1993. Homozygous deletion of the human insulin receptor gene results in leprechaunism. Nat. Genet. **5:** 71–73.
22. PSIACHOU, H. *et al.* 1993. Leprechaunism and homozygous nonsense mutation in the insulin receptor gene. Lancet **342:** 924.
23. KROOK, A., L. BRUETON & S. O'RAHILLY. 1993. Homozygous nonsense mutation in the insulin receptor gene in infant with leprechaunism. Lancet **342:** 277–278.
24. CAMACHO-HUBNER, C. *et al.* 2002. Insulin-like growth factor (IGF)-I gene deletion. Rev. Endocr. Metab. Disord. **3:** 357–361.
25. ACCILI, D. *et al.* 1996. Early neonatal death in mice homozygous for a null allele of the insulin receptor gene. Nat. Genet. **12:** 106–109.

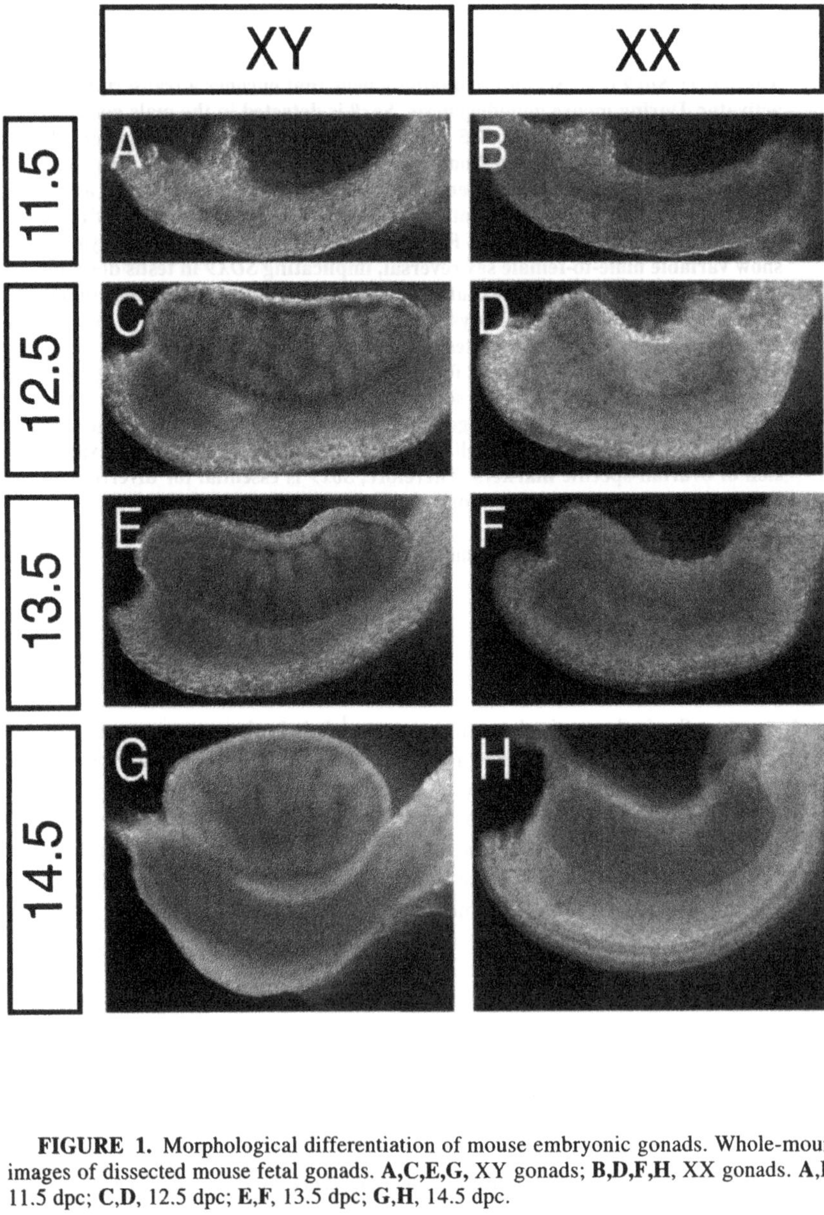

FIGURE 1. Morphological differentiation of mouse embryonic gonads. Whole-mount images of dissected mouse fetal gonads. **A,C,E,G,** XY gonads; **B,D,F,H**, XX gonads. **A**,**B**, 11.5 dpc; **C,D**, 12.5 dpc; **E,F**, 13.5 dpc; **G,H**, 14.5 dpc.

SOX9 EXPRESSION DURING MOUSE GONADOGENESIS

During mouse embryogenesis, the gonad forms on the medial surface of the mesonephros. At 11.5 days postcoitus (dpc), the fetal gonads of both XX and XY embryos are morphologically indistinguishable (FIG. 1A and B). However, 1 day later at 12.5 dpc, XY gonads can be distinguished morphologically from XX gonads by the presence of testicular cords (FIG. 1C and D). The XX gonad at 12.5 dpc does not show any overt morphological differentiation. From 12.5 dpc onwards, the testes and ovaries are easily identified by the differences in their morphologies (FIG. 1C to H).

In situ hybridization has been used to visualize *Sox9* RNA in developing mouse gonads. Using whole mount in situ hybridization, Kent and colleagues[7] reported the detection of a faint, diffuse signal at 10.5 dpc in the region of the gonad of the urogenital ridge of both XX and XY embryos. Robust signal was detected at 11.5 dpc in the XY but not the XX gonad. This XY gonad-specific pattern of expression was detected at subsequent stages of embryogenesis. Morais de Silva and colleagues[8] showed a detectable signal for *Sox9* at 10.5 dpc by whole-mount in situ hybridization in both XX and XY gonads. By 11.5 dpc, expression was only detected in the XY gonads. RNAse protection assays using urogenital ridges showed that *Sox9* transcripts were present in both XX and XY gonads at 11.0 dpc (12 somites). At 11.5 dpc there was an upregulation of *Sox9* in XY urogenital ridges, but a reduction in XX urogenital ridges. The authors suggested that some *Sox9* expression in the XX urogenital ridge could be due to nongonadal tissues. Sekido and colleagues[9] also showed by whole-mount in situ hybridization that *Sox9* expression is robust in the 11.5 dpc XY gonad but undetectable in the XX gonad. *Sox9* RNA has also been localized by in situ hybridization on sectioned material using radioactive probes.[10] At 11.5 dpc, *Sox9* transcripts were readily detected in XY gonads but were not above background in XX gonads.

Western blots using a rabbit polyclonal antibody raised against the carboxy terminus of mouse Sox9 showed expression in the urogenital ridges of 11.5 dpc XY but not XX embryos.[7] Immunohistochemistry using a rabbit polyclonal antibody raised against a C-terminal peptide of human SOX9 showed immunoreactivity in 11.5 dpc XY gonads but much less in XX gonads.[8] Sox9 protein expression was detected at 13.5 and 14.5 dpc in the fetal testis but was not detected in the fetal ovary. We used a rabbit polyclonal antibody raised against a human SOX9 synthetic peptide (Chemicon AB5535) to examine Sox9 protein localization during mouse gonadogenesis (FIG. 2). At 11.5 dpc, prior to sexual differentiation, Sox9 protein was readily detected in XY gonads but was undetectable in XX gonads (FIG. 2A and B). This male gonad-specific expression pattern was observed at subsequent stages of embryogenesis (FIG. 2C to H). Sox9 expression was restricted to Sertoli cells.

Previously, the bacterial *lacZ* gene was inserted into the *Sox9* locus by gene targeting in embryonic stem cells.[11] The insertion of *lacZ* was accompanied by deletion of a portion of exon 1 that simultaneously inactivated *Sox9*. The *lacZ* allele is equivalent to a null, and cells that transcribe the *Sox9* locus should express *lacZ* that can easily be visualized by staining for β-galactosidase (β-gal) activity. Male chimeras generated using *Sox9 lacZ/+* embryonic stem cells appeared normal and were fertile, yielding *Sox9 lacZ/+* pups that died at birth when bred with wild-type females. To examine the *Sox9* expression pattern in *Sox9 lacZ/+* mice, male chimeras were bred

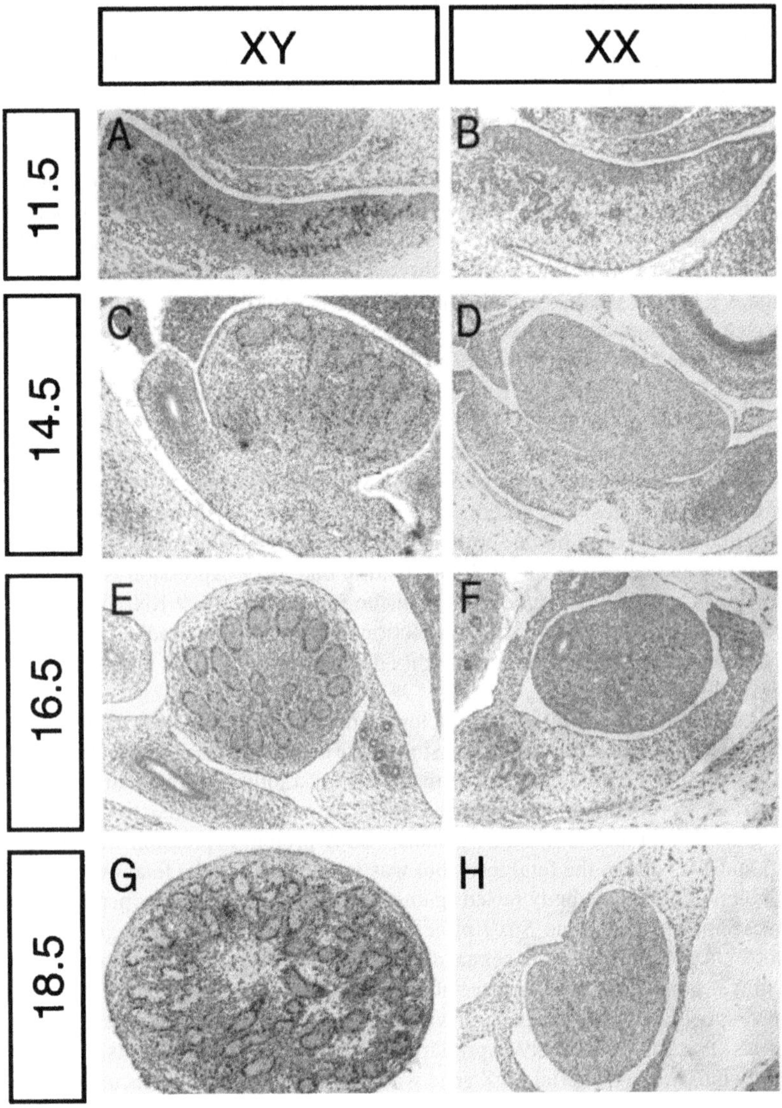

FIGURE 2. Immunohistochemical localization of Sox9 protein during mouse gonadogenesis. **A,C,E,G**, XY gonads; **B,D,F,H**, XX gonads. **A,B**, 11.5 dpc; **C,D**, 14.5 dpc; **E,F**, 16.5 dpc; **G,H**, 18.5 dpc. Sox9 protein, brown immunostaining; counterstain, hematoxylin. After 11.5 dpc, Sox9 protein expression is restricted to Sertoli cells.

with wild-type females to establish timed matings. β-gal activity was observed at 11.5 dpc in male but not female *Sox9 lacZ*/+ gonads (FIG. 3A and B). β-gal activity was also detected at 14.5 dpc in Sertoli cells of the testicular cords of the testes but not in the ovary (FIG. 3C to E). Therefore, expression of the *Sox9 lacZ* allele in the genital ridge mirrors the endogenous *Sox9* expression pattern.

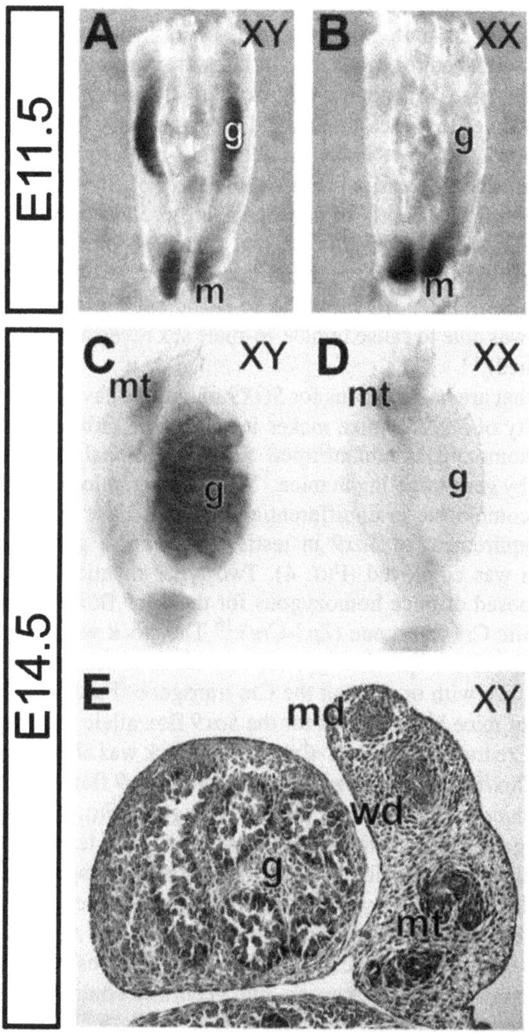

FIGURE 3. β-gal expression in *Sox9 lacZ*/+ gonads. **A,C,E**, XY; **B,D**, XX. **A,B**, 11.5 dpc; **C,D,E**, 14.5 dpc. **E**, histological section of C, counterstained with eosin. G, gonad; m, mesonephros; md, müllerian duct; mt, mesonephric tubules; wd, wolffian duct.

SOX9 IS ESSENTIAL FOR TESTIS FORMATION

Sox9$^{-/-}$ mice, generated by the cross just described, die during embryogenesis soon after 11.5 dpc because of cardiovascular defects.[14] *Sox9*$^{-/-}$ mutant embryos have gonads but, like their wild-type and heterozygous siblings, they are undifferentiated (i.e., testis differentiation has not yet occurred). To bypass this embryonic lethality, 11.5 dpc genital ridges from XY *Sox9*$^{-/-}$ and controls were isolated and cultured *in vitro* for 2–3 days.[18] In this assay, control XY genital ridges will form testicular cords and express Sertoli cell markers like anti-müllerian hormone (*Amh*). By contrast, XY *Sox9*$^{-/-}$ genital ridges do not form testicular cords or express *Amh*. Interestingly, they express female gonad-specific markers such as *Bmp2* and *follistatin*.[19] These findings demonstrate that *Sox9* is essential for testis formation and in its absence XY gonads continue an ovarian differentiation program.

Tissue-specific knockout of *Sox9* in the developing genital ridge has also been reported. In one study, an *Sf1-Cre* transgene was used to ablate *Sox9* in the XY gonad. Although Cre-mediated deletion was mosaic and variable, *Sox9* loss clearly led to defects in testicular cord formation and loss of male-specific marker expression. Recently, a *cytokeratin19-Cre* transgene was used to delete *Sox9* in developing XY gonads.[20] The *cytokeratin19-Cre* transgene is widely expressed, but Cre reporter expression was robust in 10.5 and 11.5 dpc urogenital ridges. Many conditional mutants died prior to 11.5 dpc (i.e., only about one-eighth the number of conditional mutants were recovered after 11.5 dpc). Interestingly, those XY conditional mutants that survived past 11.5 dpc showed male-to-female sex reversal, characterized by the absence of testicular cords and loss of male-specific markers, and the expression of *Wnt4* in a female-specific pattern. Some XY conditional mutants had ovotestes most probably because of mosaic Cre activity. It will be interesting to determine if *Sox9* is essential for maintaining Sertoli cell fate by conditional knockout after sex determination.

In summary, it is now clear that *Sox9* is an autosomal gene that is expressed in somatic cells of the gonad and is both necessary and sufficient for testis formation in the mouse. The male-specific expression pattern of *Sox9* in the gonads of diverse vertebrate species is consistent with the idea that *Sox9* is an evolutionarily conserved regulator of testis formation. Evidence to support this idea awaits the identification or generation of *Sox9* mutants in these species.

REFERENCES

1. SWAIN, A. & R. LOVELL-BADGE. 1999. Mammalian sex determination: a molecular drama. Genes Dev. **13:** 755–767.
2. LOVELL-BADGE, R. & E. ROBERTSON. 1990. XY female mice resulting from a heritable mutation in the primary testis-determining gene. Tdy Dev. **109:** 635–646.
3. KOOPMAN, P., J. GUBBAY, N. VIVIAN, *et al.* 1991. Male development of chromosomally female mice transgenic for Sry. Nature **351:** 117–121.
4. BOWLES, J., G. SCHEPERS, & P. KOOPMAN. 2000. Phylogeny of the SOX family of developmental transcription factors based on sequence and structural indicators. Dev. Biol. **227:** 239–255.
5. FOSTER, J.W., M.A. DOMINGUEZ-STEGLICH, S. GUIOLI, *et al.* 1994. Campomelic dysplasia and autosomal sex reversal caused by mutations in an SRY-related gene. Nature **372:** 525–530.

6. WAGNER, T., J. WIRTH, J. MEYER, *et al.* 1994. Autosomal sex reversal and campomelic dysplasia are caused by mutations in and around the SRY-related gene SOX9. Cell **79:** 1111–1120.
7. KENT, J., S.C. WHEATLEY, J.E. ANDREWS, *et al.* 1996. A male-specific role for SOX9 in vertebrate sex determination. Development **122:** 2813–2822.
8. MORAIS DA SILVA, S., A. HACKER, V. HARLEY, *et al.* 1996. Sox9 expression during gonadal development implies a conserved role for the gene in testis differentiation in mammals and birds. Nat. Genet. **14:** 62–68.
9. SEKIDO, R., I. BAR, V. NARVAEZ, *et al.* 2004. SOX9 is up-regulated by the transient expression of SRY specifically in Sertoli cell precursors. Dev. Biol. **274:** 271–279.
10. BISHOP, C.E., D.J. WHITWORTH., Y. QIN, *et al.* 2000. A transgenic insertion upstream of Sox9 is associated with dominant XX sex reversal in the mouse. Nat. Genet. **26:** 490–494.
11. BI, W., J.M. DENG, Z. ZHANG, *et al.* 1999. Sox9 is required for cartilage formation. Nat. Genet. **22:** 85–89.
12. BI, W., W. HUANG., D.J. WHITWORTH, Z. ZHANG, *et al.* 2001. Haploinsufficiency of Sox9 results in defective precartilaginous mesenchymal condensations and premature skeletal mineralization. Proc. Natl. Acad. Sci. USA **98:** 6698–6703.
13. VIDAL, V.P., M.C. CHABOISSIER, D.G. DE ROOIJ & A. SCHEDL. 2001. Sox9 induces testis development in XX transgenic mice. Nat Genet. **28:** 216–217.
14. AKIYAMA, H., M.C. CHABOISSIER, J.F. MARTIN, *et al.* 2002. The transcription factor Sox9 has essential roles in successive steps of the chondrocyte differentiation pathway and is required for expression of Sox5 and Sox6. Genes Dev. **16:** 2813–2828.
15. KIST, R., H. SCHREWE, R. BALLING & G. SCHERER. 2002. Conditional inactivation of Sox9: a mouse model for campomelic dysplasia. Genesis **32:** 121–123.
16. DE VRIES, W.N., L.T. BINNS, K.S. FANCHER, *et al.* 2000. Expression of Cre recombinase in mouse oocytes: a means to study maternal effect genes. Genesis **26:** 110–112.
17. O'GORMAN, S., N.A. DAGENAIS, M. QIAN & Y. MARCHUK. 1997. Protamine-Cre recombinase transgenes efficiently recombine target sequences in the male germ line of mice, but not in embryonic stem cells. Proc. Natl. Acad. Sci. USA **94:** 14602–14607.
18. CHABOISSIER, M.C., A. KOBAYASHI, V.I. VIDAL, *et al.* 2004. Functional analysis of Sox8 and Sox9 during sex determination in the mouse. Development **131:** 1891–1901.
19. YAO, H.H., M.M. MATZUK, C.J. JORGEZ, *et al.* 2004. Follistatin operates downstream of Wnt4 in mammalian ovary organogenesis. Dev. Dyn. **230:** 210–215.
20. BARRIONUEVO, F., S. BAGHERI-FAM, J. KLATTIG, *et al.* 2005. Homozygous inactivation of Sox9 causes complete XY sex reversal in mice. Biol. Reprod. [Epub ahead of print].

Seminiferous Cord Formation and Germ-Cell Programming

Epigenetic Transgenerational Actions of Endocrine Disruptors

MICHAEL K. SKINNER AND MATTHEW D. ANWAY

Center for Reproductive Biology, School of Molecular Biosciences, Washington State University, Pullman, Washington 99164-4231, USA

ABSTRACT: The molecular and cellular control of embryonic testis development was investigated through an analysis of the embryonic testis transcriptome to identify potential regulatory factors for male sex determination and testis morphogenesis. One critical factor identified is neurotropin 3 (NT3). At the onset of male sex determination, Sertoli cells initiate differentiation and express NT3 to act as a chemotactic factor for mesonephros cells to migrate and associate with Sertoli-germ cell aggregates to promote cord formation. Promoter analysis suggests that NT3 may be an initial downstream gene to SRY and helps promote testis morphogenesis. Endocrine disruptors were used to potentially interfere with embryonic testis development and further investigate this biological process. The estrogenic pesticide methoxychlor and antiandrogenic fungicide vinclozolin were used. Previous studies have shown that methoxychlor and vinclozolin both interfere with embryonic testis cord formation and cause increased spermatogenic cell apoptosis in the adult testis. Interestingly, transient *in vivo* exposure to endocrine disruptors at the time of male sex determination caused a transgenerational phenotype (F1-F4) of spermatogenic cell apoptosis and subfertility. This apparent epigenetic mechanism involves altered DNA methylation and permanent re-programming of the male germ-line. A series of genes with altered DNA methylation and imprinting are being identified. Observations reviewed demonstrate that a transient embryonic *in utero* exposure to an endocrine disruptor influences the embryonic testis transcriptome and through epigenetic effects (e.g., DNA methylation) results in abnormal germ-cell differentiation that subsequently influences adult spermatogenic capacity and male fertility, and that this phenotype is transgenerational through the germ-line. The novel observations of transgenerational epigenetic endocrine disruptor actions on male reproduction critically impact the potential hazards of these compounds as environmental toxins. The literature reviewed provides insight into the molecular and cellular control of embryonic testis development, male sex determination, and the programming of the male germ-line.

Address for correspondence: Michael K. Skinner, Center for Reproductive Biology, School of Molecular Biosciences, Washington State University, Pullman, WA 99164-4231. Voice: 509-335-1524; fax: 509-335-2176.

Skinner@mail.wsu.edu

Ann. N.Y. Acad. Sci. 1061: 18–32 (2005). © 2005 New York Academy of Sciences.
doi: 10.1196/annals.1336.004

KEYWORDS: endocrine disruptors; vinclozolin; methoxychlor; testis; sertoli; spermatogenesis; male fertility; epigenetic; transgenerational; sex determination; review

INTRODUCTION

Transgenerational Epigenetic Phenomena

Transgenerational effects of irradiation, chemical treatments (e.g., chemotherapy), and environmental toxins such as endocrine disruptors have been observed over the last decade. Most transgenerational observations are simply the effects of the agent on the gestating mother and subsequent actions on the offspring associated with the F1 generation.[1–3] Transgenerational effects for multiple generations have not been as thoroughly studied and will require transmission through the germ-line. In the context of the current review, transgenerational denotes germ-line transmission to multiple generations, minimally the F2 generation. The ability of an external agent to induce a transgenerational effect is through an epigenetic phenomenon involving DNA methylation or stable chromosomal alterations.[4–6] Transgenerational effects of irradiation were the first to be identified, and some have been shown to be transmitted through the germ-line to multiple generations.[4,5,7] These are often associated with mutagenesis and tumor formations in subsequent generations. The treatment of cancer with chemotherapeutics also has been shown to cause transgenerational effects,[8–10] but the impact on multiple generations has not been thoroughly investigated. Recently, nutritional effects on the F1 generation have been observed.[11] Environmental toxins such as endocrine disruptors have also been shown to influence the F1 generation after parental exposure,[9,12–14] but studies have not demonstrated transgenerational effects on multiple generations. However, the potential impact of such transgenerational effects of endocrine disruptors has been discussed.[15]

Epigenetic alterations that lead to transgenerational transmission of specific genetic traits or molecular events (e.g., imprinting) were recently identified.[16–19] These observations have led to the conclusion that re-programming through an altered methylation state of the germ-line is responsible. The impact this has on human health and evolutionary importance is significant.[16,17] Recent investigations of the DNA methylation state of primordial germ cells (PGCs) have indicated that as PGCs migrate down the genital ridge, de-methylation (i.e., erasure of methylation) begins and upon colonization in the early gonad complete de-methylation is achieved.[20–22] This has primarily been observed through analysis of specific imprinted genes.[23] During the period of sex determination in the gonad, germ cells undergo re-methylation involving sex-specific determination of the germ cells. Although de-methylation may not require the gonad somatic cells,[21] re-methylation of the germ-line appears to be dependent on association with the somatic cells in the gonads.[20,22] Due to this unique property of germ cells to undergo de-methylation and re-methylation during the period of sex determination in the developing gonad, the ability of an environmental agent such as an endocrine disruptor to influence through an epigenetic process the germ-line is postulated. This epigenetic effect on the germ-line could re-program the germ cell through an event such as altered DNA

genic metabolite of methoxychlor, HPTE, has differential effects on ER-α and ER-β, being an ER-α agonist and ER-β antagonist.[76,77] Other methoxychlor metabolites also have differential effects on ER-α and ER-β.[77] Therefore, in examining the actions of methoxychlor or HPTE, thought must be given to ER agonist and antagonist activity as well as antiandrogenic activities. Consideration of these differential activities is critical in elucidating the mechanisms of action of endocrine disruptors such as methoxychlor. Previously, methoxychlor metabolites were shown to act differentially on the ER from different species.[78] The effects of methoxychlor during embryonic or early postnatal periods can influence reproductive functions at later adult periods.[79–81] Neonatal exposure to methoxychlor can influence pregnancy,[82,83] ovarian and hypothalamic function,[84,85] reproductive behavior,[86] prostate development,[87] thymus development,[88] and testis development.[89] Therefore, transient embryonic exposure to an endocrine disruptor can reprogram and/or imprint effects that become manifest in the adult on reproductive physiology. One study has shown that the effects on a gestating mother may influence subsequent pregnancies as well.[90] In recent studies methoxychlor was utilized as a model endocrine disruptor to investigate the reproductive toxicology of this class of compound,[29,30] and spermatogenic effects were found in adults after transient embryonic exposure.

Testis Cell Biology and Development

The adult testis is a complex organ that is composed of seminiferous tubules, which are enclosed by a surrounding interstitium. The seminiferous tubules are the site of spermatogenesis, where germ cells develop into spermatozoa in close interaction with Sertoli cells. The Sertoli cells[91] form the seminiferous tubules and provide the cytoarchitectural arrangements for the developing germinal cells.[92] Tight junctional complexes between the Sertoli cells contribute to the maintenance of a blood-testis barrier[93] and create a unique environment within the tubule.[94,95] The majority of Sertoli secretory products[96–101] are hormonally regulated and provide useful markers of Sertoli cell differentiation. Surrounding the Sertoli cells are a layer of peritubular myoid cells that function in contraction of the tubule. The peritubular cell surrounds and forms the exterior wall of the seminiferous tubule. Peritubular cells are mesenchymally derived cells that secrete fibronectin[102] and several extracellular matrix components.[103] Both the peritubular and the Sertoli cells form the basement membrane surrounding the seminiferous tubule, and their interactions are important in germ-cell development. The interstitial space around the seminiferous tubules contains another somatic cell type, the Leydig cell, which is responsible for testosterone production. Leydig cells have a major influence on spermatogenesis through the actions of testosterone on both the seminiferous tubule and the pituitary. Although the Leydig cell has numerous secretory products,[104] the ability of the cell to produce androgen to act on the seminiferous tubules is the most significant secretory product of the cell. Leydig cell androgen production can be directly influenced by the actions of the endocrine disruptor methoxychlor and its metabolite HPTE.[105] Therefore, interaction of all three somatic cells, Sertoli, peritubular, and Leydig, is important for regulation of normal spermatogenic function in the testis.[104]

The process of testis morphogenesis occurs late in embryonic development (embryonic day 13 [E13] in the rat). Prior to morphogenesis of the testis the migration of primordial germ cells occurs, first from the yolk sac to the hindgut and then from

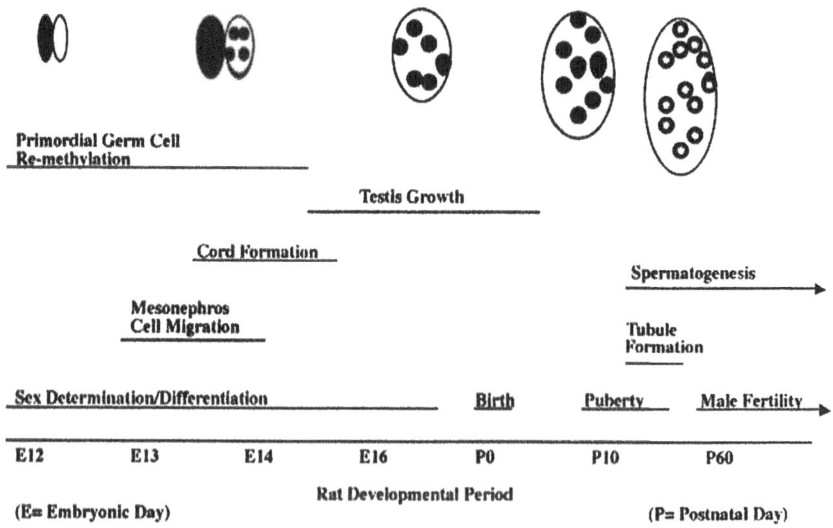

FIGURE 1. Testis development.

the hindgut to the genital ridge (FIG. 1). The gonad is bipotential after germ cell migration and morphologically can be distinguished from the adjoining mesonephros (E12 in rat), but it cannot be identified as an ovary or a testis. Two events occur early on E13 to alter the bipotential gonad. First, Sertoli cells, which are proposed to be the first cells in the testis to differentiate, aggregate around primordial germ cells.[106,107] Secondly, migration of mesenchymal cells occurs from the adjoining mesonephros into the developing gonad to surround the Sertoli cell–germinal cell aggregates. The migrating population of cells are speculated to be pre-peritubular cells.[108,109] The mechanism for this migration is a chemotactic signal from the testis to promote cell migration.[110] Therefore, during early testis development, Sertoli-peritubular cell interactions promote cord formation. The cords neonatally develop into seminiferous cords and at the onset of puberty develop into seminiferous tubules (FIG. 1). Seminiferous cords form as the Sertoli cell–primordial germ-cell aggregates become more organized and are fully surrounded by the migrated mesonephros mesenchymal cells (i.e., pre-peritubular cells). The formation of the seminiferous cords (E14 in the rat) is a critical event in the morphogenesis of the testis because it is the first indication of male sex differentiation. During the process of cord formation, the Sertoli cells undergo morphological changes including: a change in expression of mesenchymal to epithelial cell markers (cytokeratin),[111] a change in expression of cytokeratin 19 to cytokeratin 18 (cytokeratin 19 expressed in the ovary),[112] and expression of MIS, which inhibits the development of the müllerian duct, the precursor of the female uterus, cervix, fallopian tubes, and upper vagina.[113,114] Outside of the seminiferous cords the peritubular layer of cells become identifiable from the interstitium or Leydig cells at E15,[115] and 3βHSD production is detected after E15.[116] This is important because the production of testosterone and androgens by the Leydig cells has been demonstrated to stabilize the wolffian duct derivatives for normal male duct development.[117] Therefore, appropriate differentiation of somatic cell types in the testis around the time of cord formation is crucial not only for normal development of the testis, but also for the continued presence of the wolffian duct.

Primordial germ cells form aggregates with Sertoli cells prior to cord formation[106,107] and then are localized within the seminiferous cords as testis morphogenesis is initiated (FIG. 1). The germ cells undergo rapid mitosis until the late stages of embryonic development at which time they become quiescent. After birth in the rodent, germ-cell mitosis resumes, and during the onset of puberty and formation of the seminiferous tubules spermatogenesis is initiated (FIG. 1). Germ cells throughout development are in close association with somatic cells (i.e., Sertoli cells). Alteration of somatic cell differentiation could indirectly affect germ-cell development as well as directly affecting the germ cells. Recent studies examined the effects of endocrine disruptors on testis development with a focus on delayed effects on adult spermatogenic cell viability and development (i.e., spermatogenesis).[29–31]

The testis transcriptome (i.e., global gene expression profile) changes during testis development due to differentiation and growth of a variety of different cell types.[118] These changes in the testis transcriptome reflect critical regulatory genes and gene families required to promote normal testis function and development. A variety of functionally related genes such as transcription factors, signal transduction genes, cell cycle genes, cell survival genes, and growth factors will be involved in testis development and part of the transcriptome. Recent observations that endocrine disruptors (i.e., methoxychlor and vinclozolin) affect the testis during development are in part mediated through alterations in the testis transcriptome. The ability of the endocrine disruptor to alter the testis transcriptome and specific genes and gene families is in part one of the mechanisms used to alter fetal and adult testis function and development. Examples of gene families shown to influence embryonic testis development include the epidermal growth factor (EGF) family,[119–121] transforming growth factor-β (TGF-β) family, [122] and neurotropin growth factor family.[123,124]

A previous study demonstrated that a null mutation in a specific gene expressed in the embryonic testis can give the same phenotype as that seen with endocrine disruptor treatment. A mouse knockout model with a null mutation in the trk C receptor for neurotropin 3 (NT3) was used.[124] NT3 is essential for cell migration from the mesonephros and testis morphogenesis. The trk C knockout mouse was analyzed for similar phenotypes as those found with methoxychlor-treated gestating mothers. For trk C knockout animals that survived birth and puberty, testis histology showed a dramatic increase in spermatogenic cell apoptosis[124] and a similar testis phenotype.[30] Therefore, the absence of NT3 actions during embryonic testis development caused a similar phenotype as that of the endocrine disruptor-treated gestating mothers' male offspring.

These previous studies support the concept that altered expression of critical genes in the embryonic testis can cause abnormal testis development and a phenotype similar to that seen with endocrine disruptor treatment of the embryonic testis. This suggests that the affects of endocrine disruptors on the testis transcriptome will likely in part promote the epigenetic effects of methoxychlor on the germ-line and be the causal factor for the transgenerational effects observed.[29]

EPIGENETIC TRANSGENERATIONAL ACTION OF ENDOCRINE DISRUPTORS ON MALE FERTILITY

A recent study using methoxychlor and vinclozolin treatment of gestating mothers showed no major effect on sex determination or gross testis histology throughout development.[29] Animals were exposed *in utero* (E8–E15) and then spermatogenic defects were observed in the pubertal and adult F1 generation (FIG. 1). Treated animals have an increase in spermatogenic cell apoptosis. Similar results are observed at postnatal day 20 or day 60 for both endocrine disruptors. In addition to this decreased spermatogenic cell survival, sperm motility and morphology were also found to be impaired.[29–31] Interestingly, animals exposed *in utero* at E15–E20 (i.e., around birth [E20]) to the same dose of endocrine disruptor had no spermatogenic cell defects.[30,31] Therefore, only exposure during the E8–E15 period of development affected later adult spermatogenesis, and exposure after E15 had no effect on testis development.[29] This correlates with critical processes such as germ cell re-methylation, cord formation, and sex determination during the E10–E15 period compared to the growth phase after E15 (FIG. 1). In previous studies of endocrine disruptor actions on the testis using treatment regimens after E15, effects were not seen because the important E10–E15 period was missed. Due to the observation that neither methoxychlor nor vinclozolin transient exposure between E15 and E20 had any effect on postnatal/adult spermatogenesis, the effects observed from E8–E15 are not due to artifacts such as postnatal exposure from milk transfer from exposed mothers during weaning or bedding contamination from the exposed mother because either E8–E15 or E15–E20 would have the same potential transfer. In addition, no effect on total body weight was observed from either endocrine disruptor treatment so that the effects observed are not due to the effects on growth rates. Therefore, the endocrine disruptor effects observed on the F1 generation are due to local effects on testis development and germ-cell maturation during the E8–E15 period and are not a toxic effect in later postnatal development. None of the transgenerational effects described were due to endocrine disruptor toxicity because none of the subsequent gestating mothers or pups were ever exposed; only the F0 gestating mother was exposed. Therefore, the spermatogenic defect observed is due to the effects of the endocrine disruptor on transient embryonic testis development at the time of sex determination and not to exposure artifacts or nonspecific toxicity.[29–31]

Interestingly, the same increased apoptosis rate in spermatogenesis was identified in the F2 generation as the F1 generation.[29] A three- to fivefold increase in apoptosis was observed in P20, P60, and P120 male rats. Epididymal sperm were analyzed from P60 and P120 rats, and an approximately 20% decrease in sperm concentration and forward movement was also observed.[24] Similar observations were observed with the F2 generation of vinclozolin-treated animals at both P20 and P60 ages of development.

Analysis of the F3 and F4 generations demonstrated similar effects on spermatogenesis and sperm with a smaller effect in sperm motility with methoxychlor-treated animals.[24] Therefore, exposure of a gestating mother at the critical time of sex determination and testis morphogenesis (i.e., cord formation) appears to cause a germ-line effect of decreased spermatogenic capacity and decreased sperm viability that is transgenerational in the male. This experiment has now been repeated with numerous different gestating mothers with similar results.[29] An outcross of an affected treated male (F2 generation) with a wild-type untreated female (VOC) also had reduced sperm motility[29] and increased spermatogenic cell apoptosis, whereas a reverse outcross (RVOC) of a treated female (F2 generation) with a wild-type male had no effect.[29] It is currently speculated that transient exposure at the critical time of testis development is crucial and that a phenomenon such as an imprinted methylation state of the germ-line may be involved. Although genetic mutagenesis (i.e., DNA sequence change) could be a factor, the phenotype observed was present in most males from all generations, suggesting a frequency not possible for random mutagenesis, but instead more indicative of an epigenetic factor such as methylation. This intriguing observation of a transgenerational effect of an endocrine disruptor through the male germ-line will be examined mechanistically in the future.

SUMMARY

The observations that an environmental toxin (e.g., endocrine disruptor) can have an epigenetic effect on the germ-line and cause a transgenerational effect on male reproduction have a significant impact on our understanding of the potential hazards of these compounds in human health as well as that of all other mammalian species. These studies establish a novel mechanism of action not previously appreciated on how environmental toxins may act on a gestating mother to influence her grandchildren and subsequent generations. Elucidation of this phenomenon will allow us to better understand the true hazards of environmental toxins, identify the specific causal agents, and develop appropriate preventive and therapeutic approaches. Independent of the specific compound or agent of interest, the establishment of this potential mechanism of action is critical to our insight into the effects of environmental factors that influence embryonic development and adult reproduction. Although the dose of endocrine disruptors used in a recent study[29] is higher than anticipated environmental exposures, the observation that the germ-line can be reprogrammed and induce transgenerational effects is a critical phenomenon to establish and elucidate. The specific toxicology of individual environmental toxins will need to be established.

ACKNOWLEDGMENTS

Thanks to Jill Griffin for assistance in the preparation of the manuscript.

REFERENCES

1. Tsui, M. T. & W.X. Wang. 2004. Maternal transfer efficiency and transgenerational toxicity of methylmercury in *Daphnia magna*. Environ. Toxicol. Chem. **23:** 1504–1511.
2. Shimada, A. & A. Shima. 2004. Transgenerational genomic instability as revealed by a somatic mutation assay using the medaka fish. Mutat. Res. **552:** 119–124.
3. Nomura, T. *et al.* 2004. Transgenerational transmission of radiation- and chemically induced tumors and congenital anomalies in mice: studies of their possible relationship to induced chromosomal and molecular changes. Cytogenet. Genome Res. **104:** 252–260.

4. Hoyes, K.P. *et al.* 2001. Transgenerational effects of preconception paternal contamination with (55)Fe. Radiat. Res. **156:** 488–494.
5. Mohr, U. *et al.* 1999. Possible carcinogenic effects of X-rays in a transgenerational study with CBA mice. Carcinogenesis **20:** 325–332.
6. Dubrova, Y.E. 2003. Radiation-induced transgenerational instability. Oncogene **22:** 7087–7093.
7. Barber, R. *et al.* 2002. Elevated mutation rates in the germ line of first- and second-generation offspring of irradiated male mice. Proc. Natl. Acad. Sci. USA **99:** 6877–6882.
8. Fujii, T. 1997. Transgenerational effects of maternal exposure to chemicals on the functional development of the brain in the offspring. Cancer Causes Control **8:** 524–528.
9. Foran, C.M., B.N. Peterson & W.H. Benson. 2002. Transgenerational and developmental exposure of Japanese medaka (*Oryzias latipes*) to ethinylestradiol results in endocrine and reproductive differences in the response to ethinylestradiol as adults. Toxicol. Sci. **68:** 389–402.
10. Cheng, R.Y. *et al.* 2004. Epigenetic and gene expression changes related to transgenerational carcinogenesis. Mol. Carcinog. **40:** 1–11.
11. Martin, R.M. *et al.* 2004. Parents' growth in childhood and the birth weight of their offspring. Epidemiology **15:** 308–316.
12. Kang, I.J. *et al.* 2002. Effects of bisphenol a on the reproduction of Japanese medaka (*Oryzias latipes*). Environ. Toxicol. Chem. **21:** 2394–2400.
13. Schwaiger, J. *et al.* 2002. How estrogenic is nonylphenol? A transgenerational study using rainbow trout (*Oncorhynchus mykiss*) as a test organism. Aquat. Toxicol. **59:** 177–189.
14. Parks, L.G. *et al.* 2000. The plasticizer diethylhexyl phthalate induces malformations by decreasing fetal testosterone synthesis during sexual differentiation in the male rat. Toxicol. Sci. **58:** 339–349.
15. DeRosa, C. *et al.* 1998. Environmental exposures that affect the endocrine system: public health implications. J. Toxicol. Environ. Health B Crit. Rev. **1:** 3–26.
16. Roemer, I. *et al.* 1997. Epigenetic inheritance in the mouse. Curr. Biol. **7:** 277–280.
17. Rakyan, V. & E. Whitelaw. 2003. Transgenerational epigenetic inheritance. Curr. Biol. **13:** R6.
18. Rakyan, V.K. *et al.* 2003. Transgenerational inheritance of epigenetic states at the murine Axin(Fu) allele occurs after maternal and paternal transmission. Proc. Natl. Acad. Sci. USA **100:** 2538–2543.
19. Lane, N. *et al.* 2003. Resistance of IAPs to methylation reprogramming may provide a mechanism for epigenetic inheritance in the mouse. Genesis **35:** 88–93.
20. Lee, J. *et al.* 2002. Erasing genomic imprinting memory in mouse clone embryos produced from day 11.5 primordial germ cells. Development **129:** 1807–1817.
21. Sato, S. *et al.* 2003. Erasure of methylation imprinting of Igf2r during mouse primordial germ-cell development. Mol. Reprod. Dev. **65:** 41–50.
22. Hajkova, P. *et al.* 2002. Epigenetic reprogramming in mouse primordial germ cells. Mech. Dev. **117:** 15–23.
23. Durcova-Hills, G., J. Ainscough & A. McLaren. 2001. Pluripotential stem cells derived from migrating primordial germ cells. Differentiation **68:** 220–226.
24. Anway, M.D. *et al.* 2005. Epigenetic transgenerational actions of endocrine disruptors and male fertility. Science **308:** 1466–1469.
25. Carlsen, E. *et al.* 1992. Evidence for decreasing quality of semen during past 50 years. Br. Med. J. **305:** 609–613.
26. Sharpe, R.M. *et al.* 1995. Gestational and lactational exposure of rats to xenoestrogens results in reduced testicular size and sperm production. Environ. Health Perspect. **103:** 1136–1143.
27. Facemire, C.F., T.S. Gross & L.J. Guillette, Jr. 1995. Reproductive impairment in the Florida panther: nature or nurture? Environ. Health Perspect. **103** (Suppl 4)**:** 79–86.
28. Kelce, W.R. *et al.* 1997. Vinclozolin and p,p'-DDE alter androgen-dependent gene expression: in vivo confirmation of an androgen receptor-mediated mechanism. Toxicol Appl. Pharmacol. **142:** 192–200.

29. ANWAY, M.D. *et al.* 2005. Epigenetic transgenerational actions of endocrine disruptors on male fertility. Science **308:** 466–469.
30. CUPP, A.S. *et al.* 2003. Effect of transient embryonic in vivo exposure to the endocrine disruptor methoxychlor on embryonic and postnatal testis development. J. Androl. **24:** 736–745.
31. UZUMCU, M., H. SUZUKI & M.K. SKINNER. 2004. Effect of the anti-androgenic endocrine disruptor vinclozolin on embryonic testis cord formation and postnatal testis development and function. Reprod. Toxicol. **18:** 765–774. .
32. CUMMINGS, A.M. 1997. Methoxychlor as a model for environmental estrogens. Crit. Rev. Toxicol. **27:** 367–379.
33. PETTERSSON, K. & J.A. GUSTAFSSON. 2001. Role of estrogen receptor beta in estrogen action. Annu. Rev. Physiol. **63:** 165–192.
34. LEE, K.H. *et al.* 2000. Estrogen receptor alpha has a functional role in the mouse rete testis and efferent ductules. Biol. Reprod. **63:** 1873–1880.
35. BRANDENBERGER, A.W. *et al.* 1997. Tissue distribution of estrogen receptors alpha (ER-alpha) and beta (ER-beta) mRNA in the midgestational human fetus. J. Clin. Endocrinol. Metab. **82:** 3509–3512.
36. SAUNDERS, P.T. *et al.* 1997. Expression of oestrogen receptor beta (ER beta) in multiple rat tissues visualised by immunohistochemistry. J. Endocrinol. **154:** R13–16.
37. GOYAL, H.O. *et al.* 1997. Immunolocalization of androgen receptor and estrogen receptor in the developing testis and excurrent ducts of goats. Anat. Rec. **249:** 54–62.
38. LI, H. *et al.* 1997. Regulation of rat testis gonocyte proliferation by platelet-derived growth factor and estradiol: identification of signaling mechanisms involved. Endocrinology **138:** 1289–1298.
39. EDDY, E.M. *et al.* 1996. Targeted disruption of the estrogen receptor gene in male mice causes alteration of spermatogenesis and infertility. Endocrinology **137:** 4796–4805.
40. SMITH, E.P. *et al.* 1994. Estrogen resistance caused by a mutation in the estrogen-receptor gene in a man. N. Engl. J. Med. **331:** 1056–1061.
41. COUSE, J.F. *et al.* 1999. Postnatal sex reversal of the ovaries in mice lacking estrogen receptors alpha and beta. Science **286:** 2328–2331.
42. TENA-SEMPERE, M. *et al.* 2000. Neonatal exposure to estrogen differentially alters estrogen receptor alpha and beta mRNA expression in rat testis during postnatal development. J. Endocrinol. **165:** 345–357.
43. TENA-SEMPERE, M. *et al.* 2001. Neonatal imprinting and regulation of estrogen receptor alpha and beta mRNA expression by estrogen in the pituitary and hypothalamus of the male rat. Neuroendocrinology **73:** 12–25.
44. ATANASSOVA, N. *et al.* 1999. Permanent effects of neonatal estrogen exposure in rats on reproductive hormone levels, Sertoli cell number, and the efficiency of spermatogenesis in adulthood. Endocrinology **140:** 5364–5373.
45. NIELSEN, M. *et al.* 2000. Ontogeny of oestrogen receptor alpha in gonads and sex ducts of fetal and newborn mice. J. Reprod. Fertil. **118:** 195–204.
46. EBLING, F.J. *et al.* 2000. Estrogenic induction of spermatogenesis in the hypogonadal mouse. Endocrinology **141:** 2861–2869.
47. PRINS, G.S. & L. BIRCH. 1995. The developmental pattern of androgen receptor expression in rat prostate lobes is altered after neonatal exposure to estrogen. Endocrinology **136:** 1303–1314.
48. MCKINNELL, C. *et al.* 2001. Suppression of androgen action and the induction of gross abnormalities of the reproductive tract in male rats treated neonatally with diethylstilbestrol. J. Androl. **22:** 323–338.
49. WILSON, C.M. & M.J. MCPHAUL. 1996. A and B forms of the androgen receptor are expressed in a variety of human tissues. Mol. Cell Endocrinol. **120:** 51–57.
50. MAJDIC, G., M.R. MILLAR & P.T. SAUNDERS. 1995. Immunolocalisation of androgen receptor to interstitial cells in fetal rat testes and to mesenchymal and epithelial cells of associated ducts. J. Endocrinol. **147:** 285–293.
51. SILVERSIDES, D.W., C.A. PRICE & G.M. COOKE. 1995. Effects of short-term exposure to hydroxyflutamide in utero on the development of the reproductive tract in male mice. Can. J. Physiol. Pharmacol. **73:** 1582–1588.

52. NAMBU, A. & Y. KUMAMOTO. 1995. [Studies of spermatogenic damages induced by anti-cancer agent and anti-androgenic agents in rat testes.] Nippon Hinyokika Gakkai Zasshi **86:** 1221–1230.
53. GUPTA, C., A. CHANDORKAR & A.P. NGUYEN. 1996. Activation of androgen receptor in epidermal growth factor modulation of fetal mouse sexual differentiation. Mol. Cell Endocrinol. **123:** 89–95.
54. TURNER, K.J. *et al.* 2002. Effects of in utero exposure to the organophosphate insecticide fenitrothion on androgen-dependent reproductive development in the Crl:CD(SD)BR rat. Toxicol. Sci. **68:** 174–183.
55. HOTCHKISS, A.K. *et al.* 2002. Androgens and environmental antiandrogens affect reproductive development and play behavior in the Sprague-Dawley rat. Environ. Health Perspect. **110** (Suppl 3)**:** 435–439.
56. POTHULURI, J.V. *et al.* 2000. Biotransformation of vinclozolin by the fungus *Cunninghamella elegans*. J. Agric. Food Chem. **48:** 6138–6148.
57. KELCE, W.R. *et al.* 1994. Environmental hormone disruptors: evidence that vinclozolin developmental toxicity is mediated by antiandrogenic metabolites. Toxicol. Appl. Pharmacol. **126:** 276–285.
58. KELCE, W.R., L.E. GRAY & E.M. WILSON. 1998. Antiandrogens as environmental endocrine disruptors. Reprod. Fertil. Dev. **10:** 105–111.
59. GRAY, L.E., Jr. *et al.* 1999. Environmental antiandrogens: low doses of the fungicide vinclozolin alter sexual differentiation of the male rat. Toxicol. Ind. Health **15:** 48–64.
60. MONOSSON, E. *et al.* 1999. Peripubertal exposure to the antiandrogenic fungicide, vinclozolin, delays puberty, inhibits the development of androgen-dependent tissues, and alters androgen receptor function in the male rat. Toxicol. Ind. Health **15:** 65–79.
61. YU, W.J. *et al.* 2004. Reproductive disorders in pubertal and adult phase of the male rats exposed to vinclozolin during puberty. J. Vet. Med. Sci. **66:** 847–853.
62. GRAY, L.E., JR. *et al.* 1999. Administration of potentially antiandrogenic pesticides (procymidone, linuron, iprodione, chlozolinate, p,p'-DDE, and ketoconazole) and toxic substances (dibutyl- and diethylhexyl phthalate, PCB 169, and ethane dimethane sulphonate) during sexual differentiation produces diverse profiles of reproductive malformations in the male rat. Toxicol. Ind. Health **15:** 94–118.
63. HELLWIG, J. *et al.* 2000. Pre- and postnatal oral toxicity of vinclozolin in Wistar and Long-Evans rats. Regul. Toxicol. Pharmacol. **32:** 42–50.
64. MOORMAN, W.J. *et al.* 2000. Male adolescent exposure to endocrine-disrupting pesticides: vinclozolin exposure in peripubertal rabbits. Andrologia **32:** 285–293.
65. SHONO, T. *et al.* 2004. The effect of a prenatal androgen disruptor, vinclozolin, on gubernacular migration and testicular descent in rats. J. Pediatr. Surg. **39:** 213–216.
66. WOLF, C.J. *et al.* 2000. Characterization of the period of sensitivity of fetal male sexual development to vinclozolin. Toxicol. Sci. **55:** 152–161.
67. KELCE, W.R. *et al.* 1995. Persistent DDT metabolite p,p'-DDE is a potent androgen receptor antagonist. Nature **375:** 581–585.
68. WOLF, C.J., G.A. LEBLANC & L.E. GRAY, JR. 2004. Interactive effects of vinclozolin and testosterone propionate on pregnancy and sexual differentiation of the male and female SD rat. Toxicol. Sci. **78:** 135–143.
69. KAPOOR, I.P. *et al.* 1970. Comparative metabolism of methoxychlor, methiochlor, and DDT in mouse, insects, and in a model ecosystem. J. Agric. Food Chem. **18:** 1145–1152.
70. DEHAL, S.S. & D. KUPFER. 1994. Metabolism of the proestrogenic pesticide methoxychlor by hepatic P450 monooxygenases in rats and humans. Dual pathways involving novel ortho ring-hydroxylation by CYP2B. Drug Metab. Dispos. **22:** 937–946.
71. KUPFER, D., W. H. BULGER & A. D. THEOHARIDES. 1990. Metabolism of methoxychlor by hepatic P-450 monooxygenases in rat and human. 1. Characterization of a novel catechol metabolite. Chem. Res. Toxicol. **3:** 8–16.
72. BULGER, W.H. & D. KUPFER. 1985. Estrogenic activity of pesticides and other xenobiotic on the uterus and male reproductive tract. *In* Endocrine Toxicology. J.A. Thomas, K.S. Korach & J.A. McLachlan, Eds. :1–33. Raven Press. New York.

73. LAMOUREUX, C.H. & V.J. FEIL. 1980. Gas chromatographic and mass spectrometric characterization of impurities in technical methoxychlor. Anal. Chem. **63:** 1007–1037.
74. WEST, P.R. *et al.* 1982. High performance liquid chromatographic analysis of impurities and degradation products of methoxychlor. Anal. Chem. **65:** 1457–1470.
75. EROSCHENKO, V.P., A.W. ROURKE & W.F. SIMS. 1996. Estradiol or methoxychlor stimulates estrogen receptor (ER) expression in uteri. Reprod. Toxicol. **10:** 265–271.
76. GAIDO, K.W. *et al.* 1999. Differential interaction of the methoxychlor metabolite 2,2-bis-(*p*-hydroxyphenyl)-1,1,1-trichloroethane with estrogen receptors alpha and beta. Endocrinology **140:** 5746–5753.
77. GAIDO, K.W. *et al.* 2000. Interaction of methoxychlor and related compounds with estrogen receptor alpha and beta, and androgen receptor: structure-activity studies. Mol. Pharmacol. **58:** 852–858.
78. MATTHEWS, J. *et al.* 2000. Differential estrogen receptor binding of estrogenic substances: a species comparison. J. Steroid Biochem. Mol. Biol. **74:** 223–234.
79. CHAPIN, R.E. *et al.* 1997. The effects of perinatal/juvenile methoxychlor exposure on adult rat nervous, immune, and reproductive system function. Fundam. Appl. Toxicol. **40:** 138–157.
80. COOKE, P.S. & V.P. EROSCHENKO. 1990. Inhibitory effects of technical grade methoxychlor on development of neonatal male mouse reproductive organs. Biol. Reprod. **42:** 585–596.
81. SUZUKI, M. *et al.* 2004. Effects of methoxychlor exposure during perinatal period on reproductive function after maturation in rats. J. Reprod. Dev. **50:** 455–461.
82. SWARTZ, W.J. & V.P. EROSCHENKO. 1998. Neonatal exposure to technical methoxychlor alters pregnancy outcome in female mice. Reprod. Toxicol. **12:** 565–573.
83. ALWORTH, L.C. *et al.* 2002. Uterine responsiveness to estradiol and DNA methylation are altered by fetal exposure to diethylstilbestrol and methoxychlor in CD-1 mice: effects of low versus high doses. Toxicol. Appl. Pharmacol. **183:** 10–22.
84. EROSCHENKO, V.P., A.A. ABUEL-ATTA & M.S. GROBER. 1995. Neonatal exposures to technical methoxychlor alters ovaries in adult mice. Reprod. Toxicol. **9:** 379–387.
85. BORGEEST, C. *et al.* 2002. Methoxychlor may cause ovarian follicular atresia and proliferation of the ovarian epithelium in the mouse. Toxicol. Sci. **68:** 473–478.
86. PALANZA, P. *et al.* 1999. Prenatal exposure to endocrine disrupting chemicals: effects on behavioral development. Neurosci. Biobehav. Rev. **23:** 1011–1027.
87. STOKER, T.E., C.L. ROBINETTE & R.L. COOPER. 1999. Perinatal exposure to estrogenic compounds and the subsequent effects on the prostate of the adult rat: evaluation of inflammation in the ventral and lateral lobes. Reprod. Toxicol. **13:** 463–472.
88. TAKEUCHI, Y. *et al.* 2002. Thymic atrophy induced by methoxychlor in rat pups. Toxicol. Lett. **135:** 199–207.
89. LATCHOUMYCANDANE, C. & P.P. MATHUR. 2002. Effect of methoxychlor on the antioxidant system in mitochondrial and microsome-rich fractions of rat testis. Toxicology **176:** 67–75.
90. SWARTZ, W.J. & M. CORKERN. 1992. Effects of methoxychlor treatment of pregnant mice on female offspring of the treated and subsequent pregnancies. Reprod. Toxicol. **6:** 431–437.
91. SERTOLI, E. 1865. On the existence of special branched cells in the seminiferous tubule of the human testes. Morgangni **7:** 31–39.
92. FAWCETT, D.W. 1975. The ultrastructure and functions of the Sertoli cell. *In* Handbook of Physiology, Vol. V(7). R.O. Greep & E.W. Hamilton, Eds. :22–55. Am. Physiol. Soc. Washington, D.C.
93. SETCHELL, B.P. & G.M.H. WAITES. 1975. *In* Handbook of Physiology, Vol. V (7). R.O. Greep & E.W. Hamilton, Eds. :143–172. Am. Physiol. Soc. Washingto, D.C.
94. WAITES, G.M. & R.T. GLADWELL. 1982. Physiological significance of fluid secretion in the testis and blood-testis barrier. Physiol. Rev. **62:** 624–671.
95. RUSSELL, L.D. *et al.* 1983. Three-dimensional reconstruction of a rat stage V Sertoli cell. III. A study of specific cellular relationships. Am. J. Anat. **167:** 181–192.
96. DORRINGTON, J.H., I.B. FRITZ & D.T. ARMSTRONG. 1978. Control of testicular estrogen synthesis. Biol. Reprod. **18:** 55–64.

97. ROBINSON, R. & I.B. FRITZ. 1981. Metabolism of glucose by Sertoli cells in culture. Biol. Reprod. **24:** 1032–1041.
98. LACROIX, M., F.E. SMITH & I.B. FRITZ. 1977. Secretion of plasminogen activator by Sertoli cell enriched cultures. Mol. Cell Endocrinol. **9:** 227–236.
99. SKINNER, M.K. & M.D. GRISWOLD. 1983. Sertoli cells synthesize and secrete a ceruloplasmin-like protein. Biol. Reprod. **28:** 1225–1229.
100. SKINNER, M.K. & M.D. GRISWOLD. 1980. Sertoli cells synthesize and secrete transferrin–like protein. J. Biol. Chem. **255:** 9523–9525.
101. GRISWOLD, M.D. 1988. Protein secretions of Sertoli cells. Int. Rev. Cytol. **110:** 133–156.
102. TUNG, P.S., M.K. SKINNER & I.B. FRITZ. 1984. Fibronectin synthesis is a marker for peritubular cell contaminants in Sertoli cell-enriched cultures. Biol. Reprod. **30:** 199–211.
103. SKINNER, M.K., W.L. COSAND & M.D. GRISWOLD. 1984. Purification and characterization of testicular transferrin secreted by rat Sertoli cells. Biochem. J. **218:** 313-320.
104. SKINNER, M.K. 1991. Cell-cell interactions in the testis. Endocr. Rev. **12:** 45–77.
105. AKINGBEMI, B.T. *et al.* 2000. A metabolite of methoxychlor, 2,2-bis(*p*-hydroxyphenyl)-1,1, 1-trichloroethane, reduces testosterone biosynthesis in rat Leydig cells through suppression of steady-state messenger ribonucleic acid levels of the cholesterol side-chain cleavage enzyme. Biol. Reprod. **62:** 571–578.
106. KIERSZENBAUM, A.L. 1994. Mammalian spermatogenesis *in vivo* and *in vitro*: a partnership of spermatogenic and somatic cell lineages. Endocr. Rev. **15:** 116–134.
107. JOST, A., S. MAGRE & R. AGELOPOULOU. 1981. Early stages of testicular differentiation in the rat. Hum. Genet. **58:** 59–63.
108. MAGRE, S. & A. JOST. 1980. The initial phases of testicular organogenesis in the rat. An electron microscopy study. Arch. Anat. Microsc. Morphol. Exp. **69:** 297–318.
109. BUEHR, M., S. GU & A. MCLAREN. 1993. Mesonephric contribution to testis differentiation in the fetal mouse. Development **117:** 273–281.
110. MCLAREN, A. 1991. Development of the mammalian gonad: the fate of the supporting cell lineage. Bioessays **13:** 151–156.
111. FROJDMAN, K. *et al.* 1992. Intermediate filaments and epithelial differentiation of male rat embryonic gonad. Differentiation **50:** 113–123.
112. FRIDMACHER, V. *et al.* 1995. Switch in the expression of the K19/K18 keratin genes as a very early evidence of testicular differentiation in the rat. Mech. Dev. **52:** 199–207.
113. BLANCHARD, M.G. & N. JOSSO. 1974. Source of the anti-Mullerian hormone synthesized by the fetal testis: Mullerian-inhibiting activity of fetal bovine Sertoli cells in tissue culture. Pediatr. Res. **8:** 968–971.
114. GREENE, R.R. 1944. Embyology of sexual structure and hermaphroditism. J. Clin. Endorcrinology **4:** 335–348.
115. MAGRE, S. & A. JOST. 1991. Sertoli cells and testicular differentiation in the rat fetus. J. Electron Microsc. Tech. **19:** 172–188.
116. ORTH, J.M. *et al.* 1983. Environmental stress alters the developmental pattern of delta 5-3 beta-hydroxysteroid dehydrogenase activity in Leydig cells of fetal rats: a quantitative cytochemical study. Biol. Reprod. **28:** 625–631.
117. BLOCH, E., M. LEW & M. KLEIN. 1971. Studies on the inhibition of fetal androgen formation. Inhibition of testosterone synthesis in rat and rabbit fetal testes with observations on reproductive tract development. Endocrinology **89:** 16–31.
118. SMALL, C.L. *et al.* 2005. Profiling gene expression during the differentiation and development of the murine embryonic gonad. Biol. Reprod. **72:** 492–501.
119. GUPTA, C. 1996. The role of epidermal growth factor receptor (EGFR) in male reproductive tract differentiation: stimulation of EGFR expression and inhibition of Wolffian duct differentiation with anti-EGFR antibody. Endocrinology **137:** 905–910.
120. GUPTA, C., S. SIEGEL & D. ELLIS. 1991. The role of EGF in testosterone-induced reproductive tract differentiation. Dev. Biol. **146:** 106–116.
121. LEVINE, E. *et al.* 2000. Role of transforming growth factor-alpha and the epidermal growth factor receptor in embryonic rat testis development. Biol. Reprod. **62:** 477–490.

122. DESAI, K.V. & P. KONDAIAH. 2000. Androgen ablation results in differential regulation of transforming growth factor-beta isoforms in rat male accessory sex organs and epididymis. J. Mol. Endocrinol. **24:** 253–260.
123. LEVINE, E., A.S. CUPP & M.K. SKINNER. 2000. Role of neurotropins in rat embryonic testis morphogenesis (cord formation). Biol. Reprod. **62:** 132–142.
124. CUPP, A.S., L. TESSAROLLO & M.K. SKINNER. 2002. Testis developmental phenotypes in neurotropin receptor trkA and trkC null mutations: role in formation of seminiferous cords and germ cell survival. Biol. Reprod. **66:** 1838–1845.

In Vitro Generation of Germ Cells

New Techniques to Solve Current Issues

JOSHUA M. BAUGHMAN[a,b] AND NIELS GEIJSEN[b]

[a]*Graduate Program of Biological and Biomedical Sciences, Harvard Medical School, Boston, Massachusetts 02115, USA*

[b]*Harvard Medical School, Center for Regenerative Medicine and Technology, Massachusetts General Hospital, Boston, Massachusetts 02114, USA*

ABSTRACT: Primordial germ cells comprise a privileged cellular class within the embryo charged with the elite task of maintaining species longevity. While in lower organisms germ-cell fate is determined by the allocation of germ plasm, mammalian germ-line differentiation requires extracellular signals that converge upon the proximal epiblast. Studies using mutant mice or explanted embryos have identified some of the factors controlling primordial germ-cell specification, such as members of the BMP family, but considerable gaps still exist in our understanding of the complete signaling network. Comprehensive investigations of mammalian germ-line specification have been hampered by the inaccessibility of this cell population in the early embryo. Recently, however, several labs including our own have derived primordial germ cells from embryonic stem cells *in vitro,* thus providing a powerful new technique for the study of germ cells. In this review the different methods used for the *in vitro* generation of germ cells and how these techniques may be improved and applied to further advance our knowledge of germ-cell biology are discussed.

KEYWORDS: primordial germ cells; embryonic stem cells; differentiation; bone morphogenic proteins; germ-line; mammals

INTRODUCTION

Embryonic stem (ES) cells are derived from the inner cell mass of a blastocyst-stage embryo and are capable of maintaining pluripotency even after extensive *in vitro* culture. Although first isolated nearly 25 years ago, ES cells were recently catapulted to the medical, ethical, and political forefronts after their successful derivation from human blastocysts.[1–3] In combination with somatic cell nuclear transfer, human ES cells offer significant promise for the development of cell therapies to treat degenerative disease.

Through a variety of *in vitro* methods, ES cells can be induced to faithfully recapitulate several aspects of differentiation and developmental timing of the early

Address for correspondence: Niels Geijsen, Center for Regenerative Medicine and Technology, Massachusetts General Hospital, CPZN-4264, 185 Cambridge Street, Boston, MA 02114. Voice: 617-724-6013; fax: 617-724-2662.
ngeijsen@partners.org

**Ann. N.Y. Acad. Sci. 1061: 33–40 (2005). © 2005 New York Academy of Sciences.
doi: 10.1196/annals.1336.005**

embryo. Therefore, in addition to their medical applications, ES cell differentiation offers unique insight into areas of mammalian development otherwise inaccessible to experimentation. To date, spontaneous and directed differentiation of ES cells has led to the derivation of descendant cell types from all three major germ layers including neural, hematopoietic, vascular, pancreatic, muscular, and cardiac lineages.[4]

Surprisingly, despite harboring a well-known totipotency and germ-line transmission when injected into host blastocysts, ES cells have only recently been shown to differentiate into primordial germ cells (PGCs) *in vitro*.[5–8] This significant advance presents new opportunities to study early PGC development while providing prospective medical applications for easily deriving oocytes and treating infertility. In this review we describe our current understanding of murine PGC specification. We briefly outline recent experiments that have demonstrated *in vitro* derivation of PGCs and discuss the potential use of these techniques for advancing our basic understanding of PGC commitment, differentiation, and epigenetic reprogramming.

PRIMORDIAL GERM-CELL DEVELOPMENT IN THE EARLY MURINE EMBRYO

PGCs were first identified morphologically in 1954, but we still do not completely understand the necessary signals required for germ-line commitment.[9] PGC development in mammals differs from that of lower organisms whose germ cells are preformed by cytoplasmic partitioning within the oocyte.[10] Instead, mammalian PGCs become committed to their germ-line fate during early gastrulation and are first visible at the base of the incipient allantois within the posterior extraembryonic mesoderm as approximately 20–60 alkaline phosphatase (TNAP) staining cells at 7.25 days postcoitum (dpc).[9,11,12] Fate mapping has demonstrated that germ cells are derived from the proximal epiblast[11] (FIG. 1A), wherease elegant cellular transplantation studies have proven that signals emanating from the extraembryonic ectoderm are necessary for specifying the PGCs within the proximal epiblast.[13] Distal epiblast cells, normally destined for neural or surface ectoderm, can be respecified to form germ cells when placed adjacent to the extraembryonic ectoderm. Conversely, proximal epiblast cells heterotopically transplanted to distal positions do not form PGCs and instead populate the ectodermal lineages.[13] Therefore, epiblast cells represent a plastic intermediate capable of either retaining totipotency or differentiating into somatic cell types. The germ-line specification of an exclusive few epiblast cells appears to involve a distinctive niche that confers PGC fate while resisting somatic differentiation.

The first paracrine constituent of the PGC-specifying niche was recently described after finding that mice deficient in a TGF-β superfamily member, *Bmp4*, present a highly penetrant PGC-null phenotype.[14] Subsequent studies have strongly supported a role for Bmp signaling in determining germ-line fate. Bmp4 and Bmp8b act synergistically and from within the extraembryonic ectoderm to specify PGCs (FIG. 1B).[15] Bmp2, a protein highly homologous to Bmp4, is expressed from within the posterior visceral endoderm (VE) to affect PGC differentiation.[16] Finally, mice deficient in either *Smad1*, *Smad4*, or *Smad5*, the classic downstream transducers of Bmp4 activation, display low PGC numbers or fail to develop PGCs completely altogether.[17–20]

Components of the Bmp pathway are needed to direct germ-cell development, but they may not comprise the only necessary signals. Supporting this view, *Bmp4* or *Bmp8b* homozygous null mice either lack or are significantly reduced in allantoic tissue that is derived from the extraembryonic mesoderm.[21,14] Bmp proteins, thus, concurrently specify the fate of at least two proximate tissues: extraembryonic mesoderm and PGCs. It does not seem that Bmp4 acts singly and morphogenically to distinguish these fates.[14,22] It has therefore been hypothesized that Bmp proteins function to prime epiblastic cells for either an extraembryonic mesoderm or PGC fate, whereas subsequent signals, possibly propagated by indirect Bmp action on the VE, truly specify the PGC subset.[22] The nature of these putative secondary signals remains uncharacterized.

An alternative hypothesis to explain the pleiotropic actions of the Bmp proteins is that signals acting prior to Bmp induction are necessary to prime epiblast cells to respond differentially to Bmp ligands. Such priming might be accomplished by modifying receptor quantity or inducing the expression of Bmp co-receptors known to

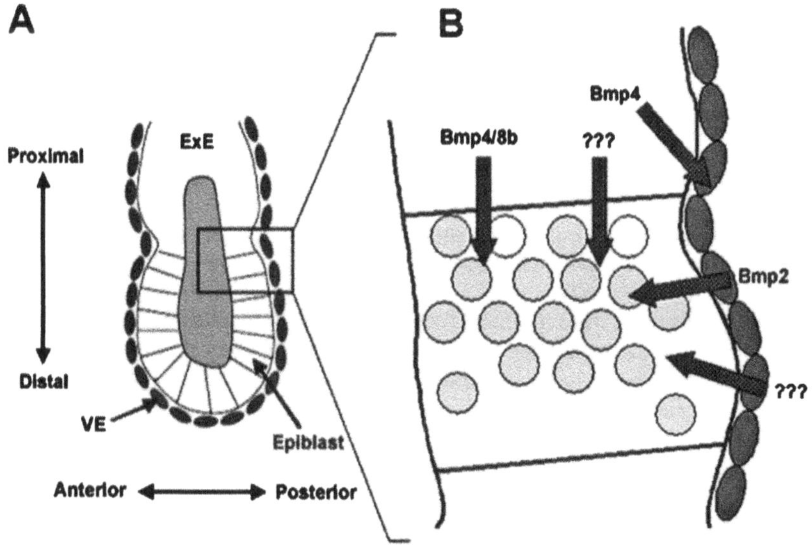

FIGURE 1. Early PGC-specifying signals in a 6.0-dpc embryo. (**A**) A 6.0-dpc pre-primitive streak stage embryo is comprised of columnar epiblast cells surrounding a hollow inner cavity (*shaded*). Proximal epiblast cells lie adjacent to the extraembryonic ectoderm (ExE). Both the ExE and epiblast are sheathed by visceral endoderm (VE). Axes of the early embryo are indicated. (**B**) Cells within the proximal epiblast can be specified to become germ cells. Bmp4 and Bmp8b are secreted from the ExE and are necessary for germ-cell development. It is unknown whether these proteins are the only necessary signals emanating from the ExE (see main text). The VE is indispensible for PGC specification. Bmp4 likely acts from the ExE to direct the secretion of other germ-line–inducing factors from within the VE. Bmp2 is expressed in the VE, but other factors originating from the VE may assist in specifying germ cells. Blimp1 is only expressed in some cells within the most proximal posterior epiblast cell layer (white cells). Fate-mapping experiments have demonstrated that these Blimp1-positive cells will contribute nearly exclusively to the germline, whereas other proximal cells (*gray*) could contribute to either extraembryonic mesoderm or PGCs.

modulate cellular response to Bmp agonists.[23] These signals could also modulate intracellular signaling pathways, such as MAPK, that are capable of modulating Smad activation.[24] Further study is needed to identify these presumed priming signals as well as their intracellular effects that confer disparate Bmp responses among cells in the proximal epiblast.

One corollary to the foregoing proposals is that cells competent to become PGCs should be molecularly distinguishable from other cell types even before they become fully committed to the germ-line during gastrulation. There has been some recent success in characterizing the distinguishing features of these PGC precursors. One group used elegant single-cell dissection, RT-PCR amplification, and differential cDNA screening to discover two novel markers of early PGC development: *stella* and *fragilis*.[25] *Fragilis* (mil-1, Ifitm3) is a member of the interferon-inducible transmembrane family (Ifitm).[25–27] Although the true functions of *fragilis* are still unclear, they may involve homotypic adhesion between PGC precursors that segregates them from surrounding tissue. To this point, the functional involvement for the Ifitm family in PGC development remains correlative to their expression pattern. Interestingly, the *fragilis* promoter harbors two interferon response elements, leading to the hypothesis that interferon may function as a secondary signal affecting PGC fate.[25] However, this hypothesis remains in question as further study has revealed that the IRE domains seem dispensable for the expression of *fragilis* in PGC precursors.[26]

Stella, a nuclear protein of unknown function, is expressed within a subset of *fragilis*-positive cells and is thought to specifically mark those cells that will become germ-line committed.[25] As with *fragilis*, however, *stella*'s role in PGC development is not clear, as first-generation homozygous mutant mice are not defective in germ-cell formation whereas second-generation mice die during early cleavage.[29] *Stella* appears to be a maternal effect gene that is a specific but possibly nonfunctional marker of early PGC cells.

Finally, Blimp1, a transcriptional repressor known to regulate B-cell terminal differentiation, was demonstrated to be essential for PGC specification.[30,31] *Blimp1* knockout mice completely lack migratory PGCs and may act by repressing the transcription of genes necessary to induce somatic cell fate such as members of the homeobox gene family.[31] *Blimp1* transcribing cells are observed as early as 6.25 dpc in only the first cell layer of the posterior proximal epiblast, suggesting that direct contact with extraembryonic ectoderm is necessary for its induction. Lineage tracing demonstrates that nearly all epiblast cells expressing *Blimp1* go on to express the PGC markers TNAP and stella. *Blimp1* is now the earliest and most specific marker known for PGC precursor cells.

These findings have significantly advanced our knowledge of the molecular identity of PGCs. However, several questions remain unanswered. For instance, what is the nature of the inductive signals that confine *Blimp1* expression to the most proximal epiblast layer and to only a subset of fragilis expressing cells?[31] Are there other markers coexpressed along with or before *Blimp1* that further characterize PGC precursors? And, what is the full set of genes targeted by *Blimp1* to affect PGC specification?

Further genomic screens will advance our knowledge of the PGC transcriptional profile. However, the limited number of PGC precursor cells in early development hinders detailed analysis of the biochemical nature of PGC inducers and downstream effectors *in vivo*. Fortunately, several groups have recently described the derivation of primordial germ cells from mouse ES cells *in vitro*. With this new *in vitro* system and the increasingly specific PGC marker set we are uniquely poised to answer the significant questions pertaining to germ-line development.

IN VITRO DERIVATION OF PGCS: NEW METHODS TO SOLVE OLD PROBLEMS

A groundbreaking study from Karen Hubner and colleagues demonstrated for the first time that ES cells can give rise to primordial germ cells *in vitro*.[5] Three other groups, including our own, subsequently confirmed and expanded on this exciting result.[6–8] There are already comprehensive reviews describing these studies,[32,33] and we therefore focus in this review on ways *in vitro* PGC derivation might be applied to study early PGC specification.

Two labs utilized embryoid bodies (EBs) to differentiate mouse ES cells into PGCs.[6,7] EBs are formed when ES cells are cultured in suspension and allowed to differentiate within spherical aggregates without pluripotency-maintaining factors. They possess the unique advantage of providing intimate intercellular contacts that emulate *in vivo* developmental niches.[4] EB-derived PGCs express a number of germ-cell markers including *stella*, *fragilis*, *Mvh1*, *SSEA1*, and *Oct4*.[7] They functionally mimic *in vivo* PGCs by responding to Bmp4 stimulation, erasing parental imprints and acquiring the ability to colonize murine testis and produce functional sperm.[6,7] We have found that a rare population of EB-derived PGCs can undergo meiosis to produce round spermatids that, when injected into a murine oocyte, are capable of producing blastocyst embryos.[7] Therefore, EBs seem to recapitulate the embryonic PGC niche in which the germ cells express both molecular and functional markers of germ-line development.

EBs are easily produced and offer an attractive alternative to *in vivo* genetic manipulation. They are amenable to new technologies such as retroviral or lentiviral RNAi libraries that may be employed to efficiently probe for factors necessary in germ-line commitment.[34,35] Additionally, small molecule effectors, neutralizing antibodies, or recombinant soluble proteins can easily be added to the culture medium to affect the EB environment and PGC development. These large-scale screening methods are not currently possible *in vivo* and they may significantly advance germ-cell study by yielding new candidate genes and pathways involved in germ-line specification.

EBs offer the advantage of a three-dimensional environment but may also confound results because of the complexity of signals and cell types they produce. A two-dimensional PGC differentiation protocol would reduce such inherent complexity. With such a system one can specifically control factors involved in PGC differentiation to produce cells homogeneously staged at different points in germ-cell development. Once implemented, this technique would allow for genomic scale expression studies or chromatin immunoprecipitations to elucidate the full marker profile and downstream effectors during PGC maturation. We could deduce the minimal set of factors sufficient to induce PGC development. And finally, it would provide enough material to study in a controlled manner the biochemical nature of PGC specification, imprint erasure, X-inactivation/reactivation, and other germ-line specific mechanisms.

22. De Sousa Lopes, S.M., B.A. Roelen, R.M. Monteiro, *et al.* 2004. BMP signaling mediated by ALK2 in the visceral endoderm is necessary for the generation of primordial germ cells in the mouse embryo. Genes Dev. **18:**1838–1849.
23. Xia, Y., Y. Sidis, A. Mukherjee, *et al.* 2005. Localization and action of Dragon (RGMb), a novel BMP co-receptor, throughout the reproductive axis. Endocrinology. Epub ahead of print, May 12, 2005.
24. Aubin, J., A. Davy & P. Soriano. 2004. *In vivo* convergence of BMP and MAPK signaling pathways: impact of differential Smad1 phosphorylation on development and homeostasis. Genes Dev. **18:** 1482–1494.
25. Saitou, M., S.C. Barton & M.A. Surani. 2002. A molecular programme for the specification of germ cell fate in mice. Nature **418:** 293–300.
26. Tanaka, S.S. & Y. Matsui. 2002. Developmentally regulated expression of mil-1 and mil-2, mouse interferon-induced transmembrane protein like genes during formation and differentiation of primordial germ cell. Mech. Dev. **119** (Suppl 1): S261–S267.
27. Lange, U.C., M. Saitou, P.S. Western, *et al.* 2003. The fragilis interferon-inducible gene family of transmembrane proteins is associated with germ cell specification in mice. BMC Dev. Biol. **3:** 1.
28. Tanaka, S.S., G. Nagamatsu, Y. Tokitake, *et al.* 2004. Regulation of expression of mouse interferon-induced transmembrane protein like gene-3, Ifitm (mil-1, fragilis), in germ cells. Dev. Dyn. **230:** 651–659.
29. Bortvin, A., M. Goodheart, M. Liao & D.C. Page. 2004. Dppa3/Pgc7/stella is a maternal factor and is not required germ cell specification in mice. BMC Dev. Biol. **4:** 2.
30. Vincent, S.D., N.R. Dunn, R. Sciammas, *et al.* 2005. The zinc finger transcriptional repressor Blimp1/Prdm1 is dispensable for early axis formation but is required for specification of primordial germ cells in the mouse. Development **132:** 1315–1325.
31. Ohinata, Y., B. Payer, D.O. O'Carroll, *et al.* 2005.Nature June 5th (ePub ahead of print).
32. West, J. & G.Q. Daley. 2004. In vitro gametogenesis from embryonic stem cells. Curr. Opin. Cell. Biol. **16:** 688–692.
33. Lacham-Kaplan, O. 2004. *In vivo* and *in vitro* differentiation of male germ cells in the mouse. Reproduction **128:**147–152.
34. Berns, K., E.M. Hijmans, J. Mullenders, *et al.* 2004. A large-scale RNAi screen in human cells identifies new components of the p53 pathway. Nature **428:** 431–437.
35. Paddison, P.J., J.M. Silva, D.S. Conklin, *et al.* 2004. A resource for large-scale RNA-interference-based screens in mammals. Nature **428:** 427–431.
36. Qi, X., T.G. Li, J. Hao, *et al.* 2004. BMP4 supports self-renewal of embryonic stem cells by inhibiting mitogen-activated protein kinase pathways. Proc. Natl. Acad. Sci. USA **101:** 6027-6032.
37. Rathjen, J., J.A. Lake, M.D. Bettess, *et al.* 1999. Formation of a primitive ectoderm like cell populaion, EPL cells, from ES cells in response to biologically derived factors. J. Cell. Sci. **112:** 601–612.
38. Rathjen, J., B.P. Haines, K.M. Hudson, *et al.* 2002. Directed differentiation of pluripotent cells to neural lineages: homogeneous formation and differentiation of a neurectoderm population. Development **129:** 2649–2661.

Structure-Function Relationships of the Luteinizing Hormone Receptor

DAVID PUETT,[a] YONGSHENG LI,[a] KRASSIMIRA ANGELOVA,[a] GENEVA DEMARS,[a] THOMAS P. MEEHAN,[a,b] FRANCESCA FANELLI,[c] AND PREMA NARAYAN[a,d]

[a]*Department of Biochemistry & Molecular Biology, University of Georgia, Athens, Georgia 30602, USA*

[b]*Current address: Department of Molecular and Human Genetics, Baylor College of Medicine, Houston, Texas 77030, USA*

[c]*Dulbecco Telethon Institute and Department of Chemistry, University of Modena and Reggio Emilia, via Campi 183, 41100, Modena, Italy*

[d]*Current address: Department of Physiology, Southern Illinois University School of Medicine, Carbondale, Illinois 62901, USA*

ABSTRACT: Of the 800-900 genes in the human genome that appear to encode G-protein-coupled receptors (GPCRs), two are known to encode receptors that bind the three heterodimeric human gonadotropins, luteinizing hormone (LH), chorionic gonadotropin (CG), and follicle-stimulating hormone (FSH). LH and CG bind to a common receptor, LHR, and FSH binds to a paralogous receptor. These GPCRs contain a relatively large ectodomain (ECD), responsible for high-affinity ligand binding, and a transmembrane portion, as in the other GPCRs. The ECD contains nine leucine-rich repeats capped by N-terminal and C-terminal cysteine-rich regions. The overall goal of this research is to elucidate the molecular mechanisms by which CG and LH bind to and activate LHR and the latter, in turn, activates Gsα. A combination of molecular modeling and site-directed mutagenesis, coupled with binding and signaling studies in transiently transfected HEK 293 cells expressing wild-type and mutant forms of LHR, has been used to develop and test models for the LHR ECD, the CG-LHR ECD complex, and the structural changes in the transmembrane helices and intracellular loops, particularly loop 2, that accompany receptor activation. In addition, a single-chain CG-LHR complex was designed in which a fusion protein of the two subunits of human CG was linked to full-length LHR. This ligand-receptor complex was shown to be constitutively active in cellular models and in transgenic mice, the latter of which exhibit precocious puberty. From a combination of molecular modeling, site-directed mutagenesis, genetic/protein engineering, and receptor characterization in cellular and animal models, considerable insight is being developed on the mechanisms of normal and aberrant activation of LHR.

KEYWORDS: luteinizing hormone receptor; G-protein-coupled receptor; heptahelical receptor; receptor ectodomain; receptor activation

Address for correspondence: Dr. David Puett, Department of Biochemistry & Molecular Biology, B129 Fred C. Davison Life Sciences Complex, University of Georgia, 120 Green St., Athens, GA 30602-7229, USA. Voice: 706-542-1676; fax: 706-542-0182.
puett@bmb.uga.edu

Ann. N.Y. Acad. Sci. 1061: 41–54 (2005).
doi: 10.1196/annals.1336.006

INTRODUCTION

The largest gene family in the human genome is that of the G-protein–coupled receptors (GPCRs), with over 800 identified genes that can be grouped into five main families.[1] These cell surface-associated heptahelical proteins mediate the action of a variety of extracellular signals into intracellular second messengers, thus regulating numerous physiological processes. The three genes encoding the glycoprotein hormone receptors, along with several other GPCRs, form a cluster within the rhodopsin gene family,[1] the proteins of which contain several conserved structural features and bind four glycoprotein hormones. The luteinizing hormone receptor (LHR) binds pituitary-derived LH and placental-derived chorionic gonadotropin (CG); the follicle-stimulating hormone and thyroid-stimulating hormone receptors (FSHR and TSHR) bind their respective pituitary hormones, FSH and TSH.[2–5]

The human (h) LHR gene, some 70 kbp in length and located on chromosome 2p21, contains 11 exons and 10 introns.[2–4] Exons 1–10 and a portion of exon 11-encode a relatively large ectodomain (ECD), responsible for high-affinity, specific ligand binding that contains nine leucine-rich repeats (LRRs), each about 25 amino acid residues in length, and is capped by N-terminal and C-terminal cysteine-rich regions.[2–5] The ECDs of FSHR and TSHR are similar, although that of TSHR contains an additional 50 amino acid residues in the C-terminal region not present in LHR and FSHR.[5] The three glycoprotein hormone receptors constitute a subclass of the LRR-containing family of human GPCRs whose other members bind small peptide and unidentified ligands, suggesting that similar LRR structural motifs may recognize considerably different types of ligands. The remainder of exon 11 in LHR encodes protein with the standard features of all GPCRs, including seven transmembrane helices, three extracellular loops, three intracellular loops, and an intracellular C-terminal tail. In addition, a palmitoylation site(s), Cys-Cys, in the C-terminal tail of hLHR leads to a fourth intracellular loop.[6,7]

The four hormones that bind to and activate the three glycoprotein hormone receptors are heterodimers, containing a common α-subunit and a hormone-specific β-subunit.[8] Crystallographic structures are available for aglycoforms and asialoforms of hCG[9–11] and hFSH.[12] The subunits in the heterodimers are highly elongated and intertwined; moreover, a portion of the β-subunit wraps around the α-subunit, forming a "seat-belt" that is stabilized by formation of a disulfide bond in the β-subunit. Overall, the structures of hCG and hFSH are quite similar, although subtle differences in the localized conformations of the portions of α and β forming the seat-belt region may be important in cognate receptor binding.

The only crystallographic structure available for a GPCR is that of rhodopsin.[13] Thus, this structure has served as the template for homology modeling of many of the other GPCRs. A crystallographic structure of an hFSH-hFSHR ECD complex was recently reported,[14] providing considerable insight into the nature of the sites on hormone and receptor that comprise the large contact surface area.

This review focuses on structure-function relationships of LHR in cellular and animal systems. Emphasis is given to the LHR ECD, the mechanism of mutation-induced receptor activation at the transmembrane level, and the coupling of LHR to Gs. Lastly, the results of studies based on protein engineering and engineered mutagenesis of LHR to give constitution activation in cellular models and in transgenic mice are also reviewed.

MOLECULAR MODELING OF THE LHR ECD

This laboratory[15] and others generated models for the LRR region of the LHR ECD[16–18] using the structure of the ribonuclease inhibitor[19] as a template. LRRs ranging from 9–14 were proposed by various groups, but the original suggestions of 9 now appear most accurate. Recent studies in this laboratory have utilized homology modeling based on the threading program PROSPECT[20] and the SCOP40 database.[21] Before threading, however, the sequence was subjected to PSI-blast to identify the conserved residues in various species, and then the sequence profile was included in the threading.

Two proteins, the Nogo receptor (NogoR)[22] and glycoprotein Ib-α,[23,24] yielded significant z-scores. These proteins, like the LHR ECD, are characterized by the LxxLxLxxNxxLPxxaFxxLxx LRR motif, where L denotes Leu, Ile, or Val (generally), N represents Asn (and sometimes Ser on Thr), P is Pro, "a" refers to a small side chain, and F is Phe. In addition, NogoR and glycoprotein Ib-α, like the LHR ECD, contain NCR and CCR flanking regions. NogoR and glycoprotein Ib-α contain 8 and 7 LRRs, respectively, and the sequence of the last two LRRs of NogoR were used to model the ninth LRR of the LHR ECD.

The MODELLER program[25] was used to generate the structure shown in FIGURE 1 (drawn with Molscript[26]) for the LHR ECD. This improved model is less curved than the original model[15] and is more complete in containing proposed conformations for both the NCR and the CCR. The former is predicted to contain two antiparallel β strands, and the latter contains α helices and a parallel β strand aligned to the ninth β strand in the LRR.

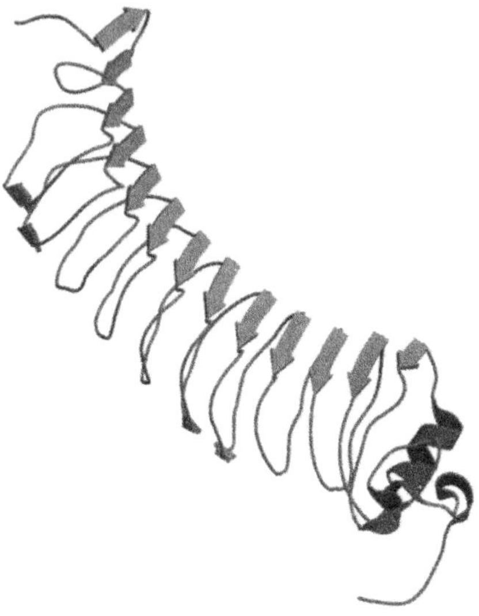

FIGURE 1. Ribbon representation of the structural model of the LHR ECD drawn with Molscript.[26]

C

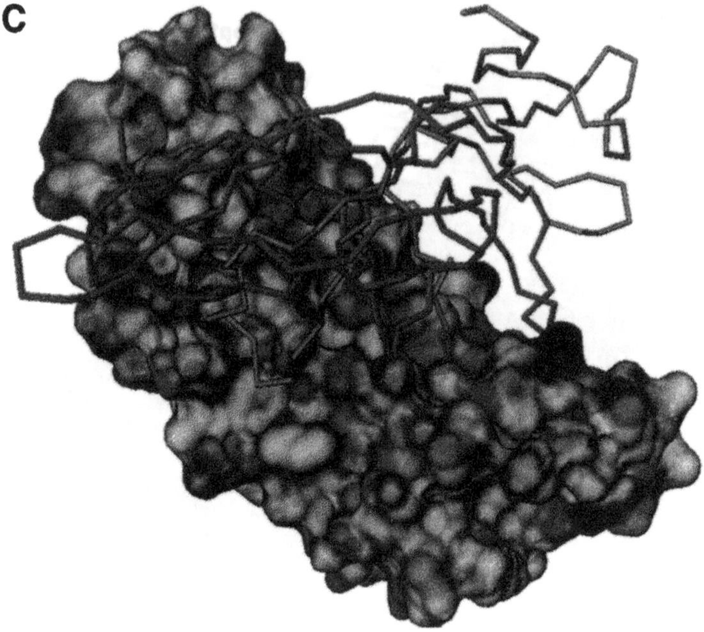

FIGURE 3 — *continued.*

el with the crystal structure of the FSH-FSHR ECD complex[14] (FIG. 3C) showed good similarity for the overall orientation and binding surface between FSH and FSHR ECD. However, the difference at high resolution is apparent, indicating that caution must be exercised in modeling complex structures such as this. Further biological experiments will be helpful to improve the hCG-LHR ECD complex model and elucidate the mechanism of hCG stimulation on LHR.

PROPOSED MECHANISM OF LHR ACTIVATION

The detailed molecular mechanisms by which GPCRs undergo transmembrane helix reorientation in the process of receptor activation, both ligand-mediated and mutation-induced, are largely unknown, but notable advances have been made in recent years with LHR. Both *ab initio* and homology modeling of the LHR transmembrane region have provided considerable insight into the nature of the changes accompanying mutation-induced activation.[2–4, 28–30]

Using the crystal structure of rhodopsin[13] as a template, molecular dynamics simulations were conducted on all characterized naturally occurring and engineered gain-of-function LHR mutants. The results have provided novel suggestions into some of the key distinctions between the inactive and active form(s) of the receptor. For example, a network of H-bonds between highly conserved amino acid residues

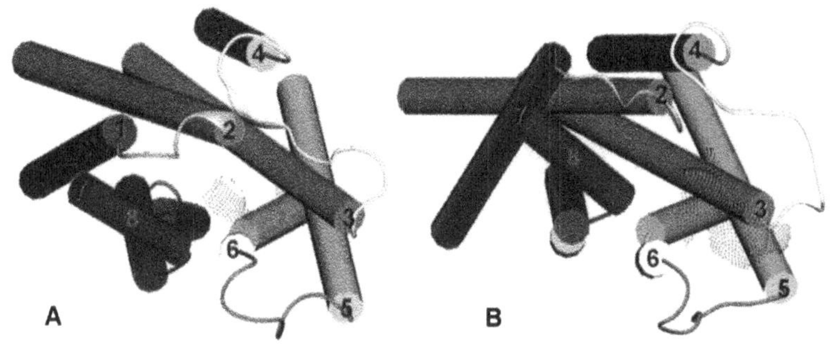

FIGURE 4. Average minimized structures of wild-type LHR (**A**) and its 6D556G constitutively active mutant (**B**). The structures are shown from the intracellular side in a direction perpendicular to the membrane surface. The seven helices are represented by cylinders, whereas the three intracellular loops are represented by thin ribbons. The extracellular loops are not shown. Helices 1, 2, 3, 4, 5, 6, and 7 are respectively colored in blue, orange, green, pink, yellow, light blue, and violet; the intracellular loops, 1, 2, and 3, are, respectively, colored in light green, white, and purple. Helix 8 of the C-terminal tail is colored in violet. The composite SASs for wild-type and mutant LHR, computed over the amino acids 3R442, 3T445, 3I446, and 6K561, are also shown, represented by gray dots. The SAS index is 28 A^2 for the WT and 110 A^2 for the 6D556G active mutant.[27]

in the cytosolic halves of several transmembrane helices, notably 1, 2, 3, 6, and 7, and charge-reinforced H-bonding interactions between the cytosolic extensions of transmembrane helices 3 and 6 serve to stabilize LHR in the inactive state. Specific interhelical H-bonding interactions include (helix number-amino acid residue-residue number): IN355 with 2D383 and 7N597, 2D383 with 7S594, 3S431 with 7N593, and 6D556 with 7N593; in addition, important charge reinforced H-bonding in the cytosolic extensions of transmembrane helices 3 and 6 include: 3R442 with the adjacent 3E441 [both in the (E/D)R(Y/W) motif] and 6D542 (FIG. 4).

Detailed analysis of all known LHR gain-of-function mutants arising from mutations in the transmembrane region led to the identification of a single parameter, the solvent-accessible surface area (SAS) computed over several amino acid residues in the cytosolic extensions of transmembrane helices 3 and 6: 3R442, 3T445, 3I446, and 6K541. SAS values <50 A^2 correspond to nonconstitutively active mutants, whereas values >50 A^2 correlate with constitutive activation.

This threshold of about 50 A^2 for SAS distinguishes a "closed" from an "open" form of the receptor, the latter presumably being the conformation that binds and activates Gsα. This distinction between nonconstitutively active and constitutively active forms of LHR has been found to hold for all single, double, and triple mutants reported, regardless of where, in the transmembrane region, the mutation(s) occurs. In addition to the cooperative network of polar amino acid residues maintaining the inactive state of the receptor, there are numerous hydrophobic interactions that provide interhelical stability. Current research is now directed towards identifying the new interactions that occur upon receptor activation and, importantly, towards delineating similarities and differences in the mode of receptor activation by ligand and by mutation(s).

G-PROTEIN-COUPLING TO ACTIVATED LHR

Following activation by hormone or by mutation, LHR interacts with a heterotrimeric G-protein ($\alpha\gamma\beta$), generally Gs, that leads to increased intracellular biosynthesis of cAMP.[4] Although the overall scheme of activated receptor binding Gsα and effecting a GDP-GTP exchange on Gsα, followed by dissociation of α from the GPCR and from $\beta\gamma$, thus allowing GTP-Gsα to bind to and activate adenylyl cyclase is reasonably well understood,[31] the molecular interactions and mechanisms associated with LHR remain to be elucidated. There is ample evidence that intracellular loops 2 and 3 are involved in this process,[32–37] and recent research from our laboratory based on Ala-scanning mutagenesis has identified several amino acid residues in the two loops that appear critical for signaling.

These studies were conducted by expressing wild-type and mutant forms of LHR in transiently transfected HEK 293 cells and measuring receptor expression levels (B_{max}), hormone affinities (IC_{50} or K_d), and signaling parameters (basal cAMP, EC_{50}, and R_{max} [maximal cAMP in response to saturating hCG]) associated with each. Expression of the results as the coupling efficiency of each mutant, Q, normalized to that of wild-type LHR, provides insight into specific amino acid residues in the two intracellular loops that are involved with Gsα binding and activation. The parameter Q is given by the following equation, [38] valid for nonconstitutively active mutants.

$$Q = (1/2)\ [1 + (K_d/EC_{50})]/(R_{max}/B_{max}).$$

Because these Ala mutants in loops 2 and 3 bind hCG with comparable affinities to those of wild-type LHR, the ratio, Qr = Qm/Qwt, that is, the coupling efficiency of a given LHR mutant relative to that of wild-type LHR, indicates signaling comparable to (Qr = 1.0), greater than (Qr >1.0), or less than (Qr <1.0) that of wild-type LHR. Analysis of each non-Ala residue in loop 2 and many of the non-Ala residues in loop 3 has enabled us to map key receptor side chains that are important in Gsα binding and/or activation.

Only two amino acid residues were identified that exhibited Qr values >1.0, W465 and K477, both in loop 2, indicating that these side chains in wild-type LHR are somewhat inhibitory to ligand-activated receptor productively coupling to G-protein. Amino acid residues in loop 2 exhibiting Qr values <0.5, that is, indicative of reduced LHR binding/activation of Gsα, are: I460, T461, R464, I468, I472, and L478 (FIG. 5); several other residues exhibited intermediate Qr values, albeit significantly different from those of wild-type LHR: H466, T467, and Y470. Of the amino acid residues characterized in loop 3, none was identified with Qr <0.5. Several residues, however, exhibited Qr values significantly different from those of wild-type LHR but between 0.5 and 1.0: I549, Y550, E557, and D564.

Thus, relative to loop 2, amino acid residues in loop 3 appear to be much less involved in G-protein–coupling. Current research is now directed towards delineating the detailed interactions of activated LHR with Gsα, with initial emphasis on the six amino acid residues identified in loop 2 that influence the coupling efficiency. Interestingly, the predictions of the importance of the cytosolic extension of transmembrane helices 3 and 6 in leading to mutation-induced receptor activation (see above) are confirmed by these experimental data based on hCG-induced receptor activation.

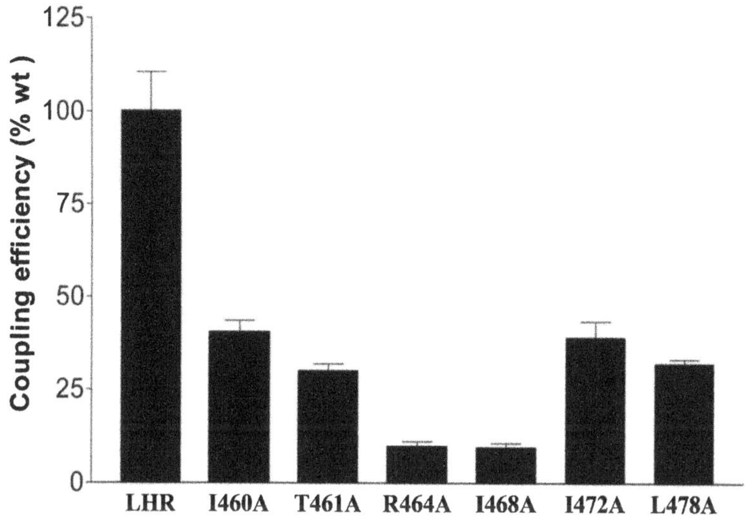

FIGURE 5. Coupling efficiency (Q) of individual amino acid residues in intracellular loop 2 that appear important in productive Gsα coupling. Q is defined in the text.

EXPERIMENTAL MODELS FOR CONSTITUTIVE ACTIVATION OF LHR

Like other members of the GPCRs, LHR can be constitutively activated by single base changes in its gene that result in replacement of specific amino acid residues. The molecular basis underlying gain-of-function mutations was presented in a previous section; this section discusses cell and animal models available to study constitutive LHR activation. A number of naturally occurring gain-of-function mutations have been identified in LHR and are responsible for certain forms of familial and sporadic precocious puberty.[39] The most common locus for such activating mutations is at position D578 in helix 6, but many other sites have also been mapped.[39]

FIGURE 6 shows increased basal levels of cAMP and inositol phosphates (IPs) in stably transfected HEK 293 cells expressing two activating mutations of LHR, D578H and D578Y. Although both mutants bind hCG with comparable affinity to that of wild-type LHR, the latter is responsive to hCG with increased signaling, whereas the former is not.

Using protein engineering, we have developed a gonadotropin-mediated constitutively active LHR. For this, fusion proteins of hCG in two orientations, N-β–α-C (YHR) and N-α-β-C (YHR3), were joined to full-length LHR with a Factor Xa cleavage site and the hCGβ peptide serving as a linker (FIG. 7).[40,41] When expressed in HEK 293 cells, these yoked hormone-receptor complexes resulted in constitutive activation (FIG. 8), and the receptor, not surprisingly, is unresponsive to exogenous hCG.

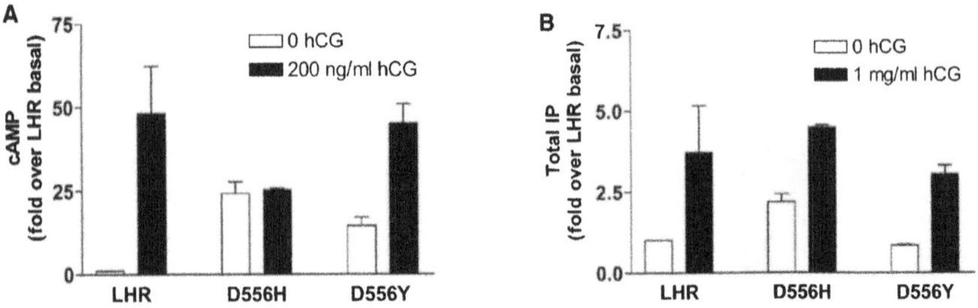

FIGURE 6. Mutation-induced constitutive activation of two LHR mutants, D578H and D578Y, showing increases in basal cAMP (**A**) and total IP (**B**) in HEK 293 cells expressing wild-type and mutant LHRs. The results are presented as fold-increase over wild-type LHR under basal conditions,that is, no added hCG.

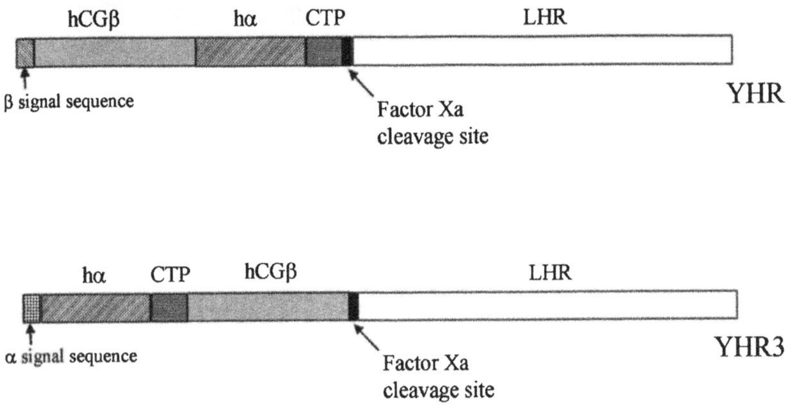

FIGURE 7. Schematics of single-chain hCG in two orientations fused to full-length LHR with a Factor Xa cleavage site and the hCGβ CTP as linker.

Transgenic mice were generated to study the effects of one of the YHRs, namely, the N-β-α-(spacer)-LHR-C form.[42] Using a 6-kb mouse inhibin α-subunit promoter and the bovine growth hormone polyadenylation signal, the transgene was microinjected into (C57BL/6xSJL) fertilized eggs, and founder mice were bred to C57BL/6 mice. RT-PCR on total RNA from several tissues demonstrated robust gonadal expression as well as expression in brain and low levels in heart and the adrenals.[42] Results of testicular levels of testosterone and serum concentrations of LH and FSH are summarized in TABLE 1 for transgenic (YHR^+) and wild-type control male mice at 3, 5, 8, and 12 weeks of age. Testosterone levels are significantly higher than control levels in mice at 3 and 5 weeks of age, and, not surprisingly, serum LH and FSH levels are lower. These differences between transgenic and wild-type controls are not observed at 8 and 12 weeks of age, with the exception of serum FSH, which remains lower in the transgenic animals. Alterations in the hormonal profiles are manifested in increased seminal vesicle (SV) weights at 3 and 5 weeks of age and in reduced

TABLE 1. Hormonal profiles, SV weights, and testicular weights of wild-type (WT) and transgenic male mice expressing YHR (YHR$^+$)

Age (wk)	Genotype	Testicular testosterone (ng/g testis)	Serum LH (ng/ml)	Serum FSH (ng/ml)	SV weight (mg)	Testis weight (mg)
3	WT	1.5 ± 0.4	0.50 ± 0.18	19.1 ± 2.0	3.1 ± 0.4	54.2 ± 4.1
	YHR$^+$	18.0 ± 2.9^b	0.1 ± 0^{b,c}	1.0 ± 0.1^b	5.8 ± 0.6*	35.1 ± 1.7^b
5	WT	4.8 ± 0.9	0.41 ± 0.22	33.5 ± 3.3	16.4 ± 3.0	120.0 ± 7.9
	YHR$^+$	69.9 ± 5.1^b	0.13 ± 0.02^b	2.8 ± 0.2^b	31.1 ± 5.1^b	65.2 ± 2.2^b
8	WT	100.8 ± 49.1	0.26 ± 0.03	38.7 ± 2.3	113.3 ± 5.3	177.9 ± 4.0
	YHR$^+$	72.6 ± 11.9	0.35 ± 0.05	16.6 ± 3.8^b	99.0 ± 10.8	106.7 ± 4.0^b
12	WT	78.1 ± 50.5	0.91 ± 0.50	42.6 ± 3.1	191.6 ± 10.2	200.1 ± 2.4
	YHR$^+$	102.3 ± 47.5	0.26 ± 0.03	19.6 ± 1.5^b	143.5 ± 5.6^b	156.1 ± 3.7^b

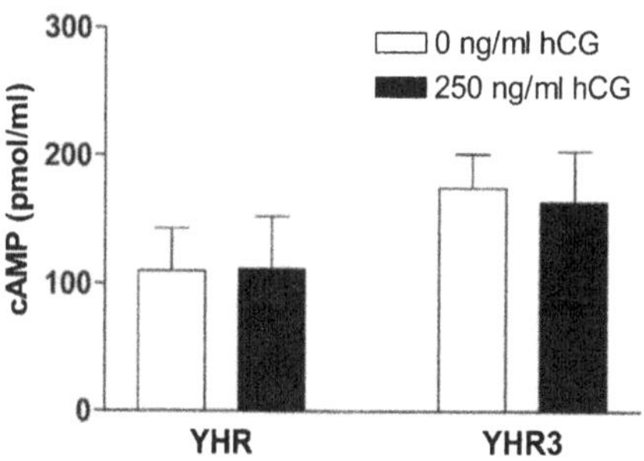

FIGURE 8. Constitutive activation of YHR and YHR3 expressed in HEK 293 cells. Values are given for cAMP under basal conditions, that is, with no exogenous hCG, and in the presence of a saturating concentration of hCG.

testicular weights (TABLE 1). In 8-week-old mice, SV weights are the same between transgenic and control animals, whereas at 12 weeks of age SV weight is less in transgenic than in control mice. The weights of the testes remain depressed in the transgenic animals at all ages investigated, this being attributed to a reduction in tubule size.[42]

This is the first animal model for constitutive LHR activation. One difference between this mouse model and young boys harboring activating LHR mutations is that precocious puberty leads to Leydig cell hyperplasia in boys but not in male mice. It will be of interest to determine if mouse models of naturally activating mutations produce similar phenotypes to those of humans or mice expressing YHR. Of interest in this context is the suggestion that constitutively active germ-line mutations of LHR, resulting in Leydig cell hyperplasia, result in constitutive activation of the Gs pathway, whereas somatic mutations, found in Leydig cell neoplasias, appear to constitutively activate both Gs and Gq pathways. Such differential activation of the G-proteins, however, was not found to be the case with mouse Leydig tumor cells.[44]

CONCLUSIONS

Using a combination of computational and experimental approaches, considerable insight is being developed into the molecular mechanisms by which hCG binds to LHR, leading to receptor activation and subsequent intracellular signaling. There are now available reasonable working models for: (1) the structure of the LHR ECD and its mode of binding hCG, and (2) the molecular changes in the transmembrane portion of LHR accompanying mutation-induced activation. Major problems to be addressed include: (1) the mechanism by which the hormone-ECD complex interacts with and activates the transmembrane portion of LHR, (2) elucidation of the similarities and differences between ligand-induced and mutation-induced receptor activation, and (3) the mode of activated LHR binding to and activating Gsα.

ACKNOWLEDGMENT

This work was supported in part by National Institutes of Health Grants DK33973 and DK69711.

REFERENCES

1. FREDRIKSSON, R., M.C. LAGERSTROM, L.-G. LUNDIN & H.B. SCHIOTH. 2003. The G-protein-coupled receptors in the human genome form five main families. Phylogenetic analysis, paralogon groups, and fingerprints. Mol. Pharmacol. **63:** 1256–1272.
2. FANELLI, F., A.P.N. THEMMEN & D. PUETT. 2001. Lutropin receptor function: insights from natural, engineered, and computer-simulated mutations IUBMB Life **51:** 149–155.
3. FANELLI, F. & D. PUETT. 2002. Structural aspects of luteinizing hormone receptor. Endocrine **18:** 285–293.
4. ASCOLI, M., F. FANELLI & D.L. SEGALOFF. 2002. The lutropin/choriogonadotropin receptor, a 2002 perspective. Endocrine Rev. **23:** 141–174.
5. VASSART, G., L. PARDO & S. COSTAGLIOLA. 2004. A molecular dissection of the glycoprotein hormone receptors. Trends Biochem. Sci. **29:** 119–126.
6. KAWATE, N. & K.J.M. MENON. 1994. Palmitoylation of luteinizing hormone/human choriogonadotropin receptors in transfected cells. Abolition of palmitoylation by mutation of Cys-621 and Cys-622 residues in the cytoplasmic tail increases ligand-induced internalization of the receptor. J. Biol. Chem. **269:** 30651–30658.
7. ZHU, H., H. WANG & M. ASCOLI. 1995. The lutropin/choriogonadotropin receptor is palmitoylated at intracellular cysteine residues. Mol. Endocrinol. **9:** 141–150.
8. HEARN, M.T.W. & P.T. GOMME. 2000. Molecular architecture and biorecognition processes of the cystine knot protein superfamily. Part I. The glycoprotein hormones. J. Mol. Recog. **13:** 223–278.
9. LAPTHORN, A.J., D.C. HARRIS, A. LITTLEJOHN, *et al.* 1994. Crystal structure of human chorionic gonadotropin. Nature **369:** 455–461.
10. WU, H., J.W. LUSTBADER, Y. LIU, *et al.* 1994. Structure of human chorionic gonadotropin at 2.6A resolution from MAD analysis of the selenomethionyl protein. Structure **2:** 545–558.
11. TEGONI, M., S. SPINELLI, M. VERHOEYEN, *et al.* 1999. Crystal structure of a ternary complex between human chorionic gonadotropin (hCG) and two Fv fragments specific for the α and β-subunits. J. Mol. Biol. **289:** 1375–1385.
12. FOX, K.M., J.A. DIAS & P. VAN ROEY. 2001. Three-dimensional structure of human follicle stimulating hormone. Mol. Endocrinol. **15:** 378–389.
13. PALCZEWSKI, K., T. KUMASALEA, T. HORI, *et al.* 2000. Crystal structure of rhodopsin: a G protein-coupled receptor. Science **289:** 739–745.

14. FAN, Q.R. & W.A. HENDRICKSON. 2005. Structure of human follicle-stimulating hormone in complex with its receptor. Nature **433:** 269–277.
15. BHOWMICK, N., J. HUANG, D. PUETT, *et al.* 1996. Determination of residues important in hormone binding to the extracellular domain of the luteinizing hormone/chorionic gonadotropin receptor by site-directed mutagenesis and modeling. Mol. Endocrinol. **10:** 1147–1159.
16. MOYLE, W.R., R.K. CAMPBELL, S.N. VENKATESWARA, *et al.* 1995. Model of human chorionic gonadotropin and lutropin receptor that explains signal transduction of the glycoprotein hormones. J. Biol. Chem. **270:** 20020–20031.
17. KAJAVA, A.V., G. VASSART & S.J. WODAK. 1995. Modeling of the three-dimensional structure of proteins with the typical leucine-rich repeats. Structure **3:** 867–877.
18. JIANG, X., M. DREANO, D.R. BUCKLER, *et al.* 1995. Structural predictions for the ligand-binding region of glycoprotein hormone receptors and the nature of hormone-receptor interactions. Structure **3:** 1341–1353.
19. KOBE, B. & J. DISENHOFER. 1993. Crystal structure of porcine ribonuclease inhibitor, a protein with leucine-rich repeats. Nature **366:** 751–756.
20. KIM, D., D. XU, J.T. GUO, *et al.* 2003. PROSPECT II: protein structure prediction program for genome-scale applications. Protein Engr. **16:** 641–650.
21. CHANDONIA, J.M., G. HON, N.S. WALKER, *et al.* 2004. The ASTRAL Compendium in 2004. Nucleic Acids Res. **32:** D189–D192.
22. HE, X.L., J.F. BAZAN, G. MCDERMOTT, *et al.* 2003. Structure of the Nogo receptor ectodomain: a recognition module implicated in myelin inhibition. Neuron **38:** 177–185.
23. EMSLEY, J., M. CRUZ, R. HANDIN & R. LIDDINGTON. 1998. Crystal structure of the von Willebrand Factor A1 domain and implications for the binding of platelet glycoprotein Ib. J. Biol. Chem. **273:** 10396–10401.
24. HUIZINGA, E.G., S. TSUJI, R.A.P. ROMIJN, *et al.* 2002. Structures of glycoprotein Ibα and its complex with von Willebrand Factor A1 domain. Science **297:** 1176–1179.
25. FISER, A. & A. SALI. 2003. Modeller: generation and refinement of homology-based protein structure models. Methods Enzymol. **374:** 461–491.
26. KRAULIS, P.J. 1991. MOLSCRIPT: a program to produce both detailed and schematic plots of protein structures. J. Appl. Crystallog. **24:** 946–950.
27. GABB, H.A., R.M. JACKSON & M.J. STERNBERG. 1997. Modelling protein docking using shape complementarity, electrostatics and biochemical information. J. Mol. Biol. **272:** 106–120.
28. FANELLI, F. 2000. Theoretical study on mutation-induced activation of the luteinizing hormone receptor. J. Mol. Biol. **296:** 1333–1351.
29. ANGELOVA, K., F. FANELLI & D. PUETT. 2002. A model for constitutive lutropin receptor activation based on molecular simulation and engineered mutations in transmembrane helices 6 and 7. J. Biol. Chem. **277:** 32202–32213.
30. FANELLI, F., M. VERHOEF-POST, M. TIMMERMAN, *et al.* 2004. Insight into mutation-induced activation of the luteinizing hormone receptor: molecular simulations predict the functional behavior of engineered mutants at M398. Mol. Endocrinol. **18:** 1499–1508.
31. CABRERA-VERA, T.M., J. VANHAUWE, T.O. THOMAS, *et al.* 2003. Insights into G protein structure, function, and regulation. Endocrine Rev. **24:** 765–781.
32. BLIN, N., J. YUN & J. WESS. 1995. Mapping of single amino acid residues required for selective activation of $G_{q/11}$ by the m3 muscarinic acetylcholine receptor. J. Biol. Chem. **270:** 17741–17748.
33. GREASLEY, P.J., A.F. FANELLI, A. SCHEER, *et al.* 2001. Mutational and computational analysis of the α_{1b}-adrenergic receptor. J. Biol. Chem. **276:** 46485–46494.
34. FERNANDEZ, L.M. & D. PUETT. 1997. Evidence for an important functional role of intracellular loop II of the lutropin receptor. Mol. Cell. Endocrinol. **128:** 161–169.
35. SCHULZ, A., T. SCHONENBERG, R. PASCHKE, *et al.* 1999. Role of the third intracellular loop for the activation of gonadotropin receptors. Mol. Endocrinol. **13:** 181–190.
36. CHAZENBALK, G.D., Y. NAGAYAMA, D. RUSSO, *et al.* 1990. Functional analysis of the cytoplasmic domains of the human thyrotropin receptor by site-directed mutagenesis. J. Biol. Chem. **265:** 20970–20975.

37. TIMOSSI, C., D. MALDONADO, A. VIZCAINO, *et al.* 2002. Structural determinants in the second intracellular loop of the human follicle-stimulating hormone receptor are involved in G_s protein activation. Mol. Cell. Endocrinol. **189:** 157–168.
38. BALLESTEROS, J., S. KITANOVIC, F. GUARNIERI, *et al.* 1998. Microdomains in G-protein-coupled receptors. J. Biol. Chem. **273:** 10445–10453.
39. THEMMEN, A.P.N. & I.T. HUHTANIEMI. 2000. Mutations of gonadotropins and gonadotropin receptors: elucidating the physiology and pathophysiology of pituitary-gonadal function. Endocrine Rev. **21:** 551–583.
40. WU, C., P. NARAYAN & D. PUETT. 1996. Protein engineering of a novel constitutively active hormone-receptor complex. J. Biol. Chem. **271:** 31638–31642.
41. NARAYAN, P., J. GRAY & D. PUETT. 2000. A biologically active single chain human chorionic gonadotropin analog with altered receptor binding properties. Endocrinology **141:** 67–71.
42. MEEHAN, T.P., B.G. HARMON, M.E. OVERCAST, *et al.* 2005. Gonadal defects and hormonal alterations in transgenic mice expressing a single chain human chorionic gonadotropin-lutropin receptor complex. J. Mol. Endocrinol. **34:** 489–503.
43. SHENKER, A. 2002. Activating mutations of the lutropin choriogonadotropin receptor in precocious puberty. Rec. Channels **8:** 3–18.
44. HIRAKAWA, T. & M. ASCOLI. 2003. A constitutively active somatic mutation of the human lutropin receptor found in Leydig cell tumors activates the same families of G proteins as germ line mutations associated with Leydig cell hyperplasia. Endocrinology **144:** 3872–3878.

in adrenal and gonadal development.[3] Absence of adrenal steroids in SF-1 null mice resulted in their death shortly after birth, and the lack of gonadal steroids during embryonic development resulted in male-to-female sex reversal. The mice were also characterized by the absence of a ventromedial nucleus of the hypothalamus and impaired expression of genes within gonadotropes of the pituitary.

SF-1 expression sites correlate with their steroidogenic function. SF-1 is expressed by all primary steroidogenic cells, such as cortical cells of the adrenal and Leydig and theca cells of the gonads, and by tissues such as the ventromedial nucleus of the hypothalamus and gonadotropes of the pituitary.[4] Its expression is also age dependent and demonstrates developmental regulation.[5] Onset of SF-1 expression occurs at e9.5 within the cells of the urogenital ridge that gives rise to gonads and adrenal glands. After sex determination is initiated at e12.5, SF-1 expression is downregulated in the ovary and persists within the testes, where it is thought to contribute to sexual differentiation in the male.

However, because the gonads fail to develop past e10 in knockout mice, it has been difficult to evaluate SF-1's role in the later events of testes development and function. In order to make a correlation between physiological processes and molecular pathways regulating SF-1 function, we developed a transgenic approach that employs a yeast artificial chromosome (YAC) containing the *Ftz-F1* together with a significant amount of the 5′ and 3′ flanking sequence. Several lines of transgenic mice, varying in transgene size, were generated and characterized for expression profiles. Subsequently, selected transgenes were introduced into the SF-1 null mouse background to obtain transgene positive (tg/$^{+}$), SF-1 negative (SF-1$^{-/-}$) rescue mice (SF-1$^{-/-}$;tg/$^{+}$). Characterization of the rescue mice revealed differences in SF-1 expression profiles and rescue abilities between different lines. The studies established a region encompassing *Ftz-F1* that is required for proper expression of SF-1 and revealed a role for SF-1 in Leydig cell differentiation.

METHODS

Rat Ftz-F1 Containing YAC. DNA pools of a rat YAC library constructed by the Whitehead Institute for Biomedical Research and MIT Center for Genome Research were screened for the presence of rat *Ftz-F1*. One of the identified *Ftz-F1*-containing YACs, clone 385G2, was purchased from Invitrogen Life Technology. YAC clone ySLH1 was generated by subcloning 500-kb Ftz-F1 containing NotI fragment, as described earlier.[6]

Generation and Screening of Transgenic Mice. YAC DNA was purified for pronuclear microinjection as reported.[7] Isolated DNA, at a concentration of 1–2 ng/L, was used to generate transgenic mice (C57B6/SJL mouse strain) through the Transgenic and Gene-Targeting Institutional Facility at the University of Kansas Medical Center. Three separated YAC transgenic mouse lines, 7, 14, and 15, were produced. Animals were genotyped for the presence of the YAC transgene by PCR of tail DNA, as described.[6] All animal manipulations were conducted according to the Guide for Care and Use of Experimental Animals.

YAC Rescue of the SF1 Knockout Mice. Heterozygous breeder pairs containing a targeted mutation that disrupts the function of SF-1 (Nr5a1tm1Klp, formerly Ftzf1tm1Klp)[3] were purchased from The Jackson Laboratory (Bar Harbor, Maine).

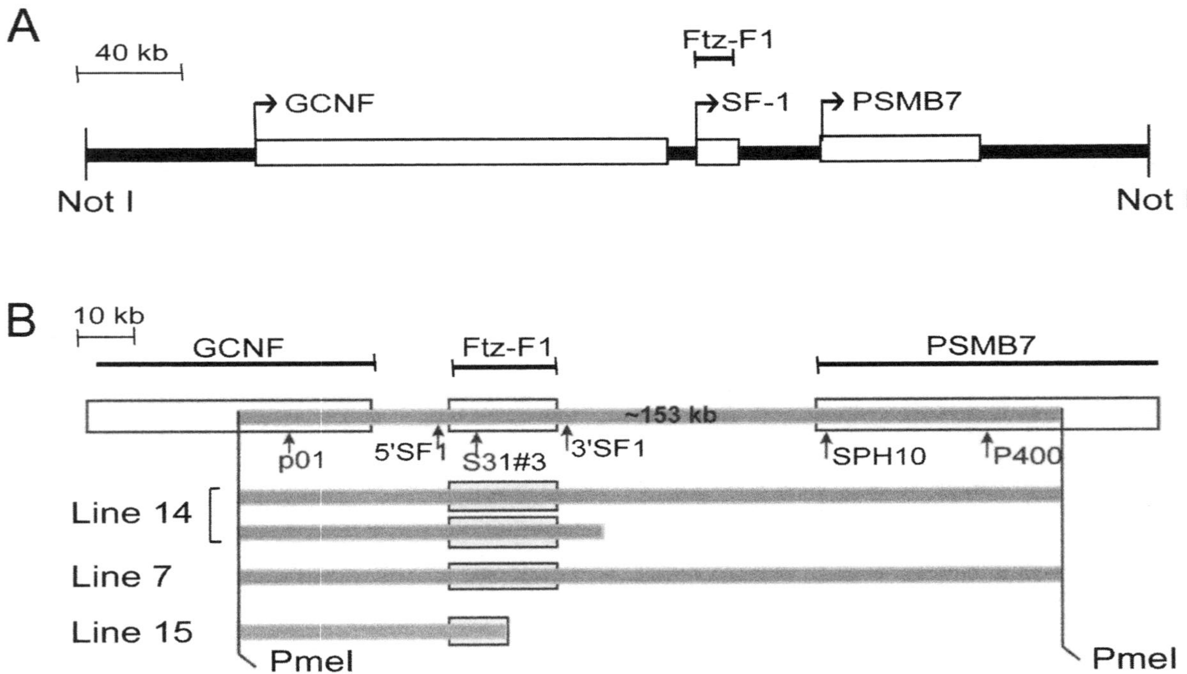

FIGURE 1. Structural characterization of the *Ftz-F1*-containing YAC, ySLH1. (**A**) Schematic depiction of ySLH1. The gene encoding SF-1, *Ftz-F1*, is centrally located and flanked by genes for proteosome protein PSMB7 and the orphan nuclear receptor GCNF. Genes are represented by rectangles and the NotI restriction endonuclease used in cloning ySHL1 noted. (**B**) Schematic representation of the YAC transgenes mapped within transgenic lines 7, 14, and 15. Positions of genes (*white bars*), Pme I restriction sites, and hybridization probes used in mapping analysis (p01, 5′SF1, S31#3, 3′SF1, SPH10, P400) are indicated on *top*. *Gray bars* denote the sequences integrated into each line, with the *top one* marking the ~153-kb fragment established as sufficient for SF-1 expression.

Transgenic Rescue of SF-1-Null Mice

TATIANA KARPOVA, R.R.M. MARAN, JEREMY PRESLEY, SERGE P. SCHERRER, LOVELLA TEJADA, AND LESLIE L. HECKERT

Departments of Molecular and Integrative Physiology, University of Kansas Medical Center, Kansas City, Kansas 66160, USA

ABSTRACT: Steroidogenic factor 1 (SF-1, Nr5a1, and Ad4bp) is an orphan nuclear receptor required for adrenal and gonad development and endocrine regulation. To extend our understanding of SF-1 function and the mechanisms controlling its expression, a transgenic rescue strategy was employed to locate important transcriptional control regions and to reveal functional roles of the protein. A rat yeast artificial chromosome containing *Ftz-F1*, the gene encoding SF-1, was used to generate mice with different transgenes that varied in size. Rat SF-1 mRNA expression was assayed to assess each transgene's targeting ability. SF-1-deficient/transgene-positive (SF-1$^{-/-}$; tg/$^{+}$) "rescue" mice were then generated and the animals' developmental and reproductive status was evaluated. The results identified differences in expression patterns and rescue abilities that provided insight into SF-1 transcriptional control and function. Comparing transgene maps and mRNA profiles placed critical transcriptional elements for pituitary and hypothalamic expression to a region 3' to intron 4, whereas examination of rescued mice revealed that an ~153-kb region of the Ftz-F1 locus recapitulates most or all activity ascribed to the endogenous allele. A second line of rescued mice was hypomorphic, with males showing defects in androgen-dependent tissues due to abnormal Leydig cell differentiation. Histological analysis of embryonic (e14.5) and adult testes from these mice implicated SF-1 in roles that are distinct in fetal and adult Leydig cells.

KEYWORDS: SF-1; testes; adrenal; YAC transgenic mice; steroidogenesis

INTRODUCTION

SF-1 is an orphan nuclear receptor encoded by the *Ftz-F1*, which is evolutionarily conserved among vertebrates and invertebrates. Structural analysis of the mouse *Ftz-F1* gene indicated four distinct transcripts, ELP1, ELP2, ELP3, and SF-1, which differ by their inclusion of alternative exons and use of different promoters.[1] The most investigated transcript in humans and rodents is the SF-1, which regulates endocrine homeostasis via its role as a transcription factor for genes encoding various steroid-producing enzymes ($P450_{SCC}$, P450arom, StAR, and αGSU) as well as steroidogenesis-regulating hormones (LHβ, GnRH, and inhibin α) and their cognate receptors (GnRHR and FSHR).[2] An essential role for SF1 in embryonic development was uncovered after generations of SF-1 knockout mice showed deficiencies

Address for correspondence: Leslie. L. Heckert, Department of Molecular and Integrative Physiology, University of Kansas Medical Center, 3901 Rainbow Blvd., Kansas City, KS 66160. Voice: 913-588-7488; fax: 913-588-7430.
lheckert@kumc.edu

**Ann. N.Y. Acad. Sci. 1061: 55–64 (2005). © 2005 New York Academy of Sciences.
doi: 10.1196/annals.1336.007**

SF-$1^{+/-}$ mice were mated to transgenic mice to generate SF-1 heterozygous transgene-positive animals (SF-$1^{+/-}$;tg/$^{+}$). These animals were bred to generate rescue mice that were homozygous null for SF-1 and positive for the YAC transgene (SF-$1^{-/-}$; tg/$^{+}$). Rescue animals containing only one YAC transgene were selected for characterization. Animals were genotyped using DNA isolated from either tail biopsy (postnatal) or isolated yolk sac (embryo), as described.[6]

Generation of Liver Plugs from Transgenic Mice and Mapping of Endogenous Transgenes. High molecular weight mouse genomic DNA was isolated from transgenic mice and analyzed by pulse field gel electrophoresis and Southern blot hybridization, as described.[8] Agarose plugs containing high molecular weight DNA were prepared from single-cell liver suspensions at a concentration of 3×10^7 cells/ml and digested with restriction enzyme Pme I. Digested DNA was fractionated by PFGE and analyzed by Southern blot analysis with different probes spanning ySLH1 (*P*.01; 5′SF1; S31#3; 3′SF1; SPH10; P400) (FIG. 1B).

RT-PCR. Transgenic mice and negative littermates were sacrificed and total RNA isolated from various tissues, as described elsewhere.[9] Complementary DNA was generated and used as a template in PCR with primers specific for either rat or mouse SF-1 cDNAs or ribosomal protein L7, as a control for cDNA synthesis. Specifics of the methods are described elsewhere.[10]

Immunohistochemistry. Whole embryos or testes from adult animals were rapidly dissected and fixed in freshly prepared 4% paraformaldehyde solution prepared with phosphate-buffered saline (pH 7.4), followed by dehydration and paraffin embedding, according to standard procedures. Sections 5 μ were immunofluorescently labeled for SF-1, using a 1:2000 dilution of polyclonal rabbit primary antibody (generously provided by K. Morohashi), or for $P450_{SCC}$ (generously provided by M.J. Soares), using 1:200 dilution of rabbit antiserum. TRITC-conjugated goat anti-rabbit secondary antibody (The Jackson Laboratory, Bar Harbor, Maine) was used to visualize SF-1 and $P450_{SCC}$ signals. For SF-1 analysis, hypothalamus, pituitary, adrenal glands, and gonads were also collected and processed as just described.

RESULTS

To evaluate transcriptional regulation of SF-1 *in vivo*, we employed YACs as transgenes and monitored transgene expression in the embryo and postnatal animals. Several *Ftz-F1*-containing rat YACs were identified, and one was chosen for further characterization (ySLH1). Within ySLH1, *Ftz-F1* is centrally located and flanked by genes encoding a second orphan nuclear receptor, germ-cell nuclear factor (Gcnf), and proteasome (prosome, macropain) subunit beta type 7 (Psmb7, FIG. 1A). Three lines of transgenic mice carrying the YAC were generated, and structural characterization of the integrated transgenes was performed prior to functional analysis. Notably, fragmentation of YAC DNA during integration provided several lines with different transgene sizes (FIG. 1B). Lines 7 and 14 contained a fragment of approximately 153 kb, with breakpoints not fully defined but located near the outer regions specified by the two Pme I sites, whereas line 15 contained a transgene that lost a significant amount of *Ftz-F1* encoding sequence terminated at the end of intron 4 (FIG. 1B). Two transgene segments containing a full copy of *Ftz-F1* were integrated in line 14, whereas only one full copy of the gene was found in line 7.

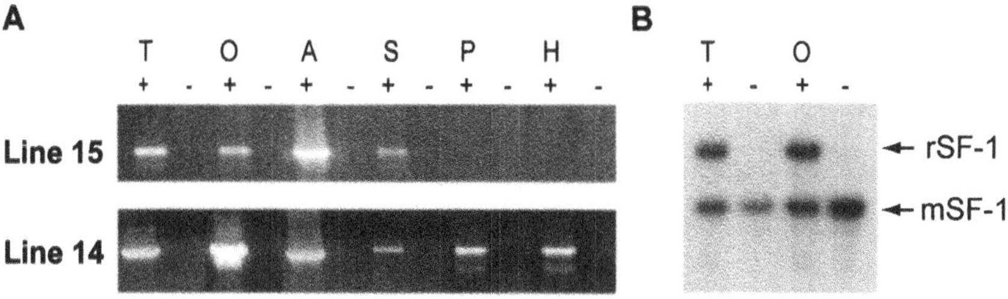

C

Line #	Expressing Tissues
7	Adrenal Ovary Testis Spleen Pituitary VMH
14	Adrenal Ovary Testis Spleen Pituitary VMH
15	Adrenal Ovary Testis Spleen

FIGURE 2. Transgene expression profile for lines 7, 14, and 15. (**A**) RNA isolated from transgene positive (+) and negative (–) tissues was analyzed for rat SF-1 mRNA in lines 14 and 15, by RT-PCR. Tissues are: testis (T), ovary (O), adrenal (A), spleen (S), pituitary (P), and hypothalamus (H). (**B**) Ribonuclease protection analysis from line 14 mice, showing levels of transgenic SF-1 (rSF-1) and endogenous SF-1 (mSF-1) mRNAs in testes and ovaries transgenic mice (+) and their negative (–) littermates. (**C**) Summary table for transgene expression in lines 7, 14, and 15. Tissues positive for transgene expression are indicated in the column next to their respective lines.

For each line the expression profile for SF-1 was compared with that for endogenous SF-1. RT-PCR analysis was performed using rat-specific primers that detect only transgenic, not endogenous, SF-1 transcripts. RT-PCR and RPA analysis in lines 7 and 14 revealed expression of the rat transgene in all SF-1–positive tissues such as testes, ovaries, adrenal, pituitary, hypothalamus, and spleen, and no expression was observed in SF-1–negative tissues (FIG. 2 and data not shown). By contrast, no transgene expression was observed in the VMH or pituitary in line 15 (FIG. 2), suggesting that sequences lost from the transgene's 3′ are needed to fulfill this expression pattern.

Although RT-PCR and RPA assays indicated correct tissue-specific expression of the transgenes in lines 7 and 14, temporal and cellular regulatory profiles and physiological changes of SF-1 could not be accurately measured in these mice. Therefore, to extend characterization of transgene function, lines 7 and 14 transgenes were introduced into a SF-1 null background and rescue mice, which lacked endogenous SF-1 but contained the transgene, were obtained. The line 14 transgene was able to completely rescue the null phenotype, as $SF1^{-/-}$;tg/+ mice were healthy with normally developed adrenal glands and gonads (FIG. 3A and data not shown). Reproductive performance of both sexes among line 14 rescue mice was indistinguishable from that of WT animals, including similar litter size and litter numbers within the reproductive lifetime period (TABLE 1). Further examination of transgene function was performed using immunodetection of transgenic SF-1protein within steroidogenic

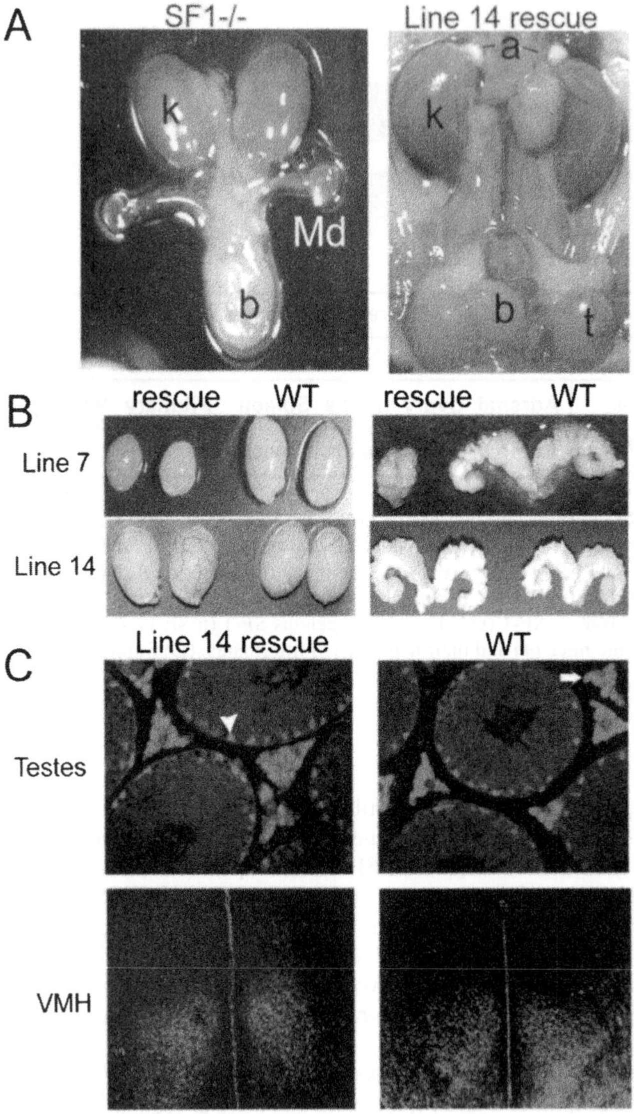

FIGURE 3. SF-1 function in rescue mice. (**A**) Normal development of adrenal glands (a) and testes (t) in line 14 rescue mice (*right*) compared with absence of adrenal glands and gonads and persistence of müllerian duct in SF-1 null mice (*left*). (**B**) Anatomical evaluation of lines 7 and 14 rescued mice revealed differences in the ability of their transgenes to rescue SF-1 function in the null mice. Significantly smaller testes and seminal vesicles size were registered in line 7 rescue mice compared with wild-type (WT) littermates, whereas testes and seminal vesicle weights were equivalent in WT and line 14 rescue animals. (**C**) Immunodetection of transgenic SF-1 in line 14 rescue mice and endogenous SF-1 in WT mice. Similar SF-1 expression was observed in Sertoli (*arrowhead*) and Leydig cells (*arrow*) of rescue and WT testes as well as similar expression pattern indicated in VMH of rescue and WT animals.

TABLE 1. Reproductive performance of rescue mice in line 14

Genotype	Sex	n	Litter number, observed in the group	Average litter size
Rescue	Male	19	25	8.6 ± 1.6
WT	Male	10	12	8.9 ± 2.1
Rescue	Female	22	25	8.2 ± 1.6
WT	Female	12	17	7.6 ± 2.0

tissues in rescue mice lacking endogenous protein production. Comparison of SF-1 expression between rescued and control animals revealed that SF-1 expressed from the transgene follows the same cellular profile as that of the endogenous SF-1 (FIG. 3C and data not shown).[6]

In contrast to line 14, the line 7 transgene was unable to rescue all defects of the characterized SF-1 null mice, demonstrating reduced reproductive success in females and complete infertility in males. Line 7 rescue male mice presented with reduced testis size and hypoplastic or completely undeveloped accessory sexual organs, and some were characterized with feminized external genitalia and cryptorchid testes (FIG. 3B and data not shown). Evaluation of testes at different stages of development, beginning at e14.5, revealed a delay of fetal Leydig cell differentiation. Immunohistochemical analysis of $P450_{SCC}$ at e14.5 in rescued animals showed a significant deficiency in its expression, indicating a lack of functional Leydig cells at this time point (FIG. 4A and B). Interestingly, by postnatal day 0, $P450_{SCC}$ expression was similar to that of the control animals, indicating a delay rather than a block in development of the fetal Leydig cells (FIG. 4C and D). Surprisingly, the Leydig cell population in adult rescue testes was markedly reduced compared to that of control animals, yet $P450_{SCC}$ expression in the remaining Leydig cells appeared normal (FIG. 4E and F). Thus, insufficient expression of SF-1 from the line 7 transgene created a hypomorphic phenotype with defects in both fetal and adult Leydig cells and revealed distinct roles for SF-1 in the establishment of the fetal and adult Leydig cell population.

DISCUSSION

We have demonstrated the successful use of YAC transgenesis to evaluate structural sequence requirements for *SF-1* transcription. Our study established that ~153-kb sequence is sufficient for accurate SF-1 expression in all SF-1–positive sites. Previous transgenic studies devoted to the identification of such sequence requirements were unable to direct transgenic SF-1 expression to all appropriate steroidogenic tissues. One transgenic mouse study, using a 674-bp SF-1 promoter directing a Lac-Z reporter, demonstrated an expression profile limited to the gonad.[11] Another mouse study employed a 50-kb region of the *Ftz-F1* locus and demonstrated correct expression in most, but not all, SF-1–positive tissues,[12] further supporting a role for distal regulatory regions in the multilevel complex profile of SF-1 expression. In an attempt to include all required regulatory sequences, we employed a YAC transgenic system, which easily accommodates large amounts of genomic sequence. The study

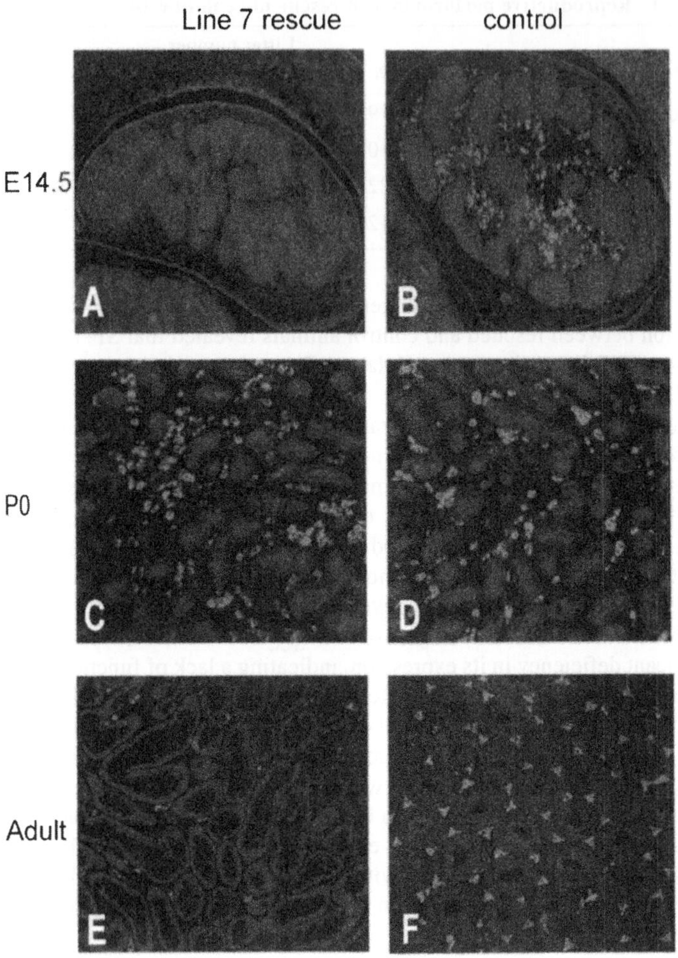

FIGURE 4. $P450_{SCC}$ expression in Leydig cells of line 7 rescue mice. Immunohistochemistry was used to evaluate $P450_{SCC}$ expression in line 7 (**A,C,E**) and control mice (**B,D,F**) at embryonic day 14.5 (**A,B**), postnatal day 0 (**C,D**), and adult ages (**E,F**). $P450_{SCC}$ was used as a marker for the Leydig cell population.

revealed a genomic region that recapitulates SF-1 expression to all appropriate steroidogenic tissues, resolving a limitation of the previous transgenic studies. Interestingly, analysis of line 15 revealed a loss of transgene sequence that correlated to expression deficiencies. Further analysis to identify the sequences lost in line 15 determined a regulatory region responsible for hypothalamic and pituitary expression of SF-1. Comparison of the integrated transgenes in lines 7, 14, and 15 placed these elements in a region downstream of the 3′ end in intron 4.

Transfer of the YAC transgene into the SF-1 null mouse background extended analysis of SF-1 into a more physiological context and opened new possibilities to perform *in vivo* functional analyses that are limited in the knockout mice, as they lack steroidogenic organs and die shortly after birth. Thus, in the transgenic rescue model, the developmental and physiological ramifications of altered SF-1 levels can be evaluated. In the case of line 7 rescue mice, the hypomorphic phenotype allowed for *in vivo* assessment of SF-1 function in Leydig cell development. Notably, the phenotype is due to altered SF-1 activity and not to the effects of additional genes within the YAC, as no phenotypic changes were associated with a second YAC rescue model and neither were they present in transgenic mice in the wild-type background. Examination of line 7 hypomorphic mice allowed evaluation of Leydig cell development, which indicated distinct roles for SF-1 in the establishment of fetal and adult Leydig cell populations. Thus, in the SF-1 hypomorphic mice, fetal Leydig cells were developmentally delayed, whereas adult Leydig cells were either blocked at a required proliferation stage or absent altogether. The different Leydig cell defects suggest several interesting hypotheses regarding SF-1 function in Leydig cells and the relation between fetal and adult populations. Thus, the phenotype may reflect a single Leydig precursor cell that forms both adult and fetal populations, and SF-1's role in the cell's development depends on the animal's developmental stage. Alternatively, the two Leydig populations may derive from distinct precursors in which SF-1 acts differently. Additional studies are ongoing to distinguish these possibilities and further our understanding of SF-1 in Leydig cell development.

Although it is tempting to speculate that the phenotype observed is due to the intrinsic effects of SF-1 in Leydig cells, the potential role of Sertoli cells, which also express SF-1, in this process cannot be overlooked. Thus, Sertoli cells, through production of desert hedgehog, are known to direct development of fetal Leydig cells.[13] Currently, it is not known if a similar Sertoli cell/Leydig cell relationship exists in the development of adult Leydig cells. It is now critical to extend these studies and distinguish between the intrinsic (Leydig) and extrinsic (Sertoli) roles of SF-1 in Leydig cell development. Further evaluation of hypomorphic mice will likely provide the opportunity to do this.

ACKNOWLEDGMENT

This work was supported by National Institutes of Health Grant R01 HD38498 (to L.L.H.) and a Specialized Cooperative Center Program in Reproduction Research (U54 HD 33994 to Paul Terranova) from the National Institute of Child Health and Human Development.

REFERENCES

1. Ninomiya, Y. *et al.* 1995. Genomic organization and isoforms of the mouse ELP gene. J. Biochem. (Tokyo) **118:** 380–389.
2. Parker, K.L. & B.P. Schimmer. 1997. Steroidogenic factor 1. A key determinant of endocrine development and function. Endocr. Rev. **18:** 361–377.
3. Luo. X., Y. Ikeda & K.L. Parker. 1994. A cell-specific nuclear receptor is essential for adrenal and gonadal development and sexual differentiation. Cell **77:** 481–490.
4. Morohashi, K. *et al.* 1994. Functional difference between Ad4BP and ELP, and their distributions in steroidogenic tissues. Mol. Endocrinol. **8:** 643–653.

5. IKEDA, Y., W.H. INGRAHAM & K.L. PARKER. 1994. Developmental expression of mouse steroidogenic factor 1, an essential regulation of the steroid hydroxylases. Mol. Endocrinol. **8:** 654–662.
6. KARPOVA, T. *et al.* 2005. A *Ftz-F1*-containing yeast artificial chromosome recapitulates expression of steroidogenic factor 1 *in vivo.* Mol. Endocrinol. **19:** 2549–2563.
7. PETERSON, K.R. *et al.* 1995. Use of yeast artificial chromosomes (YACs) in studies of mammalian development: production of beta-globin locus YAC mice carrying human globin developmental mutants. Proc. Natl. Acad. Sci. USA **92:** 5655–5659.
8. RIETHMAN, H. *et al.* 1997. Preparation, manipulation, and mapping of HMW DNA. *In* Genome Analysis: A Laboratory Manual. 1st Ed. B. Birren & E.D. Green, Eds. :83–248. Plainview. Cold Spring Harbor Laboratory Press.
9. CHEN, J. & L. HECKERT. 2001. Dmrt1 expression is regulated by follicle-stimulating hormone and phorbol esters in postnatal Sertoli cells. Endocrinology **142:** 1167–1178
10. HARJU, S. & K. PETERSON. 2001. Sensitive ribonuclease protection assay employing glycogen as a carrier and a single inactivation/precipitation step. Biotechniques JID - 8306785 **30:** 1198–200, 1202, 1204.
11. WILHELM, D. & C. ENGLERT. 2002 The Wilms tumor suppressor WT1 regulates early gonad development by activation of Sf1. Genes Dev. **16:** 1839–1851.
12. STALLINGS, N.R. *et al.* 2002. Development of a transgenic green fluorescent protein lineage marker for steroidogenic factor 1. Mol. Endocrinol. **16:** 2360–2370.
13. PARK, S.Y. *et al.* 2005. Nuclear receptors Sf1 and Dax1 function cooperatively to mediate somatic cell differentiation during testis development. Development **132:** 2415–2423.

Estrogens in Testis Biology

JAYNE E. SIERENS, SHARON F. SNEDDON, FRANCES COLLINS, MICHAEL R. MILLAR, AND PHILIPPA T. K. SAUNDERS

MRC Human Reproductive Sciences Unit, The Queen's Medical Research Institute, Edinburgh, EH16 4TJ, UK

ABSTRACT: Levels of estrogen within the male reproductive tract are higher than in the general circulation and the aromatase enzyme is expressed in the adult testis. Estrogens such as estradiol (E2) modify cell function by binding to high-affinity estrogen receptors (ER). Two subtypes (ERα and ERβ) have been identified. Studies in animals have shown that over- or underexposure to estrogens can have an impact on testis function. For example, mice with targeted disruption of the aromatase *cyp19* gene become infertile because round spermatids fail to differentiate normally. In rodents, ERα is expressed in Leydig cells; ERα mRNA and protein are not detectable in testes from humans or primates. High levels of expression of ERα occur in the efferent ductules in rodents, primates, and the human. ERβ protein has been immunolocalized to all somatic cells and to some germ cells in these same species. Messenger RNAs for splice variant isoforms of human ERβ are expressed in human testes. Homologues of the ERβ2 variant have been cloned from primates; this isoform does not exist in rodents and does not bind E2. Full-length ERβ protein (ERß1) and ERβ2 have differential patterns of expression in human testes. In conclusion, although estogens are synthesized in the testis and it has been suggested that E2 may function as a germ cell survival factor, the mechanisms by which estrogens influence male fertility remain uncertain and rodents may be poor models in which to examine this.

KEYWORDS: Sertoli cells; estrogen; ERβ; fertility; receptor; spermatid; fetal

INTRODUCTION

Our quest to understand how estrogenic ligands alter cell function in the reproductive system of the male has largely focused upon the expression and functional activity of estrogen receptors (ERs) that belong to a superfamily of ligand-activated transcription factors.[1] The cDNA of the first ER was cloned in 1986[2] and a second ER was identified 10 years later.[3] These receptors are known as ERα (ESR1, NR3A1) and ERβ (ESR2, NR3A2) and are the products of two genes located on different chromosomes.[4] They are believed to have arisen from a single gene duplication event at least 450 million years ago.[5] The effects of estrogens can be mediated through different pathways (reviewed in Nilsson *et al.*[1] and Hall *et al.*[6]). Classically,

Address for correspondence: Professor Philippa Saunders, MRC Human Reproductive Sciences Unit, The Queen's Medical Research Institute, 47 Little France Crescent, Edinburgh, EH16 4TJ, Scotland, UK. Voice: 44-131-242-6388; fax: 44-131-242-6231.
p.saunders@ed.ac.uk

**Ann. N.Y. Acad. Sci. 1061: 65–76 (2005). © 2005 New York Academy of Sciences.
doi: 10.1196/annals.1336.008**

after ligand binding the ER undergoes a conformational change, dimerizes with a second ER, and then interacts with binding sites on DNA. The two regulatory regions that have been investigated most thoroughly are the estrogen-response elements (EREs) and AP-1 sites, both of which are found in the promoter regions of estrogen-responsive genes. In addition, steroid ligand–independent activation of ERs due to direct phosphorylation via growth factor-driven pathways has been reported.[7]

ESTROGEN RECEPTORS: STRUCTURE AND FUNCTION

ERα and ERβ bind estradiol (E2) with high affinity; both proteins contain a central DNA binding domain and a C-terminal ligand-binding domain, Two transactivation domains (AF-1 and AF-2) are located in the N- and C-terminal portions of the protein, respectively (see reviews by Nilsson *et al.*[1] and Saunders [8]). Although significant similarities do exist between the amino acid sequences of the two receptors (for example within their DNA binding domains), they exhibit different binding affinities for some ligands, notably genistein, raloxifene, and tamoxifen.[9–11] The crystal structures of the ligand-binding domains of ERα and ERβ both contain 12 helices. Structure–function analyses have demonstrated that, after binding of receptors to agonists or antagonists, the orientation of the twelfth helix that contains the AF-2 domain, influences the ability of the receptor to recruit co-activators.[12] The identification of those amino acids within the ligand-binding pockets of ERα and ERβ, which interact directly with natural and synthetic estrogens,[12] has facilitated the development of a number of ER subtype-selective ligands.[13,14] Some of these ligands are already available from commercial suppliers. FIGURE 1 shows the results of assays in which ERα- or ERβ-containing constructs were transiently transfected into cells together with a plasmid reporter construct containing a luciferase cDNA, the expression of which could be induced after binding of ligand-activated ERs to a promoter containing three copies of a consensus estrogen-response element, ERE[15] In agreement with results of other published studies, E2 acts as an agonist for both ERα and ERβ, but the synthetic ligands exhibit some ER subtype specificity. For example, PPT, a propylpyrazole triol (Tocris Cookson Inc, Ellisville, MO; catalogue No: 1426), has been reported to have a 410-fold binding affinity preference for ERα.[16] In our reporter assay, expression of the luciferase reporter was increased ~75- to 170-fold in the presence of ERα exposed to 10^{-8} to 10^{-7} M PPT, while activation was minimal when ERα was replaced by ERß (FIG. 1). In contrast, DPN (2,3-bis (4-hydroxyphenyl) propionitrile; Tocris, catalogue No: 1494) acts as an agonist on both ER subtypes, but has been reported to have a 70-fold-higher relative binding affinity for ERβ.[17] In our hands, DPN was apparently ERβ-selective at lower concentrations (10^{-12} to 10^{-8} M) but at higher concentrations (>10^{-6} M) was an effective ERα agonist (FIG. 1). The recent development of a new generation of ER-selective agonists provides an exciting avenue for expansion of studies *in vitro* and in *vivo* on the relative contribution made by ligand-activated ERα or ERβ to gene expression.[14,18]

Splice variant isoforms of the ERβ gene have been identified in the human.[19–22] the bovine,[23] and rodents.[21,24,25] One striking feature revealed by sequencing of these products is that the variant most commonly found in rodent tissues, which is

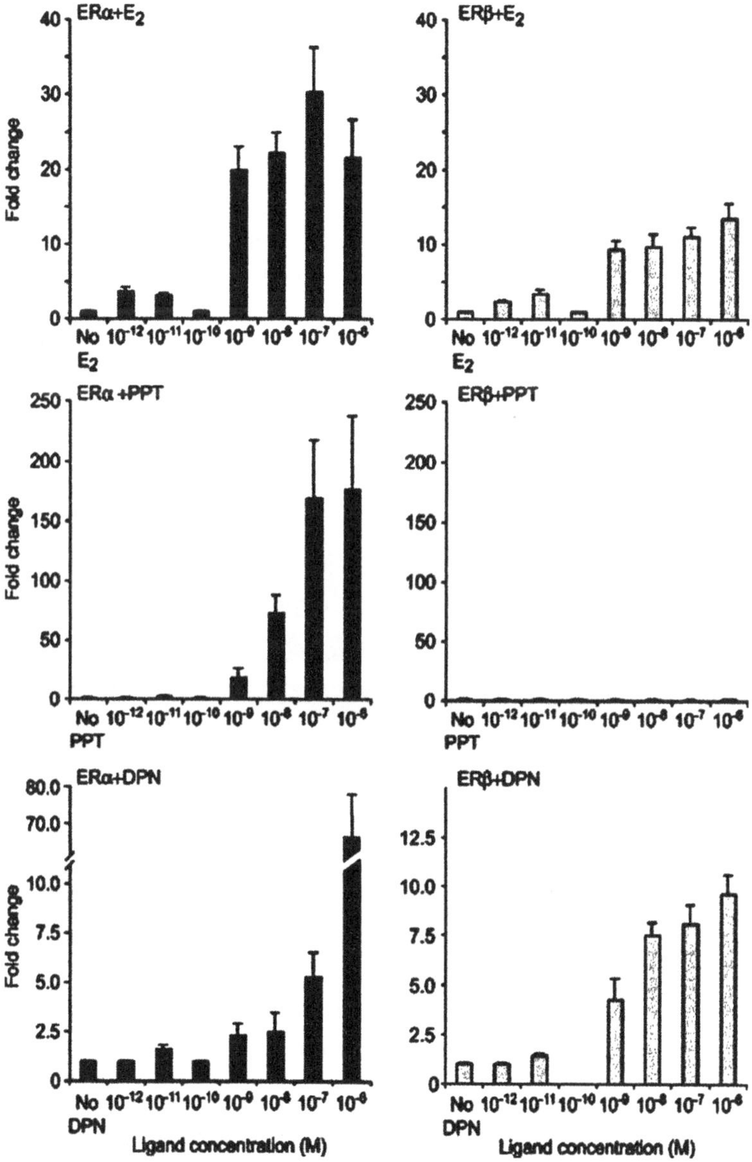

FIGURE 1. Comparison between the ability of E2 and two receptor-selective steroids (PPT, DPN) to induce expression of a 3× ERE-containing luciferase reporter construct after transient co-transfection with ERα or ERβ in Hek293 cells.[75]

formed by an insertion of 18 amino acids within the ligand-binding domain,[24,26] has not been detected in human tissue extracts.[27] Ogawa *et al.* identified an ERβ variant (hERßcx) in which the 61 amino acids usually found at the C-terminus of the wild-type ERβ protein were replaced by a unique 26-amino-acid peptide encoded by an alternative eighth exon.[19] Moore *et al.* cloned cDNAs for five hERß splice isoforms (hERß2-ß5), and the mRNAs for these isoforms have been detected in a wide range of human tissues, including the testis, ovary, and endometrium.[20,28] Examination of the sequences revealed that ERβ2 is identical to hERβcx. Recently we have discovered that mRNAs homologous to hERβ2 also exist in Old and New World primates, and we have shown that maERβ2 mRNA found in the common marmoset is expressed in a tissue-dependent pattern identical to that seen in the human (FIG. 2).[29]

In vitro studies have demonstrated that homo-dimers (ERα-ERα or ERβ-ERβ) or hetero-dimers (ERα-ERβ) can be formed when both genes are expressed in the same cell.[26,30] On the basis of studies in transfected cells and in rodent models it has been suggested that one of the functional consequences of ERβ expression may be the blunting of ERα-mediated transcriptional activity.[15,31] Gel-shift studies have shown that human ERβ variants can form hetero-dimers with full length ERα or ERβ.[20] Ogawa *et al.*[19] reported that ERβ2 cannot bind E2 and that it preferentially forms a hetero-dimer with ERα rather than with ERβ. We have used cells transfected with

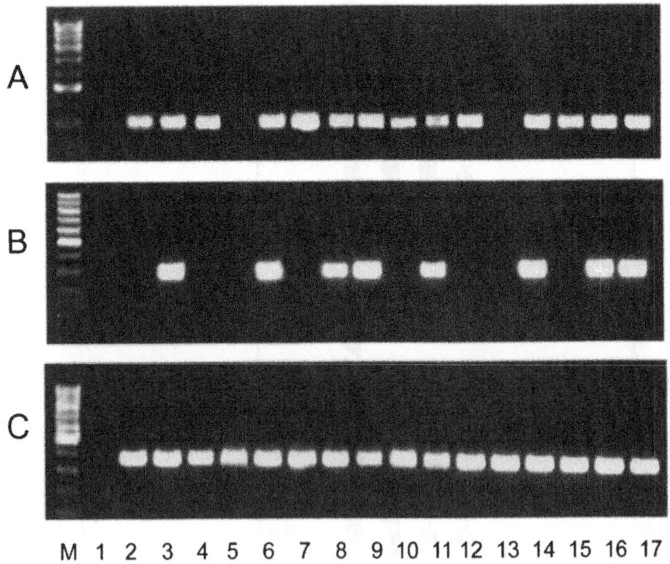

FIGURE 2. Messenger RNAs encoding full-length ERβ1 as well as the ERβ2 splice variant have an identical tissue distribution in the human and in the common marmoset monkey. (**A**) ERβ1 (240 bp); (**B**) ERβ2 (135 bp); (**C**) GAPDH (280 bp). Samples tested were (1) negative control, no tissue; (2) human brain; (3) human heart; (4) human liver; (5) human kidney; (6) human ovary; (7) human placenta; (8) human testis library (BD Biosciences); (9) human testis; (10) marmoset brain; (11) marmoset heart; (12) marmoset liver; (13) marmoset kidney; (14) marmoset ovary; (15) marmoset placenta; (16) marmoset testis; and (17) marmoset testis. A 100-bp marker was run on all gels. Data from Sierens.[75]

hER constructs tagged with fluorescent proteins (GFP or DsRed) to examine the behavior of the receptor proteins after ligand exposure. Although ERα and full-length ERβ(ERβ1) both adopt a punctate "active" subnuclear distribution within 1 h of introduction of E2, ERβ2-GFP proteins remain diffuse and are unable to activate expression of a luciferase construct containing an ERE promoter.

ARE ESTROGENS AVAILABLE WITHIN THE TESTIS?

Estrogens are synthesized from androgens by the aromatase complex, which contains the cytochrome P450 enzyme encoded by the *CYP19* gene.[32] Expression of the aromatase protein has been detected in Leydig cells and elongate spermatids in the mouse and human.[33,34] Aromatase has been demonstrated in Sertoli cells isolated from immature rat testes[35] in ejaculated spermatozoa from several species.[36,37] It has been proposed that local production of E2 might contribute to activation of genomic or non-genomic estrogen receptors on the spermatozoa themselves or in the rest of the male reproductive system. Studies using isolated cells have revealed that expression of the *cyp19* gene can be modulated by androgens, cytokines, and the transcription factors SF-1 and LRH-1.[38,39]

The estrogen sulfotransferase enzyme, which catalyzes the sulfoconjugation and inactivation of estrogens and may therefore regulate the activity of locally produced estrogens, is also expressed within the testis. The protein has been localized to Leydig cells in the human, as well as in other species, and its expression shown to be dependent upon LH and androgen.[40,41]

WHERE ARE ESTROGEN RECEPTORS EXPRESSED IN THE TESTIS?

Rodents

In the testes of adult rats and mice, ERα immunoexpression has been detected in the Leydig cells and the peritubular myoid cells, but not in the Sertoli and germ cells.[42] The cells within the epithelium of efferent ductules in both species express high levels of ERα.[42, 43] In comparison, ERß protein can be detected in multiple cell types, including the Sertoli cells, and in some but not all germ cells, as well as in epithelial and stromal cells within the male reproductive system (summarized in TABLE 1).[42–45]

Studies on the impact of estrogens on isolated testicular cell types have been hampered by the limited capacity of isolated germ cells to undergo functional maturation *in vitro* and by the loss of steroid receptors when isolated Sertoli cells are maintained in primary culture. To overcome this problem we have used the SK11 cell line that was originally derived from Sertoli cells of a 10-day-old *H2K^b^-tsA58* transgenic mouse.[46] We have demonstrated that these transformed Sertoli cells express ERβ mRNA and protein, which is consistent with the results of immunohistochemistry using fixed tissue sections from mouse testes.[47] Incubation of these cells with E2 induced expression of an ERE-luciferase reporter gene and an upregulation in expression of ERß mRNA.[47]

TABLE 1. Estrogen receptor expression in the male reproductive tract of the adult mouse

Tissue/Cell Type	ERα	ERβ
Testis		
Sertoli cells	—	++
Leydig cells	+++	+++
Germ cells	—	Variable
Efferent ductule	E+++/S–	E+++/S+++or-
Epididymis		
Initialsegment	E+++/S–	E+++/S+
Caput/corpus	E*/S–	E+++/S++
Cauda	E*/S+++	E+++/S+++
Vas deferens	E*/S+++	E+++/S+++

NOTE: Data are summarized from Zhou *et al.*[42] E = epithelial cells; S = stromal cells; E* = a very few immunopositive epithelial (clear) cells, which are present in the regions of the epididymis other than the initial segment. In the vas deferens a few ERα-positive basal cells can be detected.

Human and Primate

We have failed to detect ERα mRNA in human tissue biopsy specimens, even though the mRNA can be detected in "testicular" cDNA pools and libraries from commercial sources.[48] This is consistent with the failure to immunolocalize ERα to cells within the human testes[48,49] even though high levels of ERα are expressed in the efferent ductules.[50]

Messenger RNAs encoding ERß1, three of the hERß C-terminal splice variants (ERβ2, ERβ4, and ERβ5), as well as forms deleted in exons 5 and/or 6 have all been detected in tissue extracts of human testes.[28,48,49] Immunohistochemistry has revealed that hERβ1 and several of the ERβ variants (hERβ2, β4, β5) are all expressed.[28,48] Notably, the highest levels of immunoexpression of ERβ1 are detected in round spermatids, whereas the ERβ2 variant appeared to be most abundant in Sertoli cells.[48–50] The complete cell-specific pattern of expression of ERβ1 and ERβ2 is summarized in FIGURE 3. Differential expression of ERβ1 and ERβ2 has been demonstrated in other human tissues,[51] and these results raise questions over the results from studies conducted using human tissues where the epitope for the anti-ERβ antibody is not delineated. Proteins homologous to hERβ1 have also been immunolocalized to sections of testes from primate species and their pattern of expression appears to be identical to that seen in the human.[29]

WHAT CAN WE LEARN FROM MODELS?

Rodents

Studies on the influence of estrogens on reproductive function have also included the use of mice in which estrogen responsivness has been altered by the targeted disruption of ER receptors (ERKO, βERKO, αβERKO[52–54]). In addition, bioavailabil-

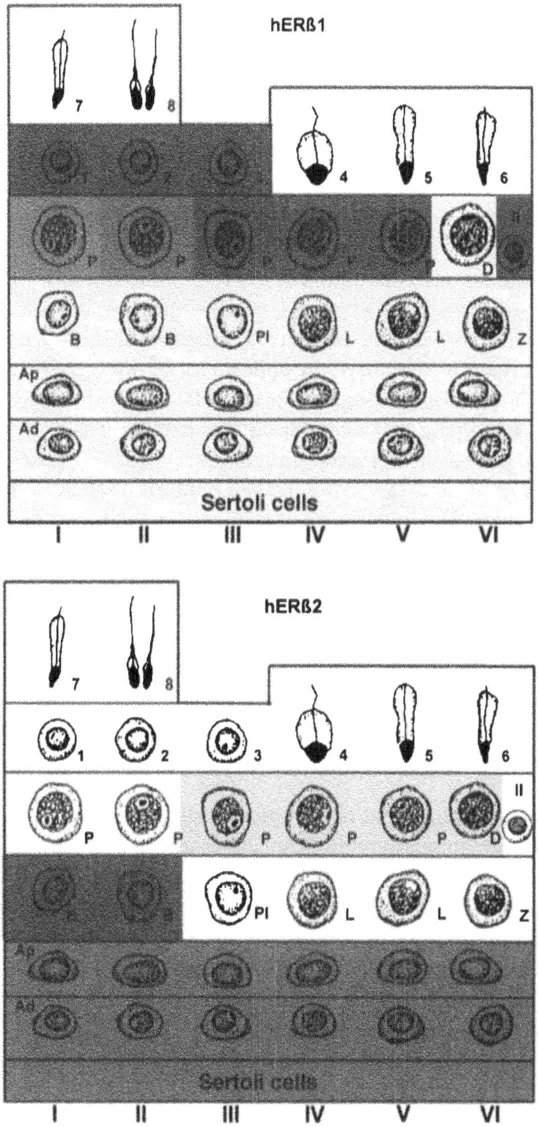

FIGURE 3. Summary of the pattern of immunoexpression of ERβ1 and ERβ2 in human testes. The diagram was redrawn from that published by Clermont.[76] The intensity of shading reflects the intensity of the immunopositive reaction within the nuclei of the different cell types (negative to ++++). (From Saunders *et al.*[48] Reproduced by permission.)

ity of estrogens has been manipulated by performing knockouts of the aromatase *cyp19* (ArKO[55]) and estrogen sulfotransferase (EST) enzymes[56] and by the administration of anti-estrogens.[57–59]

In the ERαKO, testicular function declines in adulthood, when fluid builds up within the lumens of the seminiferous tubules because the function of the efferent ductules is impaired.[60] ERα is not required directly for germ-cell maturation, and transplanted αERKO germ cells can give rise to sperm capable of fertilizing eggs, leading to the production of offspring.[61] The ERβKO males do not have any obvious disturbances in testicular or epididymal function, and the development of a prostatic phenotype has been claimed by some,[54] but disputed by others.[52] Notably, the phenotype of the ERαβKO mice is not the same as the ArKO (see below), which may be consistent with residual ER activity and/or gene activation via alternative pathways.[62]

Male mice with targeted disruption of *cyp19*[63,64] are initially fertile, but develop disruptions in testis function between 4 months and 1 year of age. In these older males spermatogenesis was arrested at early spermiogenic stages, and was characterized by an increase in apoptosis, the appearance of multinucleated cells, a significant reduction in round and elongated spermatids, but no changes in Sertoli cells or earlier germ cells. Leydig cell hyperplasia/hypertrophy was also evident, presumably as a consequence of increased circulating LH. The phenotype includes the development of a central adiposity reminiscent of that seen in patients with aromatase deficiency. The importance of estrogens in maintenance of testis function was underlined by the finding that the development of infertility could be accelerated by removing all phytoestrogens from the diet and partially prevented/delayed when they were included.[65]

Mice with targeted disruption of the gene encoding EST have also been generated. Although the testes in young animals appear normal, their Leydig cells, as they age, develop a phenotype of increased cholesterol ester accumulation and impaired steroidogenesis, which was associated with seminiferous tubule damage.[56,66] The phenotype could be induced in younger animals by the administration of exogenous E2.[66]

The functional consequences of estrogen deprivation in adulthood have also been investigated by the administration of the anti-estrogen Faslodex (ICI182870). This compound binds to both ERα and ERβ and reduces the stability of ERα *in vitro*.[67] In rats and mice long-term treatment with Faslodex (2 to 150 days) resulted in efferent duct dilation, a progressive decrease in testis weight, decreased sperm concentration, and decreased fertility.[57–59] These changes are similar to those observed in the ERαKO mice, and it is notable that they occurred after a reduction in expression of ERα but no change in expression of ERβ.[58]

Human Patients and Primates

Apart from papers describing the pattern of expression of aromatase, EST, and the ERs, information relevant to elucidating a role for estrogens in testis function in the human and in primates is limited. Although it has been reported that levels of estrogens are higher in semen of infertile males compared with fertile ones,[68] the functional significance of this finding is not clear. Recently it was reported that the frequency of the heterozygous RsaI AG genotype was three times higher in infertile men compared with fertile controls.[69] This polymorphism does not change the protein sequence, but the authors of this study suggested it might have an impact on splicing. In a small study, a difference in the incidence of an XbaI polymorphism of ERα gene between controls and azoospermic patients has been claimed.[70] Elevated levels of testosterone and gonadotrophins have been detected in patients with mutations in the *CYP19* gene,[71] and E2 treatment of one of the patients for 3 months increased his sperm count.[72] Pentikainen *et al.*[73] have reported that incubation with estrogens reduces levels of germ cell apoptosis in isolated human seminiferous tubules, which suggests that the expression of ERß1 within germ cells may play a role in germ-cell survival.

CONCLUSIONS AND FUTURE DIRECTIONS

Evidence from animal species and humans is consistent with the adult testis being an organ capable of the biosynthesis of estrogens from androgen precursors. The observation that disturbances in testis function occur both in patients with deficits in aromatase enzyme function as well as in mice with targeted disruption of the aromatase *cyp19* gene argues for an important functional role for locally produced estrogens in germ-cell maturation/male fertility. Expression of ERs in a variety of testicular cell types has been documented by a number of investigators. Differences exist between the patterns of expression of ER subtypes and variant isoforms in rodents and those that have been identified in man and in non-human primates. Future studies on the role(s) played by estrogens in testicular function are likely to benefit from the development of new ER subtype selective agonists and antagonists, the use of primate models, and access to new cell lines derived from human tissues or ES cells.[74]

REFERENCES

1. NILSSON, S. *et al.* 2001. Mechanisms of estrogen action. Physiol. Rev. **81:** 1535–1565.
2. GREEN, S. *et al.* 1986. Human oestrogen receptor cDNA: sequence, expression and homology to v-erb-A. Nature **320:** 134–139.
3. KUIPER, G.G. *et al.* 1996. Cloning of a novel receptor expressed in rat prostate and ovary. Proc. Natl. Acad. Sci. USA **93:** 5925–5930.
4. ENMARK, E. *et al.* 1997. Human estrogen receptor beta-gene structure, chromosomal localization, and expression pattern. J. Clin. Endocrinol. Metab. **82:** 4258–4265.
5. KELLEY, S.T. & V.G. THACKRAY. 1999. Phylogenetic analyses reveal ancient duplication of estrogen receptor isoforms. J. Mol. Evol. **49:** 609–614.
6. HALL, J.M., J.F. COUSE & K.S. KORACH. 2001. The multifaceted mechanisms of estradiol and estrogen receptor signaling. J. Biol. Chem. **276:** 36869–36872.
7. IGNAR-TROWBRIDGE, D.M. *et al.* 1996. Peptide growth factor cross-talk with the estrogen receptor requires the A/B domain and occurs independently of protein kinase C or estradiol. Endocrinology **137:** 1735–1744.
8. SAUNDERS, P.T. 1998. Oestrogen receptor beta (ER beta). Rev. Reprod. **3:** 164–171.
9. KUIPER, G.G.J.M. *et al.* 1997. Comparison of the ligand binding specificity and transcript tissue distribution of estrogen receptors alpha and beta. Endocrinology **138:** 863–870.
10. BARKHEM, T. *et al.* 1998. Differential response of estrogen receptor alpha and estrogen receptor beta to partial estrogen agonists/antagonists. Mol. Pharmacol. **54:** 105–112.
11. AN, J. *et al.* 2001. Estrogen receptor beta-selective transcriptional activity and recruitment of coregulators by phytoestrogens. J. Biol. Chem. **276:** 17808–17814.

12. PIKE, A.C. *et al.* 1999. Structure of the ligand-binding domain of oestrogen receptor beta in the presence of a partial agonist and a full antagonist. EMBO J. **18:** 4608–4618.
13. KATZENELLENBOGEN, B.S. *et al.* 2000. Estrogen receptors: selective ligands, partners, and distinctive pharmacology. Recent Prog. Horm. Res. **55:** 163–193.
14. HILLISCH, A. *et al.* 2004. Dissecting physiological roles of estrogen receptor α and ß with potent selective ligands from structure based design. Mol. Endocrinol. **18:** 1599–1609.
15. HALL, J.M. & D.P. MCDONNELL. 1999. The estrogen receptor beta-isoform (ERbeta) of the human estrogen receptor modulates ERalpha transcriptional activity and is a key regulator of the cellular response to estrogens and antiestrogens. Endocrinology **140:** 5566–5578.
16. STAUFFER, S.R. *et al.* 2000. Pyrazole ligands: structure-affinity/activity relationships and estrogen receptor-alpha-selective agonists. J. Med. Chem. **43:** 4934–4947.
17. MEYERS, M.J. *et al.* 2001. Estrogen receptor-beta potency-selective ligands: structure-activity relationship studies of diarylpropionitriles and their acetylene and polar analogues. J. Med. Chem. **44:** 4230–4251.
18. FRASOR, J. *et al.* 2003. Response-specific and ligand dose-dependent modulation of estrogen receptor (ER) alpha activity by ERbeta in the uterus. Endocrinology **144:** 3159–3166.
19. OGAWA, S. *et al.* 1998. Molecular cloning and characterization of human estrogen receptor betacx: a potential inhibitor ofestrogen action in human. Nucleic Acids Res. **26:** 3505–12.
20. MOORE, J.T. *et al.* 1998. Cloning and characterization of human estrogen receptor beta isoforms. Biochem. Biophys. Res. Commun. **247:** 75–78.
21. LU, B. *et al.* 1998. Estrogen receptor-beta mRNA variants in human and murine tissues. Mol. Cell Endocrinol. **138:** 199–203.
22. INOUE, S. *et al.* 2000. An estrogen receptor beta isoform that lacks exon 5 has dominant negative activity on both ERalpha and ERbeta. Biochem .Biophys. Res. Commun. **279:** 814–819.
23. WALTHER, N. *et al.* 1999. Cloning of bovine estrogen receptor beta (ERbeta): expression of novel deleted isoforms in reproductive tissues. Mol. Cell Endocrinol. **152:** 37–45.
24. CHU, S. & P.J. FULLER. 1997. Identification of a splice variant of the rat estrogen receptor beta gene. Mol. Cell Endocrinol. **132:** 195–199.
25. LU, B. *et al.* 2000. Functional characteristics of a novel murine estrogen receptor-beta isoform, estrogen receptor-beta 2. J. Mol. Endocrinol. **25:** 229–242.
26. PETTERSSON, K. *et al.* 1997. Mouse estrogen receptor beta forms estrogen response element-binding heterodimers with estrogen receptor alpha. Mol. Endocrinol. **11:** 1486–1496.
27. CHU, S. *et al.* 2000. Estrogen receptor isoform gene expression in ovarian stromal and epithelial tumors. J. Clin. Endocrinol. Metab. **85:** 1200–1205.
28. SCOBIE, G.A. *et al.* 2002. Human oestrogen receptors: differential expression of ER alpha and beta and the identification of ER beta variants. Steroids **67:** 985–992.
29. SIERENS, J.E., *et al.* 2004. Cloning of oestrogen receptor beta from Old and New World primates: identification of splice variants and functional analysis. J. Mol. Endocrinol. **32:** 703–718.
30. COWLEY, S.M. *et al.* 1997. Estrogen receptors alpha and beta form heterodimers on DNA. J. Biol. Chem. **272:** 19858–19862.
31. WEIHUA, Z. *et al.* 2000. Estrogen receptor (ER) beta, a modulator of ERalpha in the uterus. Proc. Natl. Acad. Sci. USA **97:** 5936–5941.
32. SIMPSON, E.R. & S.R. DAVIS. 2001. Minireview: aromatase and the regulation of estrogen biosynthesis-some new perspectives. Endocrinology **142:** 4589–4594.
33. NITTA, H. *et al.* 1993. Germ cells of the mouse testis express P450 aromatase. Endocrinology **132:** 1396–1401.
34. TURNER, K.J., *et al.* 2002. Development and validation of a new monoclonal antibody to mammalian aromatase. J. Endocrinol. **172:** 21–30.

35. DORRINGTON, J. H. & S.A. KHAN. 1993. Steroid production, metabolism, and release by Sertoli cells. *In* The Sertoli Cell. L. Russell, & M. Griswold, Eds.: 538–549. Cache River Press. Clearwater, FL.
36. CARREAU, S. *et al.* 1999. Sources of oestrogen in the testis and reproductive tract. Int. J. Androl. **22:** 211–223.
37. LAMBARD, S. *et al.* 2004. Human immature germ cells and ejaculated spermatozoa contain aromatase and oestrogen receptors. J. Mol. Endocrinol. **32:** 279–289.
38. BOURGUIBA, S. *et al.* 2003. Regulation of aromatase gene expression in Leydig cells and germ cells. J. Steroid Biochem. Mol. Biol. **86:** 335–343.
39. PEZZI, V. *et al.* 2004. Differential expression of steroidogenic factor-1/adrenal 4 binding protein and liver receptor homolog-1 (LRH-1)/fetoprotein transcription factor in the rat testis: LRH-1 as a potential regulator of testicular aromatase expression. Endocrinology **145:** 2186–2196.
40. MIKI, Y. *et al.* 2002. Systemic distribution of steroid sulfatase and estrogen sulfotransferase in human adult and fetal tissues. J. Clin. Endocrinol. Metab. **87:** 5760–5768.
41. QIAN, Y.M. & W.C. SONG. 1999. Regulation of estrogen sulfotransferase expression in Leydig cells by cyclic adenosine 3′,5′-monophosphate and androgen. Endocrinology **140:** 1048–1053.
42. ZHOU, Q. *et al.* 2002. Localization of androgen and estrogen receptors in adult male mouse reproductive tract. J. Androl. **23:** 870–881.
43. HESS, R.A. *et al.* 1997. Estrogen receptor (alpha and beta) expression in the excurrent ducts of the adult male rat reproductive tract. J. Androl. **18:** 602–611.
44. SAUNDERS, P.T. *et al.* 1998. Expression of oestrogen receptor beta (ER beta) occurs in multiple cell types, including some germ cells, in the rat testis. J. Endocrinol. **156:** R13–R17.
45. VAN PELT, A.M. *et al.* 1999. Ontogeny of estrogen receptor-beta expression in rat testis. Endocrinology 140: 478–483.
46. WALTHER, N. *et al.* 1996. Sertoli cell lines established from H-2Kb-tsA58 transgenic mice differentially regulate the expression of cell-specific genes. Exp. Cell Res. **225:** 411–421.
47. SNEDDON, S.F., N. WALTHER & P.T. SAUNDERS. 2005. Expression of androgen and estrogen receptors in Sertoli cells: studies using the mouse SK11 cell line. Endocrinology **146:** 5304–5312.
48. SAUNDERS, P.T. *et al.* 2002. ERbeta1 and the ERbeta2 splice variant (ERbetacx/beta2) are expressed in distinct cell populations in the adult human testis. J. Clin. Endocrinol. Metabolism **87:** 2706–2715.
49. MAKINEN, S. *et al.* 2001. Localization of oestrogen receptors alpha and beta in human testis. Mol. Hum. Reprod. **7:** 497–503.
50. SAUNDERS, P.T. *et al.* 2001. Differential expression of oestrogen receptor alpha and beta proteins in the testes and male reproductive system of human and non-human primates. Mol. Hum. Reprod. **7:** 227–236.
51. CRITCHLEY, H.O. *et al.* 2002. Wild-type estrogen receptor (ERbeta1) and the splice variant (ERbetacx/beta2) are both expressed within the human endometrium throughout the normal menstrual cycle. J. Clin. Endocrinol. Metab. **87:** 5265–5273.
52. DUPONT, S. *et al.* 2000. Effect of single and compound knockouts of estrogen receptors alpha (ERalpha) and beta (ERbeta) on mouse reproductive phenotypes. Development **127:** 4277–4291.
53. COUSE, J.F. *et al.* 1999. Postnatal sex reversal of the ovaries in mice lacking estrogen receptors alpha and beta. Science **286:** 2328–2331.
54. KREGE, J.H. *et al.* 1998. Generation and reproductive phenotypes of mice lacking estrogen receptor beta. Proc. Natl. Acad. Sci. USA **95:** 15677–15682.
55. FISHER, C.R. *et al.* 1998. Characterization of mice deficient in aromatase (ArKO) because of targeted disruption of the cyp19 gene. Proc. Natl. Acad. Sci. USA **95:** 6965–6970.
56. TONG, M.H., L.K. CHRISTENSON & W.C. SONG. 2004. Aberrant cholesterol transport and impaired steroidogenesis in Leydig cells lacking estrogen sulfotransferase. Endocrinology **145:** 2487–2497.

57. Oliveira, C.A. *et al.* 2002. ER function in the adult male rat: short- and long-term effects of the antiestrogen ICI 182,780 on the testis and efferent ductules, without changes in testosterone. Endocrinology **143:** 2399–2409.
58. Oliveira, C.A. *et al.* 2003. The antiestrogen ICI 182,780 decreases the expression of estrogen receptor-alpha but has no effect on estrogen receptor-beta and androgen receptor in rat efferent ductules. Reprod. Biol. Endocrinol. **1:** 75.
59. Cho, H.W. *et al.* 2003. The antiestrogen ICI 182,780 induces early effects on the adult male mouse reproductive tract and long-term decreased fertility without testicular atrophy. Reprod. Biol. Endocrinol. **1:** 17.
60. Hess, R.A. *et al.* 1997. A role for oestrogens in the male reproductive system. Nature **390:** 509–512.
61. Mahato, D. *et al.* 2000. Spermatogenic cells do not require estrogen receptor-alpha for development or function. Endocrinology **141:** 1273–1276.
62. Couse, J. F. *et al.* 1995. Analysis of transcription and estrogen insensitivity in the female mouse after targeted disruption of the estrogen receptor gene. MolEndocrinol. **9:** 1441–1454.
63. Robertson, K.M. *et al.* 1999. Impairment of spermatogenesis in mice lacking a functional aromatase (cyp 19) gene. Proc. Natl. Acad. Sci. USA **96:** 7986–7991.
64. Robertson, K. M. *et al.* 2001. Characterization of the fertility of male aromatase knockout mice. J. Androl. **22:** 825–830.
65. Robertson, K.M. *et al.* 2002. The phenotype of the aromatase knockout mouse reveals dietary phytoestrogens impact significantly on testis function. Endocrinology **143:** 2913–2921.
66. Qian, Y.M. *et al.* 2001. Targeted disruption of the mouse estrogen sulfotransferase gene reveals a role of estrogen metabolism in intracrine and paracrine estrogen regulation. Endocrinology **142:** 5342–5350.
67. Van Den Bemd, G.J. *et al.* 1999. Distinct effects on the conformation of estrogen receptor alpha and beta by both the antiestrogens ICI 164,384 and ICI 182,780 leading to opposite effects on receptor stability. Biochem. Biophys. Res. Commun. **261:** 1–5.
68. Bujan, L. *et al.* 1993. Increased oestradiol level in seminal plasma in infertile men. Hum. Reprod. **8:** 74–77.
69. Aschim, E.L. *et al.* 2005. The RsaI polymorphism in the estrogen receptor-beta gene is associated with male infertility. J. Clin. Endocrinol. Metab. **90:** 5343–5348.
70. Kukuvitis, A. *et al.* 2002. Association of oestrogen receptor alpha polymorphisms and androgen receptor CAG trinucleotide repeats with male infertility: a study in 109 Greek infertile men. Int. J. Androl. **25:** 149–152.
71. Simpson, E.R. 1998. Genetic mutations resulting in estrogen insufficiency in the male. Mol. Cell Endocrinol. **145:** 55–59.
72. Herrmann, B.L. *et al.* 2002. Impact of estrogen replacement therapy in a male with congenital aromatase deficiency caused by a novel mutation in the CYP19 gene. J. Clin. Endocrinol. Metab. **87:** 5476–5484.
73. Pentikainen, V. *et al.* 2000. Estradiol acts as a germ cell survival factor in the human testis in vitro. J. Clin. Endocrinol. Metab. **85:** 2057–2067.
74. Geijsen, N. *et al.* 2004. Derivation of embryonic germ cells and male gametes from embryonic stem cells. Nature **427:** 148–154.
75. Sierens, J.E. 2003. Mechanisms of Transcriptional Modulation of Gene Expression by Oestrogen Receptors. Ph.D thesis, University of Edinburgh, Edinburgh, Scotland, UK.
76. Clermont, Y. 1963. The cycle of the seminiferous epithelium in man. Am. J. Anat. **112:** 35–51.

Foxa1 and Foxa2 Interact with the Androgen Receptor to Regulate Prostate and Epididymal Genes Differentially

XIUPING YU,[a] APARNA GUPTA,[a] YONGQING WANG,[a] KICHIYA SUZUKI,[a,b] JANNI MIROSEVICH,[e] MARIE-CLAIRE ORGEBIN-CRIST,[b,c,d] AND ROBERT J. MATUSIK[a,b,d]

[a]*Department of Urologic Surgery,* [b]*Center for Reproductive Biology Research,*
[c]*Department of Obstetrics and Gynecology, and*
[d]*Department of Cell and Developmental Biology, Vanderbilt University School of Medicine, Nashville, Tennessee 37232, USA*

[e]*Molecular Medicine, Beckman Research Institute at the City of Hope, Duarte, California 91010, USA*

ABSTRACT: Previous studies from our group have shown that Foxa1 is expressed in the prostate and interacts with the androgen receptor (AR) to regulate prostate-specific genes such as prostate-specific antigen (PSA) and probasin (PB). We report here that Foxa2 but not Foxa1 is expressed in the epididymis. Further, Foxa2 interacts with the AR to regulate the mouse epididymal retinoic acid binding protein (mE-RABP) gene, an epididymis-specific gene. Binding of Foxa2 to the mE-RABP promoter was confirmed by gel-shift and chromatin immunoprecipitation (ChIP) assays. Overexpression of Foxa2 suppresses androgen activation of the mE-RABP promoter while overexpression of Foxa2 with prostate-specific promoters activates gene expression in an androgen-independent manner. GST pull-down assays determined that both Foxa1 and Foxa2 physically interact with the DNA binding domain of the AR. The interaction between Foxa proteins and AR was further confirmed by gel-shift assays where Foxa protein was recruited to AR binding oligomers even when Foxa binding sites were not present, and AR was recruited to Foxa binding oligomers even in the absence of an AR binding site. Given that Foxa1 and Foxa2 proteins are expressed differentially in the prostate and epididymis, these data suggest that the Foxa proteins have distinct effects on AR-regulated genes in different male reproductive accessory organs.

KEYWORDS: epididymis; prostate; Foxa; androgen receptor

INTRODUCTION

Forkhead box A (Foxa) proteins belong to the forkhead family of transcription factors (TF), comprising a set of three closely related members encoded by different genes. The molecular weights of Foxa1, Foxa2, and Foxa3 are 50, 47, and 42 kD, respectively.[1] This subfamily of proteins was previously called hepatocyte nuclear

Address for correspondence: Robert J. Matusik,Vanderbilt University Medical Center,A-1302 Medical Center North,Nashville, TN 37232-2765. Voice: 615-343-1902; fax: 615-322-8990
robert.matusik@vanderbilt.edu

Ann. N.Y. Acad. Sci. 1061: 77–93 (2005).
doi: 10.1196/annals.1336.009

factor 3 (HNF3α, HNF3β and HNF3δ) because they were first identified as liver-enriched factors.[2]

Foxa proteins share an almost identical DNA binding domain called winged-helix/forkhead domain.[3] When the Foxa protein is bound to DNA, the winged helix structure can displace linker histones H1 and H5 to disrupt nucleosome structures. Foxa proteins alter the highly organized and compacted chromatin into open and active conformation, so that additional TFs can be recruited.[4] Foxa proteins are thought to act as "genetic potentiators," because they can enhance the activation of genes by facilitating the binding of other TFs during development and differentiation.

Foxa proteins have been reported to regulate gene expression in endodermally derived tissues such as liver, lung, pancreas, and prostate.[5] Recent studies have shown that Foxa proteins are expressed in organs derived from all the three germ layers.[6,7,8] Studies from mouse development have shown that Foxa transcription factors play a critical role in segmentation and organ development. The first evidence of Foxa's involvement in development comes from the early expression pattern of Foxa genes in the mouse embryo, which predicts their potential roles in organ development. Of the three Foxa genes, Foxa2 is activated first at E6.5, followed by Foxa1 activation at E7-8. Foxa3 gene is activated later in development after the formation of gut.[9]

The functional significance of Foxa proteins in development was further confirmed by gene knock-out studies. A Foxa3 null genotype is not lethal in mice,[10] but Foxa1 null mice die perinatally.[11] Homozygous Foxa2-deficient mice die at embryonic day 10 to day 11 because of foregut defects, failure to form notochord, and disorganization of somites,[12] which confirms the function of Foxa2 in the development of organs that originate from both endoderm and mesoderm layers.

Although all three Foxa proteins can recognize the same DNA sequences, slight differences in binding affinity and DNA bending capacity may account for their specificity. The differential tissue distribution of Foxa proteins and co-regulators may also distinguish their function. In the prostate, Foxa1 is expressed constitutively throughout organogenesis, from early development to adult differentiated function. In contrast, Foxa2 is only expressed in developing prostate at E18 through day 1 when prostate is undergoing budding.[13] Our previous studies have shown that Foxa1 interacts with the androgen receptor (AR) and regulates prostate-specific genes and that Foxa1 is critical for prostate development.[14] In this study, we examined how Foxa proteins interact with AR to regulate prostatic and epididymis-specific genes differentially.

A prominent feature of epididymis protein-encoding genes is that they display a highly region-specific pattern of expression. This constitutes the basis for regionalization of epididymal epithelium, which has been shown to be essential for sperm maturation.[15,16] However, little is known about the mechanism underlying the region-specific gene expression. To study epididymal secretory proteins, our group has identified mE-RABP from mouse mid/distal caput epididymidis, as the ortholog for both the rat and human epididymal retinoic acid binding protein (E-RABP).[17–19] mE-RABP is specifically synthesized and secreted by the principal cells of the mid/distal caput epididymidis. The highly region-specific expression pattern of mE-RABP makes it a good model to study the molecular mechanism involved in tissue- and region-specific transcriptional regulation of epididymal genes.

To define the regulatory region of the mE-RABP gene that targets epididymis-specific gene expression, transgenic mice harboring 5 kb of the 5′-flanking region of mE-RABP promoter linked to the chloramphenicol acetyltransferase (CAT) reporter gene were generated.[20] The CAT transgene displayed faithful expression to the middle and distal caput epididymidis in exactly the same pattern as the endogenous mE-RABP gene. This indicated that the 5-kb DNA fragment contained the necessary *cis*-acting elements to drive tissue- and region-specific gene expression. Our latest study in transgenic mice further narrowed the regulatory region that would dictate epididymal and region specific expression to a 1.8-kb fragment of DNA (unpublished data). In the present study we focused on this 5′-flanking region to identify the important *cis*-DNA elements and associated transcription factors that are involved in the transcriptional regulation as well as tissue-specific expression of the mE-RABP gene.

Through 5′-deletion analysis and EMSA, a Foxa2 binding site was identified on the mE-RABP promoter. As with Foxa1, Foxa2 was also shown to interact with AR and coordinately regulate mE-RABP gene expression. In addition, our studies revealed that Foxa proteins were expressed differentially in prostate and epididymis, with Foxa1 expressed preferentially in prostate and Foxa2 in the epididymis. Further, co-transfection that would result in high levels of Foxa2 suppressed activity of the mE-RABP promoter, but activated the PSA promoter even in the absence of androgens. Since prostate cancer patients in whom androgen therapy has failed reexpress the PSA gene, we examined early and advanced-stage human prostate cancer to determine whether Foxa2 was expressed. By immunohistochemical analysis, we detected Foxa2 only in a limited set of advanced adenocarcinoma but in all cases of neuroendocrine prostate cancer, a rare and androgen-independent form of prostate cancer.[21] Although the data represent a correlation only between Foxa2 and some advanced androgen-independent prostate cancers, it is tempting to suggest that Foxa2 may play a role in prostate-cancer progression, and may represent the basic mechanism by which androgen ablation fails in some prostate cancer patients. Further, these experiments show that Foxa1 and Foxa2 function in a tissue-specific manner to regulate androgen-dependent genes.

MATERIAL AND METHODS

Constructs

A 1.8-kb mE-RABP luciferase construct was generated by PCR using high-fidelity DNA polymerase (*Pfx*, Invitrogen, Carlsbad, CA) and mE-RABP promoter-specific primers with *Mlu*I or *Hind*III site added to the 5′ end of the forward or reverse primer, respectively. The resulting PCR fragment was inserted into a pGL3-Basic vector (Promega) at *Mlu*I/*Hind*III sites. The pcDNA HNF-3 (Foxa1), pCMV-HNF-3β (Foxa2),[22] AR expression vector and various AR-GST constructs were described previously.[5] The MMTV-Luc construct was kindly provided by Dr. Darryl Granner.

Cell Culture and Transfection Assays

The immortalized mouse epididymal epithelial cell line DC2 was cultured as described previously.[23] Transient transfection assays were performed using lipofectamine 2000 (Life Technologies, Inc., Rockville, MD) according to the manufacturer's protocol. The pRL-SV40 (Promega Corp., Madison, WI) was used as an internal control to normalize transfection efficiency. The luciferase activity was determined using the dual luciferase reporter assay system (E1960; Promega Corp., Madison, WI) and LUMIstar (BMG Lab. Technologies, Inc., Durham, NC). Results were presented as luciferase/mg/min or percentage.

RT-PCR

Total RNA was extracted from dorsal/lateral prostate or epididymis using RNeasy mini kit (Qiagen, Valencia, CA), and treated with RNase free DNase (Qiagen, Valencia, CA) to eliminate genomic DNA contamination. Primers specific for Foxa1 or Foxa2 were designed to span at least one intron to ensure mRNA-specific amplification. PCR was done using primers for Foxa1 (forward 5′ TGAAGATGGA AGGGCATGAGAGCA 3′, reverse 5′ CGTGTTGGCGTAGGACATGT TGAA 3′), Foxa2 (forward 5′ ATGCTGGGAGCCGTGAAGATGGAA 3′, reverse 5′ ATTCCAGCGCCCACATAGGATGA 3′). The expected fragments for Foxa1 and Foxa2 were 200 bp and 218 bp, respectively. The PCR conditions were as follows: 94°C/4 min, and 35 cycles of 94°C/30 sec, 58°C/45 sec, and 72°C/1 min.

Immunohistochemistry

Prostate lobes were dissected from 7-week-old CD-1 mice, fixed in 10% buffered formalin, and embedded in paraffin. Epididymis was fixed in Bouin Fixation Fluid. Immunostaining for Foxa1 and Foxa2 was performed as described previously.[13] Foxa1 (C20), Foxa2 (P19), and AR (N20) antibodies were from Santa Cruz Biotechnology, Inc. (Santa Cruz, CA). Synaptophysin antibody was from BD Transduction Laboratories (San Jose, CA).

Electrophoretic Mobility Shift Assays (EMSA)

Nuclear extracts from DC2 cells were prepared as described previously.[24] *In vitro* translated Foxa1 or Foxa2 proteins were synthesized using the TNT T7 Quick Coupled Transcription/Translation System (Promega Corp., Madison, WI). GST-AR fusion proteins were purified as described previously.[25] Oligonucleotides were synthesized by Integrated DNA Technologies (Coralville, IA). Probes were end-labeled using [α-^{32}P] dATP (Amersham Biosciences, Piscataway, NJ), DNA polymerase I Klenow fragment (New England Biolabs, Beverly, MA), and oligomers for Foxa probe (5′ GCAAGATCCTGAAGGATTTGTTTATTTGATCAAAACACAGGAGGC 3′), or for ARBS3 probe (5′ CTAAGAGTACAGATCCTTACTG 3′). EMSA was performed as described previously.[5] In oligonucleotide competitions, 200- to 600-fold molar excess of cold, double-strand oligonucleotide was added to the preincubation mix. TTR competitor contains Foxa binding site (5′ TGACTAAGTCAATAATCAGAATCAG 3′); nonspecific competitor does not have Foxa binding motif (5′ CACTGCCCAGTCAAGTGTTCTTGA 3′). For super shift, 2 μL of 0.2 μg/μL antibody was used in each EMSA reaction. All super shift antibodies (Foxa1, C-20; Foxa2, P19; Foxa3, N-19) were purchased from Santa Cruz Biotechnology, Inc.

Chromatin Immunoprecipitation (ChIP)

ChIP was carried out on DC2 cells using the ChIP kit from Upstate (Milford, MA). AR or Foxa2 antibody was used to pull down the formaldehyde cross-linked protein/DNA complex, and DNA was extracted using QIAquick Spin Column (Qiagen, Valencia, CA). 3–5 μL of extracted DNA was used as template in each PCR. ChIP primer covered 200 bp of the mE-RABP promoter from −1.4 to −1.2 kb (forward 5′ AATCTTGGAGGATAGGTGTGGCAG 3′, reverse 5′ GAAGTGATCCCT-GAAGGCCAATGA 3′). A primer set recognizing far downstream of the promoter region was designed to serve as a control, which covers part of mE-RABP gene exon 3 and exon 4 (U68381, 3712-4012) and will amplify genomic DNA to produce a 300-bp fragment: forward primer 5′ CACAGCTAGGAACAGAGAGCCCTGAC 3′, reverse primer 5′ CAGGTGAGATGGGACAGGGCTTCTAT 3′.

GST Pull-Down

To study the interaction of AR with Foxa proteins, GST pull-down was performed as previously reported.[5] The glutathione agarose beads, bound to various AR-GST proteins, were incubated with synthesized Foxa1 or Foxa2 protein. The pulled-down protein beads were washed, eluted, and subjected to SDS-PAGE and Western blot.

RESULTS

Foxa Proteins Are Expressed Differentially in Prostate and Epididymis

As previously reported,[5] RT-PCR showed that Foxa1 was highly expressed in prostate (FIG. 1A). In the epididymis, both Foxa1 and Foxa2 mRNA were detectable by RT-PCR at 35 cycles but the Foxa2 levels were much higher than Foxa1. Foxa3 mRNA was not detectable in either prostate or epididymis (data not shown).

Immunohistochemical study using Foxa1 or Foxa2 antibody revealed similar expression profile as detected by RT-PCR (FIG. 1B). Foxa1 was preferentially expressed in the prostate and was mainly detected as a nuclear protein in prostatic epithelial cells while Foxa2 was found in the nucleus of the epididymis principal cell. Expression of Foxa1 in the prostate was detected in all epithelial cells, while Foxa2 displayed a region-specific expression pattern along the epididymal tube. No staining was observed in efferent ducts. No staining or weak staining was evident in the initial segment, and a gradient of more intense staining was observed from the proximal caput to the cauda with the strongest signal in cauda (data not shown).

Identification of a Foxa2 Binding Site on mE-RABP Promoter

To identify the *cis*-DNA elements and associated binding proteins involved in mE-RABP gene regulation, 5′ deletion analysis was performed starting 1.8 kb and finishing 0.5 kb upstream of the transcription start site at 100 base-pair intervals. These constructs were transfected into epididymal cells. Transfection revealed that the 1.3-kb construct gave the highest androgen response while the −1.8 to −1.4 kb region inhibited this response. EMSA has identified two AR binding sites in between −1.3 kb and −1.2 kb of the upstream region, which were denoted as ARBS-2 and ARBS-3, in accordance with the previous nomenclature reporting ARBS0 and ARBS1 in the mE-RABP promoter. A potential Foxa binding motif was predicted by a TFs search program (<http://www.cbil.upenn.edu/tess/>; FIG. 2A) upstream of ARBS2 and ARBS3. Our work on prostate-specific gene expression has identified

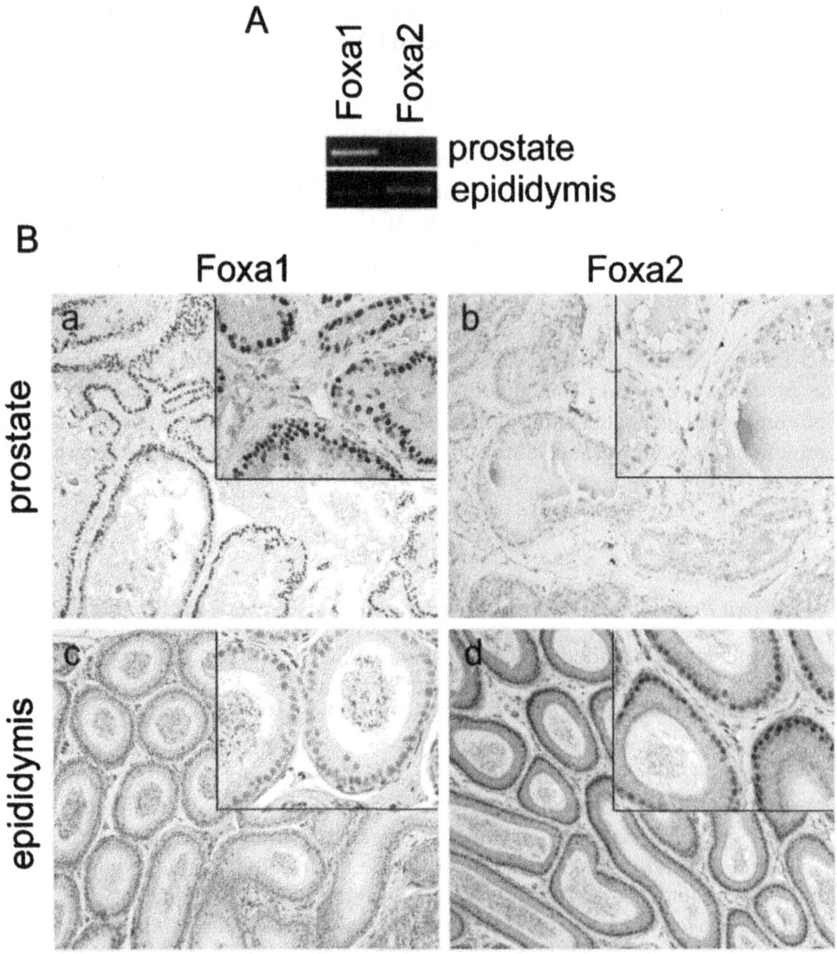

FIGURE 1. Foxa1 and Foxa2 are expressed differentially in prostate and epididymis. (**A**) RT-PCR. Foxa1 and Foxa2 expression was analyzed by RT-PCR using total RNA extracted from adult mouse prostate and epididymis. A prominent Foxa1 band was detected in prostate. In the epididymis, both Foxa1 and Foxa2 mRNA were detectable, but Foxa2 levels were much higher than Foxa1. (**B**) Immunohistochemistry analysis of Foxa proteins in prostate (**a** and **b**) and epididymis (**c** and **d**) (100 ×). Foxa1 expression was localized to the nucleus of prostate luminal epithelial cells (a, inset, 400 ×) and Foxa2 was localized to the nucleus of epididymis principal cells (**d**, inset, 400 ×).

that tissue-specific elements reside near the ARBSs. Thus, as a starting point, to identify the *cis*-acting elements required for epididymis-specific expression, we concentrated our efforts to identify TFs adjacent to the ARBS motifs. Transcription factors binding to this region will at least coordinate with AR to regulate gene expression and may be critical for epididymis-specific expression.

A fragment covering the potential Foxa binding motif (Foxa probe) was synthesized and used in EMSA to study the Foxa binding site. When the Foxa probe was incubated with nuclear extract from DC2 cells, an epididymal cell line,[23] three specific bands (FIG. 2B, labeled a, b, and c) were observed. Of these, bands a and b were completely eliminated by 300-fold excess nonlabeled self-competitor or by oligonucleotides containing the Foxa binding sequence from the transthyretin (TTR) promoter (FIG. 2B), but not by a nonpecific competitor (data not shown). Foxa2's binding to the probe was confirmed by antibody supershift (FIG. 2C, lanes 2 and 5), while antibodies to Foxa1 or Foxa3 failed to produce a supershift.

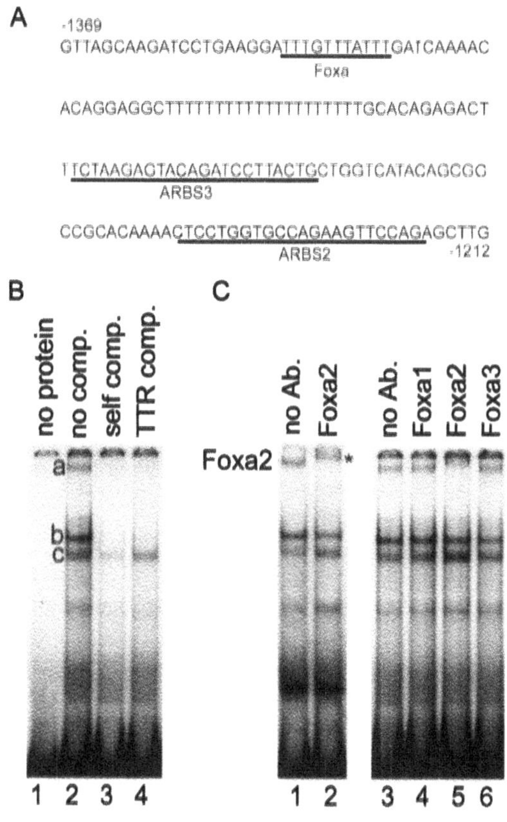

FIGURE 2. Identification of Foxa2 binding site on mE-RABP promoter. (**A**) Foxa2 binding motif exists in the vicinity of AR binding sites on mE-RABP promoter. Underlined are the Foxa binding site core sequence and AR binding sites. (**B**) EMSA. Oligomers containing potential Foxa binding motif on mE-RABP promoter were labeled as probe. Foxa binding sequence on TTR gene promoter was used as a competitor. EMSA assay was performed using nuclear extract from DC2 cells and labeled Foxa probe. Three DNA-protein complexes (*bands* **a, b, c**) were formed. 300-fold excess non-labeled self competitor (*lane* 3) or TTR competitor (*lane* 4), which contains consensus Foxa binding sites, completely eliminated *bands* **a** and **b**. (**C**) Antibody super shift. Foxa2 antibody (*lanes* 2 and 5), but not Foxa1 (*lane* 4) or Foxa3 (*lane* 6) antibody gave a supershift, as indicated by an *asterisk*.

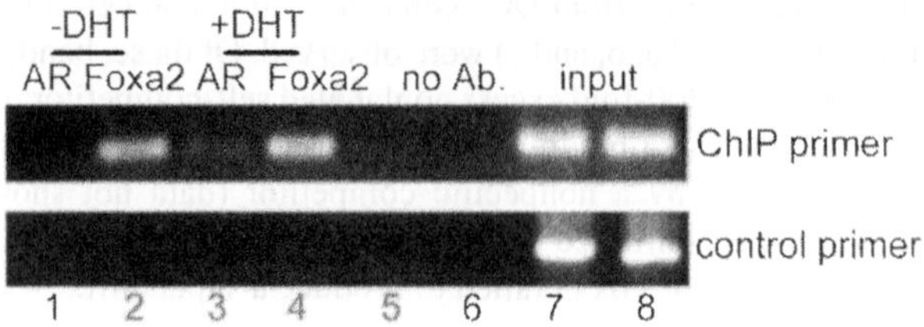

FIGURE 3. Chromatin immunoprecipitation (ChIP) assay confirmed Foxa2 binding on mE-RABP promoter in epididymal cells. ChIP assay was performed in DC2 cells treated with (*lanes* 3 and 4) or without (*lane* 1 and 2)10^{-8} M DHT. After DHT induction, AR was recruited to mE-RABP promoter (*lane* 3). In contrast, Foxa2 constitutively occupied mE-RABP promoter in the presence (*lane* 4) or absence (*lane* 2) of DHT.

Foxa2 Protein Occupies mE-RABP Promoter in Vivo

Chromatin immunoprecipitation (ChIP) was performed on DC2 cells as described in the MATERIAL AND METHODS section. Test primer pairs covered –1.2 kb to –1.4 kb of the mE-RABP promoter region where the AR and Foxa2 binding sites were identified: a control primer pair corresponding to a far downstream region of mE-RABP promoter that lacks consensus sequences for these TFs. As shown in FIGURE 3, AR was recruited to the mE-RABP promoter in the presence of DHT. In contrast, Foxa2 constantly occupied the promoter in the presence or absence of DHT. In our analysis of prostate-specific promoters, we observed a similar order of transcription factor binding to the DNA binding sequences such that Foxa1 was always bound, but AR was recruited to the promoter following DHT treatment of prostate cancer cells *in vitro*.[5]

Foxa Proteins Interact with AR

Foxa1 has been reported to interact with AR and regulate prostatic gene expression.[5] To determine whether similar protein interactions also occur between Foxa2 and AR, GST-pull down experiments were performed. *In vitro* translated full length Foxa1 or Foxa2 was incubated with various purified AR-GST fusion proteins in which different domains of AR were fused with GST. Similar to Foxa1, Foxa2 bound specifically to AR-GST protein (FIG. 4A). Further, the DNA binding domain of AR was required and sufficient for the protein/protein interaction, indicating that Foxa2 interacts with AR through DNA binding domain (FIG. 4A).

The association of Foxa proteins with AR was further confirmed by EMSA. As shown in FIGURE 4B, AR alone did not bind to Foxa specific probe, but it was recruited to the probe in the presence of either Foxa1 or Foxa2 protein. The reverse experiment was carried out using an AR binding probe (FIG. 4C), which corresponds to the ARBS3 from mE-RABP promoter. *In vitro* translated Foxa1 and Foxa2 protein did not bind to ARBS3 probe (FIG. 4C, lanes 1 and 2), but was recruited to the DNA in the presence of AR. In contrast, an *in vitro*-translated control protein was not recruited by AR to ARBS3 probe (FIG. 4D), supporting the specific interaction between AR and Foxa proteins. These EMSA results indicated that Foxa1 or Foxa2

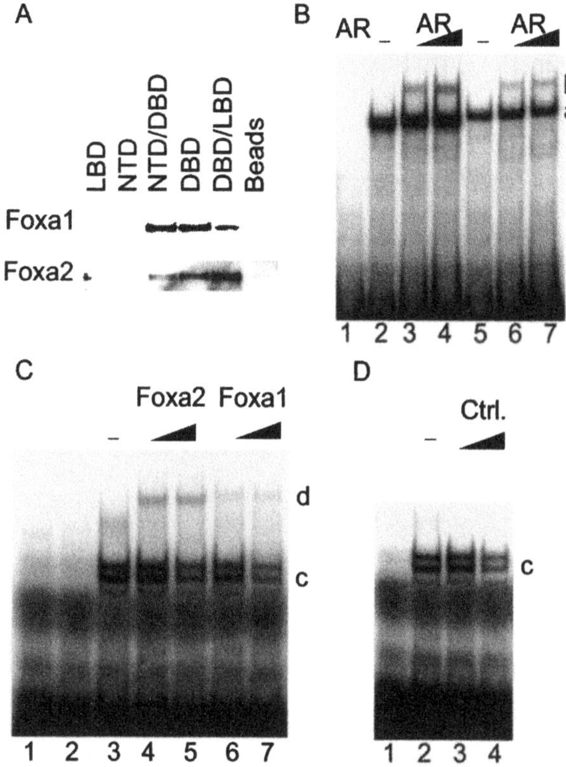

FIGURE 4. Foxa proteins interact with the AR. (**A**) GST pull-down. Different domains of AR were fused with GST and used in GST pull-down assays. LBD = ligand binding domain; NTD = N-terminal domain; DBD = DNA binding domain. As with Foxa1, Foxa2 specifically bound to DNA binding domain of AR. (**B**) EMSA. Foxa probes were incubated with AR (*lane* 1), Foxa1 (*lane* 2) or Foxa2 (*lane* 5) protein. Both Foxa1 and Foxa2 bound to the probe and formed a complex (labeled **a** in FIG. 4B). AR did not bind to this probe (*lane*1), but was recruited to the probe in the presence of Foxa1 (*lanes* 3 to 4) or Foxa2 (*lanes* 6 to 7) and formed an extra complex (labeled **b** in FIG. 4B). (**C**) EMSA. Reverse experiment was carried out using AR binding probe. *In vitro*-translated Foxa1 or Foxa2 did not bind to ARBS3 probe (*lanes* 1 and 2); AR bound to this probe and formed a complex (labeled **c** in FIG. 4C, *lanes* 3 to 7). With the addition of increasing amount of Foxa1 or Foxa2 protein, one more protein-DNA complex was formed: complex **d** in lanes 4 to 7, which consisted of DNA/AR/Foxa. (**D**) EMSA. An *in vitro* translated control protein did not bind to ARBS3 probe (*lane* 1), and was not recruited to the probe in the presence of AR (*lanes* 3 and 4).

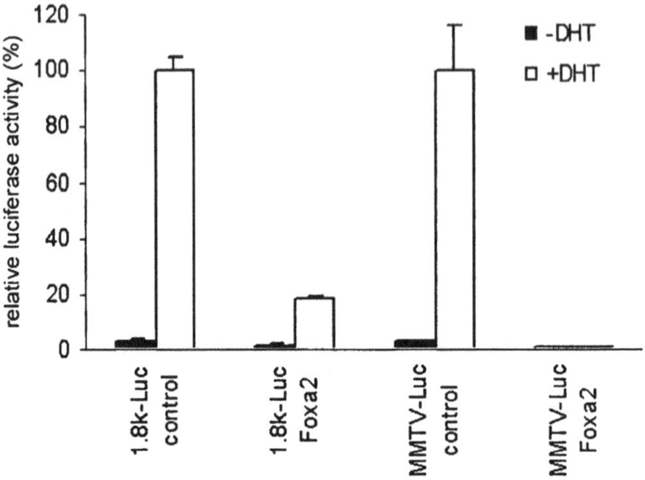

FIGURE 5. Co-transfection of Foxa2 suppressed androgen induction on mE-RABP promoter. DC2 cells were transfected with 1.8 kb mE-RABP-luciferase construct and Foxa2 expression plasmid. Co-transfection of Foxa2 reduced androgen activation of mE-RABP promoter to 20%. Similarly, co-transfection of Foxa2 also reduced androgen activation of MMTV promoter.

proteins can be recruited specifically by AR to the AR binding site through protein/protein interaction without direct binding of the forkhead proteins to the DNA.

Transfected Foxa2 Suppresses the Epididymal Promoter, but Activates the Prostatic Promoter in the Absence of Androgen

Our studies have shown that Foxa1 is critical for androgen-induced expression of the prostate-specific PSA promoter[5] and that Foxa2 can activate the PSA promoter in the absence of DHT.[21] In contrast, co-transfection of Foxa2 suppressed androgen induction on mE-RABP promoter. Consistent with a recent report,[26] suppression of activity by Foxa2 was also observed on the MMTV promoter (FIG. 5).

Foxa2 is Expressed in Prostatic Neuroendocrine Cancer

During prostate development, Foxa2 is expressed only in budding prostate between E18 to day 1. After birth, the Foxa2 expression dramatically declines.[13] In agreement with previous findings,[21] Foxa2 was found to be co-expressed with NE marker-synaptophysin both in the mouse prostate cancer TRAMP model (which develops prostate NE tumor) and in human NE cancer tissues (FIG. 6). In TRAMP mouse prostates, the NE marker synaptophysin and Foxa2 co-express in prostatic intraepithelial neoplasia, PIN (FIG. 6A prior to the appearance of NE prostate carcinoma.

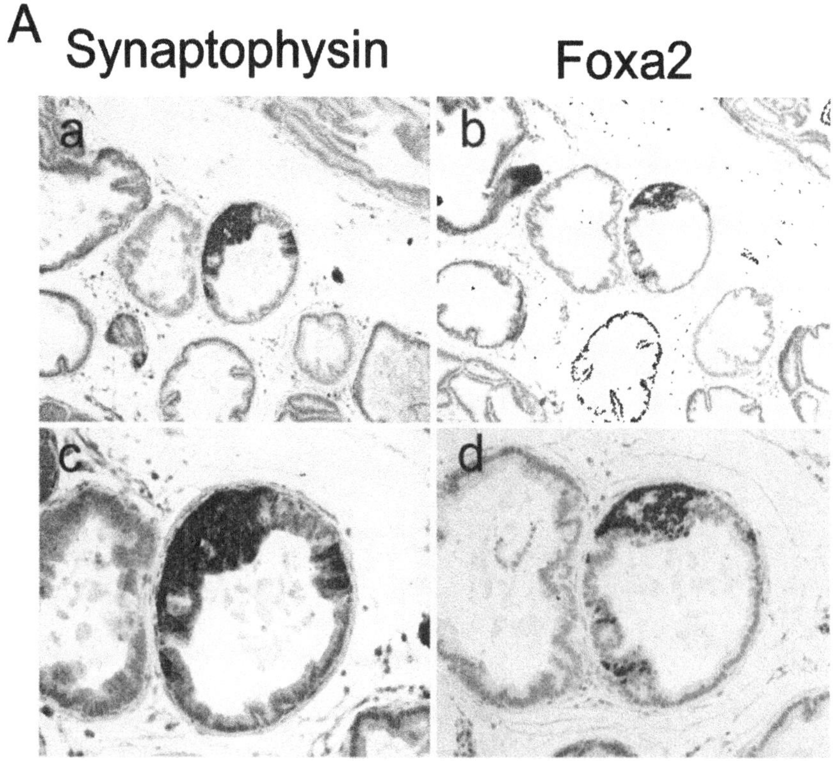

FIGURE 6A. Foxa2 is expressed in prostatic NE cancer. Foxa2 is coexpressed with NE marker in PIN in the TRAMP mouse prostate cancer model. Serial sections from TRAMP prostates were immunostained for synaptophysin and Foxa2. **a** and **b** represent 100× magnification, whereas **c** and **d** are 400× magnification, respectively. Foxa2 was detected in synaptophysin-positive cells.

Foxa2 Expression was Occasionally Observed in Glands of High-Grade Carcinomas, and in All Human Prostate NE Cancer

Tissues for this study were from human biopsy samples. Serial section immunohistochemistry demonstrated the co-expression of Foxa2 in human NE cancer (FIG. 6B). Serial sections a to d showed simultaneous expression of Foxa2 and synaptophysin. In another set of serial sections (e to h), Foxa2 expression was observed in synaptophysin-positive cells. These findings demonstrate that Foxa2 is a marker for NE cells and strongly supports the hypothesis that Foxa2 is involved in prostate NE cancer progression.[21]

DISCUSSION

Our group's studies have shown that Foxa1 is involved in androgen-regulated prostate-specific gene expression. Foxa1 is expressed constitutively in prostate, and prostates rescued from Foxa1-deficient mice displayed basal-like cell hyper-

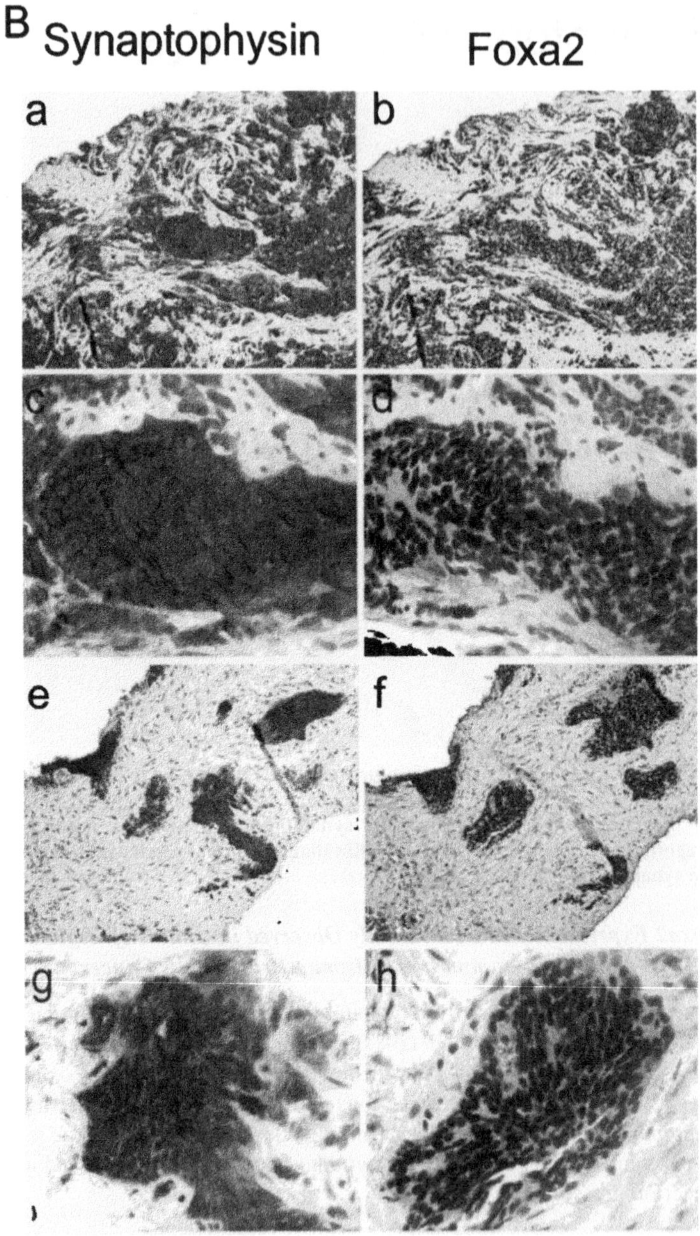

FIGURE 6B. Foxa2 is expressed in prostatic NE cancer (*continued*). Human prostate biopsy samples were used for the study. Foxa2 was coexpressed with the NE marker synaptophysin in human prostatic NE cancer specimens. **a** to **b** (100×) and **c** to **d** (400×) are from one set of serial sections; **e** to **f** (100×) and **g** to **h** (400×) are from another set. Foxa2 was coexpressed with the NE marker synaptophysin in all the human prostatic NE cancer samples examined.

plasia.[5,14] Foxa2 is only expressed in budding prostate during development between 18E to 1D and is dramatically reduced in the adult prostate tissue to barely detectable levels.[13] However, Foxa2 is expressed in normal prostate neuroendocrine (NE) cells and in NE cancer.[13,21] Although NE cells represent only a minor population of the prostate epithelium, they may contribute to the progression of prostate cancer, in that NE differentiation can be found in 10%-100% adenocarcinomas depending on the sensitivity of the technique applied, and tumors with a prominent NE component are typically hormone refractory and highly aggressive.[27]

Foxa2 was found to co-express the NE marker synaptophysin in TRAMP PIN lesions as well as NE cancer. As we have reported, Foxa2 is also expressed in a subset of high-grade primary human prostate adenocarcinomas and in all cases of human prostate NE cancer examined.[21] Because Foxa2 can bind to the Foxa1 binding site on the PSA promoter, we tested Foxa2's ability to replace Foxa1. Surprisingly, Foxa2 can convert the androgen-dependent PSA promoter to function in an androgen-independent manner.[21] This suggests that Foxa2, or a closely related forkhead protein, may represent a common mechanism by which prostate cancer can progress from androgen-dependent to androgen-independent disease.

Our data also showed that Foxa2 regulates epididymal gene expression and exhibits a regionalized expression pattern along the epididymal duct, which is consistent with recently released data showing that Foxa2 is expressed in segments 5, 7, 8, and 9 of the epididymis (<http://mrg.genetics.washington.edu>). A transcriptional profiling study revealed that more than 2,000 genes exhibit a regionalized expression pattern in epididymis.[28] Among these genes, regulatory factors, which also target segments in a region-specific manner, may account for the differentially regulated gene expression in the epididymis.

For example, the *hox* family of transcription factors has been shown to play a pivotal role during development by determining body patterning and segmentation, and these are expressed predominantly in particular segments of the epididymis.[29] Further, the functional significance of the regionalized expression of this family of transcription factors was confirmed by loss-of-function studies: *hox*a-10 and *hox*a-11 knockout mice are infertile and display disorganized epididymal structure.[30,31] These results suggest that the region-specific distribution of transcription factors can be the determining factor in segment-specific expression of genes.

Given that the Foxa2 transcription factor displays a regionalized expression pattern along the epididymal duct, and is capable of interacting with AR to regulate epididymis-specific gene expression, it is logical to suggest that Foxa2 contributes to epididymis regionalization of gene expression and may play a role in sperm maturation as well as in the development of the epididymis. This hypothesis is supported by other studies which have shown that Foxa proteins play critical roles in both segmentation and organ development. When overexpressed in posterior neural plate, the *Xenopus* XBF-1 (now termed FoxG1) suppresses endogenous neuronal differentiation and induces adjacent cell neurogenesis,[32] indicating the involvement of this forkhead protein in early embryo patterning. Overexpression of Foxa2 in midbrain/hindbrain region in transgenic mice results in disorganized development of this structure with a disruption in dorsoventral patterning and induction of a floor plate marker,[33] thus further confirming a role for Foxa2 in organ segmentation.

The epididymis is an organ that develops from the mesoderm. During development, intermediate mesoderm forms the urogenital ridge from which arises the pronephric ducts (part of urogenital ridge), Wolffian ducts, and mesonephric ducts. In male mice, at E13-14, under the stimulation of testosterone secreted by Leydig cells within the fetal testes, Wolffian ducts start differentiation and eventually develop into epididymis, vas deferens, seminal vesicle, and efferent ducts.[34] Although Foxa proteins are mainly expressed in endodermally derived organs, some studies have also showed that Foxa2 plays a key role in both endoderm and mesoderm development.[11,17] Our study adds new data suggesting that Foxa2 is involved in gene regulation in epididymis, a mesodermal organ.

Most forkhead proteins have been reported as transcriptional activators, but some members of this family have also been shown to repress transcription, such as FoxD3 and FoxG1.[35] FoxG1 suppresses target gene transcription through interaction with the transcriptional co-repressor of the Groucho/TLE family proteins.[36] In a GST-pull down assay, Foxa2 was also shown to directly interact with Groucho/TLE protein.[37] Overexpression of the TLE protein suppressed Foxa2-mediated transactivation, while co-transfection of Grg5 (a dominant negative form of TLE) enhanced Foxa2's transactivation.[37] This suggests that TLE can act as co-repressor of Foxa2 to inhibit target gene transcription.

In our study, the overexpression of Foxa2 in transient transfections suppressed mE-RABP promoter transactivation. In addition, our later studies showed that both TLE and Grg5 were highly expressed in epididymis (data not shown), indicating that TLE/Grg5 might be involved in Foxa2's regulation of epididymal genes. A balance among TLE/Grg5, Foxa2, and other TFs may define the region-specific gene expression patterns seen in the epididymis. Further experiments are required to decipher the role of TLE and Grg5 to alter Foxa2 function that is expressed in a gradient along the epididymis.

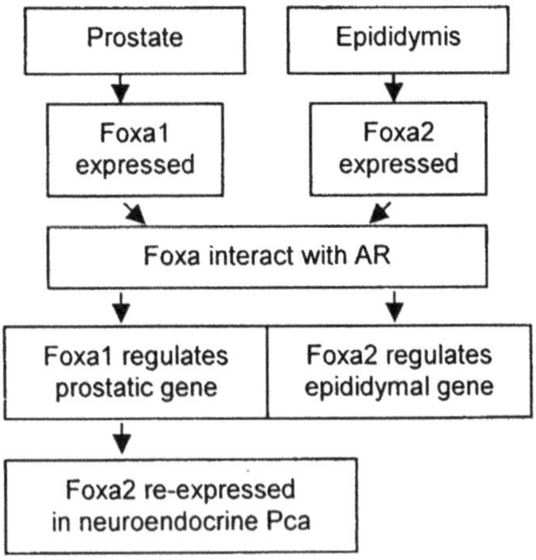

FIGURE 7. Foxa proteins differentially regulate prostatic and epididymal genes. Foxa proteins are differentially expressed in prostate and epididymis, with Foxa1 in normal prostate and Foxa2 in the epididymis. Both Foxa1 and Foxa2 interact with androgen receptor (AR) and coordinate with AR to regulate gene expression in these two organs. Foxa2 expression was occasionally observed in glands of high-grade prostate carcinomas and in all human prostate NE cancer.

In summary, our studies demonstrate that Foxa1 and Foxa2 are important transcription factors that control multiple androgen-regulated genes. As illustrated in FIGURE 7, in two organs controlled by androgens, the Foxa proteins are expressed differentially and are not exchangeable functionally. In the prostate, Foxa1 is expressed, capable of interacting with AR and regulating prostate-specific gene expression, while in the epididymis, Foxa2 is expressed and interacts with AR to regulate epididymal genes. Foxa2 is expressed only in early budding of the prostate and in prostatic NE cells. However, in addition to NE prostate cancer, Foxa2 can be expressed in late-stage prostate cancer. Our data, showing that Foxa2 can regulate androgen-dependent prostatic genes in an androgen-independent fashion, suggest a possible mechanism by which prostate cancer patients can fail androgen ablation therapy and progress to androgen independence.

ACKNOWLEDGMENTS

We thank Renjie Jin, Tom Case and Manik Paul for technical support and help. This research was supported by National Institutes of Health grants to Robert J. Matusik (R01 CA76142-06, R01, and DK55748-05, R01 AG023490-01); the Frances Williams Preston Laboratories of the T.J. Martell Foundation; and by an NIH grant to Marie-Claire Orgebin-Crist (HD36900). Janni Mirosevich is a recipient of a Department of Defense Postdoctoral Traineeship Award (W81XWH-04-1-0050).

REFERENCES

1. LAI, E., V.R. PREZIOSO, W.F. TAO, *et al.* 1991. Hepatocyte nuclear factor 3 alpha belongs to a gene family in mammals that is homologous to the Drosophila homeotic gene fork head. Genes Dev. **5:** 416–427.
2. LAI, E., V.R. PREZIOSO, E. SMITH, *et al.* 1990. HNF-3A, a hepatocyte-enriched transcription factor of novel structure is regulated transcriptionally. Genes Dev. **4:** 1427–1436.
3. KAUFMANN, E. & W. KNOCHEL. 1996. Five years on the wings of fork head. Mech. Dev. **57:** 3–20.
4. CLARK, K.L., E.D. HALAY, E. LAI & S.K. BURLEY. 1993. Co-crystal structure of the HNF-3/fork head DNA-recognition motif resembles histone H5. Nature **364:** 412–420.
5. GAO, N., J.-F ZHANG, M.A. RAO, *et al.* 2003. The role of hepatocyte nuclear factor-3{alpha} (Foxa1) and androgen receptor in transcriptional regulation of prostatic genes. Mol. Endocrinol. **17:** 1484–1507.
6. BESNARD, V., S.E. WERT, W.M. HULL & J.A. WHITSETT. 2004. Immunohistochemical localization of Foxa1 and Foxa2 in mouse embryos and adult tissues. Gene Expr. Patterns. **5:** 193–208.
7. ANG, S.L., A. WIERDA, D. WONG, *et al.* 1993. The formation and maintenance of the definitive endoderm lineage in the mouse: involvement of HNF3/forkhead proteins. Development **119:** 1301–1315.
8. LANTZ, K.A. & K.H. KAESTNER. 2005. Winged-helix transcription factors and pancreatic development. Clin. Sci. (Lond) **108:** 195–204.
9. SASAKI, H. & B.L. HOGAN. 1993. Differential expression of multiple fork head related genes during gastrulation and axial pattern formation in the mouse embryo. Development **118:** 47-59.
10. KAESTNER, K.H., H. HIEMISCH & G. SCHUTZ. 1998. Targeted disruption of the gene encoding hepatocyte nuclear factor 3gamma results in reduced transcription of hepatocyte-specific genes Mol. Cell Biol. **18:** 4245–4251.
11. KAESTNER, K.H., J. KATZ, Y. LIU, *et al.* 1999. Inactivation of the winged helix transcription factor HNF3alpha affects glucose homeostasis and islet glucagon gene expression in vivo. Genes Dev. **13:** 495–504.

12. Weinstein, D.C., A. Altaba, W.S. Chen, *et al.* 1994. The winged-helix transcription factor HNF-3 beta is required for notochord development in the mouse embryo. Cell **78:** 575–588.
13. Mirosevich, J., N. Gao & R.J. Matusik. 2005. Expression of Foxa transcription factors in the developing and adult murine prostate. Prostate **62:** 339–352.
14. Gao, N., K. Ishii, J. Mirosevich, *et al.* 2005. Forkhead box a1 regulates prostate ductal morphogenesis and promotes epithelial cell maturation. Development **132:**3431–3443.
15. Cornwall, G.A., J.J. Lareyre, R.J. Matusik, *et al.* 2002. Gene expression and epididymal function. *In* The Epididymis: From molecules to clinical practice. Robaire, B. & B.T. Hinton, Eds. 169–199. Kluwer Academic and ;Plenum Publishers. New York.
16. Sonnenberg-riethmacher, E., B.walter, D. Riethmacher, *et al.* 1996. The c-ros tyrosine kinase receptor controls regionalization and differentiation of epithelial cells in the epididymis. Genes Dev. **10:** 1184–1193.
17. Lareyre, J.J., W.-L Zheng, G.-Q. Zhao, *et al.* 1998. Molecular cloning and hormonal regulation of a murine epididymal retinoic acid-binding protein mRNA Endocrinology. **139:** 2971–2981.
18. Suzuki, k., J.j. Lareyre, D. Sanchez, *et al.* 2004. Molecular evolution of epididymal lipocalin genes localized on mouse chromosome 2. Gene **339:** 49–59.
19. Lareyre, J.J., M.G. Mattei, S. Kasper, *et al.* 1998. Structure and putative function of a murine epididymal retinoic acid- binding protein (mE-RABP). J. Reprod. Fertil. Suppl. **53:** 59–65.
20. Lareyre, J.J., T.Z. Thomas, W.L. Zheng, *et al.* 1999. A 5-kilobase pair promoter fragment of the murine epididymal retinoic acid-binding protein gene drives the tissue-specific, cell-specific, and androgen-regulated expression of a foreign gene in the epididymis of transgenic mice J. Biol. Chem. **274:** 8282–8290.
21. Mirosevich, J., N. Gao, A. Gupta, *et al.* 2005. The expression and role of Foxa proteins in prostate cancer. The Prostate, In press.
22. Wang, J.C., P.E. Stromstedt, R.M. O'brien & D.K. Granner. 1996. Hepatic nuclear factor 3 is an accessory factor required for the stimulation of phosphoenolpyruvate carboxykinase gene transcription by glucocorticoids. Mol. Endocrinol. **10:** 794–800.
23. Araki, Y., K. Suzuki, R.J. Matusik, *et al.* 2002. Immortalized epididymal cell lines from transgenic mice overexpressing temperature-sensitive simian virus 40 large T-antigen gene. J. Androl **23:** 854–869.
24. Dignam, J., R. Lebovitz & R. Roeder. 1983. Accurate transcription initiation by polymerase II in soluble extract isolated mammalian nuclei. Nucleic Acids Research **11:** 1475–1489.
25. Snoek, R., P.S. Rennie, S. Kasper, *et al.* 1996. Induction of cell-free, in vitro transcription by recombinant androgen receptor peptides. J. Steroid Biochem. Mol. Biol. **59:** 243–250.
26. Holmqvist, P.H., S. Belikov, K.S. Zaret & O. Wrange. 2005. FoxA1 binding to the MMTV LTR modulates chromatin structure and transcription. Exp. Cell Res **304:** 593–603.
27. Di Sant'agnese, P.A. 1998. Neuroendocrine differentiation in prostatic carcinoma: an update. Prostate Suppl **8:** 74–79.
28. Johnston, D.S., S.A. Jelinsky, H.J. Bang, *et al.* 2005. The mouse epididymal transcriptome: transcriptional profiling of segmental gene expression in the epididymis. Biol. Reprod. **73:** 404–413.
29. Bomgardner, D., B.T. Hinton & T.T. Turner. 2003. 5′ hox genes and meis 1, a hox-DNA binding cofactor, are expressed in the adult mouse epididymis. Biol. Reprod. **68:** 644–650.
30. Benson, G.v., H. Lim, B.C. Paria, *et al.* 1996. Mechanisms of reduced fertility in Hoxa-10 mutant mice: uterine homeosis and loss of maternal Hoxa-10 expression. Development **122:** 2687–2696.
31. Hsieh-li, H.M., D.P. Witte, M. Weinstein, *et al.* 1995. Hoxa 11 structure, extensive antisense transcription, and function in male and female fertility. Development **121:** 1373–1385.

32. BOURGUIGNON, C., J. LI & N. PAPALOPULU. 1998. XBF-1, a winged helix transcription factor with dual activity, has a role in positioning neurogenesis in Xenopus competent ectoderm. Development **125:** 4889–4900.
33. SASAKI, H. & B.L. HOGAN. 1994. HNF-3 beta as a regulator of floor plate development. Cell **76:** 103–115.
34. STAACK, A., A.A. DONJACOUR, J. BRODY, *et al.* 2003. Mouse urogenital development: a practical approach. Differentiation **71:** 402–413.
35. SUTTON, J., R. COSTA, M. KLUG, *et al.* 1996. Genesis, a winged helix transcriptional repressor with expression restricted to embryonic stem cells. J. Biol. Chem. **271:** 23126–23133.
36. YAO, J., E. LAI & S. STIFANI. 2001. The winged-helix protein brain factor 1 interacts with groucho and hes proteins to repress transcription. Mol. Cell Biol. **21:** 1962–1972.
37. WANG, J.C., M. WALTNER-LAW, K. YAMADA, *et al.* 2000. Transducin-like enhancer of split proteins, the human homologs of Drosophila groucho, interact with hepatic nuclear factor 3beta. J. Biol. Chem. **275:** 18418–18423.

Role of Glial Cell Line–Derived Neurotrophic Factor in Germ-Line Stem Cell Fate

LAURA BRAYDICH-STOLLE,[a] COURTNEY NOLAN,[a] MARTIN DYM,[b] AND MARIE-CLAUDE HOFMANN[a]

[a]*Department of Biology, University of Dayton, Dayton, Ohio 45469, USA*

[b]*Department of Cell Biology, Georgetown University Medical Center, Washington, DC 20057, USA*

ABSTRACT: The overall goal of this study is to unravel the role(s) played by glial cell line-derived neurotrophic factor (GDNF) in the fate of spermatogonial stem cells. There is great interest in the biology of spermatogonial stem cells, or A_{single} spermatogonia, because of their importance in the treatment of infertility, the development of contraceptives, and the understanding of the etiology of testicular cancer, particularly seminoma. In the mouse, spermatogonial stem cells express GFRα-1, the receptor for GDNF, and respond to this growth factor *in vivo* and *in vitro*. GDNF is produced by the adjacent Sertoli cells, which are part of the germ-line stem cell niche in vertebrates. We specifically isolated GFRα-1–positive spermatogonia using an immunomagnetic bead technique. We then stimulated the cells with 100 ng/mL of rGDNF for 10 hours; unstimulated cells served as negative controls. Microarray analysis, immunocytochemistry, and Western blotting revealed that Numb, a regulator of the Notch pathway, is upregulated by GDNF in spermatogonial stem cells. There are indications that in rats, mice, and humans, the Notch pathway promotes spermatogonial differentiation. We observed that an increase in Numb expression is concomitant with Notch degradation in these cells. Thus, through Numb, GDNF might inhibit differentiation and allows the maintenance of the stem cell pool in the mouse seminiferous epithelium.

KEYWORDS: GDNF; Notch; Numb; spermatogonia; testis stem cells

INTRODUCTION

In mammals, spermatogenesis is a tightly controlled process that takes place within the seminferous tubules of the testes and leads to sperm production. This process depends on the self-renewal and differentiation of spermatogonial stem cells (SSCs), or A_{single} spermatogonia, which are adjacent to the basement membrane of the seminiferous epithelium. Self-renewal and differentiation of the SSCs are regulated by soluble factors produced by the somatic Sertoli cells. One of these growth factors is glial cell line–derived neurotrophic factor (GDNF), which stimulates the

Address for correspondence: Marie-Claude Hofmann, Ph.D., Department of Biology, University of Dayton, 300 College Park, Dayton, OH 45469-2320. Voice: 937-229-2894; fax: 937-229-2021.
Marie-Claude.Hofmann@notes.udayton.edu

Ann. N.Y. Acad. Sci. 1061: 94–99 (2005). © 2005 New York Academy of Sciences.
doi: 10.1196/annals.1336.010

TABLE 1. Differential expression of genes influenced by GDNF and belonging to the Notch signaling pathway

Name	-Fold Expression (microarray analysis)	GenBank accession #
Notch-1	1.20	Z11886
Notch-2	1.10	D32210
Notch-3	1.10	X74760
Numb	4.44	U70674
Presenilin	−1.84	U57325

proliferation of the A_{single} spermatogonia and their direct progeny, the A_{paired} spermatogonia, *in vivo* and *in vitro.*[1,2] GDNF signals through a two-component receptor system comprising a GPI-anchored co-receptor called GFRα-1, and the Ret transmembrane receptor. Binding of GDNF to GFRα-1 and Ret activates Ret autophosphorylation and the intracellular relay of the signal.[3]

Signaling pathways leading to spermatogonial stem cell self-renewal and differentiation are poorly understood. This is due to the fact that this stem cell population is small (0.03% of the total number of germ cells) and difficult to specifically isolate. In order to elucidate which genes are differentially regulated by GDNF, we recently performed microarray analysis of pure GFRα-1–positive mouse spermatogonia isolated through the STAPUT method followed by magnetic activated cell sorting (MACS).[2] Some of the genes differentially regulated by GDNF appeared to belong to the Notch signaling pathway, as presented in TABLE 1. While GDNF does not influence the expression of Notch-1, -2, -3, or presenilin, it seems to upregulate the expression of the gene *numb*, a known repressor of Notch signaling.[4] The Notch proteins (Notch 1-4) form a family of transmembrane receptors mediating juxtacrine signaling. These receptors and their ligands have been intensively studied in *Drosophila* and *C. elegans* and more recently in vertebrates.[4] Notch signaling regulates stem cell fate choices in different tissues, generally by adjusting the cell's response to differentiation stimuli instead of specifying a cell's fate directly. Therefore, depending on the tissue and environmental cues, Notch signaling can promote stem cell self-renewal or differentiation. Once a Notch receptor is activated by its ligand, the receptor is cleaved proteolytically and an active form of the receptor is released. This active form is called the Notch intracellular domain (NICD). NICD can then enter the nucleus, where it acts as a transcription factor.[5] One modulator of Notch is Numb, a cytoplasmic protein that binds to NICD and inhibits its action by promoting either its endocytosis or ubiquitination.[6,7]

Notch receptors and their homologues are expressed in the germ line, and seem to promote germ cell proliferation in *C. elegans* and *Drosophila.*[8,9] There are reports that the Notch signaling pathway promotes spermatogonial stem cell differentiation in humans.[10] The exact function of Notch signaling in murine spermatogonial stem cells is not known, but it has been hypothesized that it could promote differentiation as well.[11,12] In the mammalian testis, the ligands for the Notch receptors, Jagged-1 and Jagged-2, are expressed by Sertoli cells. The present study investigated the influence of GDNF on the protein expression of Numb and the Notch-1 intracellular domain (NICD) in spermatogonial stem cells. Our data indicate that GDNF might induce self-renewal through the production of Numb and the degradation of Notch-1.

MATERIAL AND METHODS

Isolation and Culture of GFRα-1–Positive Cells

Testes were isolated from 60 mouse pups at 6 dpp. The tunica was removed to expose the seminiferous tubules. The tubules were digested with a sequential enzymatic treatment to yield a single cell suspension. The cells were then separated using the STAPUT method followed by magnetic activated cell sorting (MACS) using an anti-mouse GFRα-1 antibody (Santa Cruz Biotechnology, Santa Cruz, CA).[2] The GFRα-1–positive cells were seeded in microtiter plates at a concentration of 5×10^4 cells per well, in minimal media consisting of D-MEM/F12 complemented with 10% NU synthetic serum (Fisher Scientific, Pittsburgh, PA).

Cultures of Seminiferous Tubules

Testes were removed from immature and adult mice and the tunica was removed to expose the seminiferous tubules. The tubules were then digested with 1 mg/mL of collagenase and 1 μg/mL DNase to remove the basement membrane and expose the GFRα-1–positive cells. The tubules were placed in minimal media in 24-well tissue culture plates overnight and used for the experiments the next day.

GDNF Treatment

GFRα-1–Positive Spermatogonia

After STAPUT/MACS isolation, the spermatogonia were cultured for 1 day in microtiter plates containing minimal media with or without 100 ng/mL of GDNF. The next day, the cultures were fixed *in situ* with ice-cold methanol, the protein Numb revealed by immunocytochemistry and its expression evaluated directly in the culture plates using an Olympus Nikon Eclipse TE 200 inverted microscope.

Organ Cultures

The digested seminiferous tubules were treated with 100 ng/mL GDNF (R&D Systems, Minneapolis, MN) or 100 ng/mL GDNF + 5 μg/mL GDNF neutralizing antibody (R&D Systems). Untreated tubules were used as a control. Fresh treatments were applied daily for 3 days, and then proteins were isolated as described below.

Protein Isolation, Concentration, and Western Blotting

The seminiferous tubules were homogenized in a modified RIPA lysis buffer to extract the proteins. The samples were then concentrated using centrifugal concentrators from Millipore (cutoff = 30,000 Da)(Millipore, Bedford, MA). After the samples were concentrated, their protein content was determined using the Bio-Rad DC assay (BioRad Laboratories, Hercules, CA). Twenty-five μg of proteins/lane were run on a 10% Tris-HCl gel and transferred to nitrocellulose membranes. The mem-

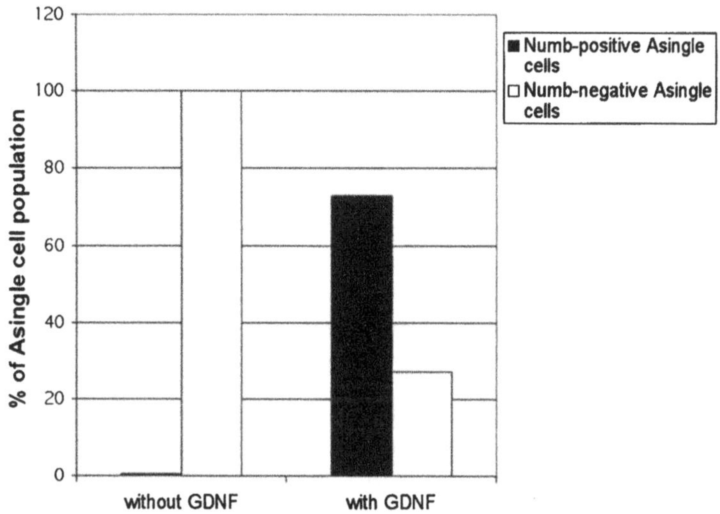

FIGURE 1. Expression of Numb in Asingle spermatogonia treated with GDNF. GFRα-1–positive spermatogonia were isolated using the STAPUT/MACS method, cultured overnight in presence or absence of 100 ng/mL GDNF and processed for Numb staining by immunocytochemistry. The percentage of positive/negative Asingle cells is shown.

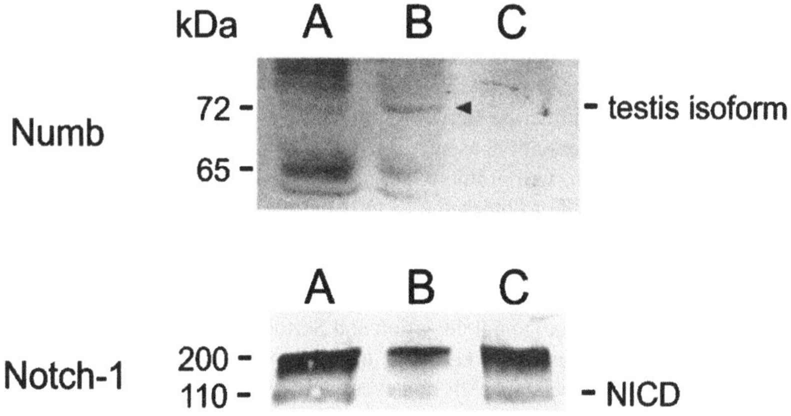

FIGURE 2. Expression of Numb and Notch-1 in seminiferous tubules treated with GDNF. Seminiferous tubules were cultured for 3 days in presence of 100 ng/mL GDNF and prepared for SDS-PAGE/Western blotting. (**A**) No GDNF; (**B**) GDNF; (**C**) GDNF + 5 μg/mL neutralizing antibody. *Upper panel*: Western blot for Numb, showing upregulation of the testis isoform of Numb by GDNF(arrowhead). *Lower panel*: Western blot for Notch-1, showing downregulation of Notch-1 and NICD by GDNF.

branes were probed with antibodies for Numb and Notch-1 (Santa Cruz Biotechnolgy), and revealed using the immunoperoxidase technique and 4-chloronaphtol. Experiments were done in triplicate. The testis isoform of the Numb protein has a molecular weight of 72 kDa, and the Notch-1 intracellular domain has a molecular weight of 120 kDa.

RESULTS AND DISCUSSION

The study of spermatogonial stem cells and their niche is of paramount importance to understand the mechanisms of self-renewal and differentiation of unipotent stem cells, and to elucidate which disruptions of specific signaling pathways lead to male infertility or testicular neoplasias. In the present study, we showed that the treatment of isolated spermatogonial stem cells and seminiferous tubules with 100 ng/mL GDNF increases the expression of the protein Numb, which confirmed our results obtained by microarray analysis and semi-quantitative RT-PCR.[2] As shown in FIGURE 1, immunocytochemistry data revealed that after overnight treatment with GDNF *in vitro*, many A_{single} spermatogonia express Numb (73%) , while the staining was not observed in absence of GDNF. Further, we specifically showed that only the testis isoform of Numb (MW = 72 kDa) is upregulated (FIG. 2, upper panel). Numb is a membrane-associated protein that functions as a determinant of cell fate during *Drosophila* and mouse development. Four specific isoforms of mammalian Numb exist, with molecular weights of 65, 66, 71, and 72 kDa that are generated by alternative splicing of the Numb mRNA. The expression of some of these isoforms is regulated in a cell type–specific manner.[13] One function of Numb is the regulation of the Notch-1 signaling pathway by promoting the endocytosis or ubiquitination of the Notch intracellular domain (NICD).[6–7] We next tested the samples for the expression of Notch-1 and the NICD protein in presence or absence of GDNF. As shown in FIGURE 2 (lower panel), the presence of GDNF lowers the amount of NICD available, in comparison to control samples without GDNF, or with GDNF treated with a neutralizing antibody. These data allow us to devise the following working model: the juxtracrine signaling initiated by the binding of Jagged-1 and Notch-1 induces differentiation, as suggested by others.[10–12] GDNF, if present at a sufficient concentration, will promote the production of Numb, which will be able to degrade enough NICD to block the differentiation process and thus favor spermatogonial stem cell self-renewal. This model is in accord with the data of the group of H. Sariola, which showed that the dosage of GDNF is important. Indeed, in GDNF heterozygous knockout mice, spermatogonial differentiation prevails over self-renewal, leading to spermatogonial depletion and Sertoli cell–only syndrome. On the other side, the overexpression of GDNF in transgenic mice induces aberrant proliferation of spermatogonial stem cells and eventually seminoma-like tumors.[14,15]

REFERENCES

1. MENG, X., M. LINDAHL, M.E. HYVONEN, *et al.* 2000. Regulation of cell fate decision of undifferentiated spermatogonia by GDNF. Science **287:** 1489–1493.
2. HOFMANN, M.C., L. BRAYDICH-STOLLE & M. DYM. 2005. Isolation of mouse germ line stem cells; influence of GDNF. Dev. Biol. **279:** 114–124.

3. JING, S., D. WEN, Y. YU, *et al.* 1996. GDNF-induced activation of the ret protein tyrosine kinase is mediated by GDNFR-alpha, a novel receptor for GDNF. Cell **85:** 1113–1124.
4. WEINMASTER, G. 1997. The ins and outs of notch signaling. Mol. Cell Neurosci. **9:** 91–102.
5. JARRIAULT, S., C. BROU, F. LOGEAT, *et al.* 1995. Signalling downstream of activated mammalian Notch. Nature. **377:** 355–358.
6. SCHWEISGUTH, F. 2004. Regulation of Notch signaling activity. Curr. Biol. **14:** R129–138.
7. MCGILL, M.A., & C.J. MCGLADE. 2003. Mammalian numb proteins promote Notch1 receptor ubiquitination and degradation of the Notch1 intracellular domain. J. Biol. Chem. **278:** 23196–23203.
8. CRITTENDEN, S.L., E.R. TROEMEL, T.C. EVANS, & J. KIMBLE. 1994. GLP-1 is localized to the mitotic region of the *C. elegans* germ line. Development **120:** 2901–2911.
9. XU, T., L.A. CARON, R.G. FEHON, & S. ARTAVANIS-TSAKONAS. 1992. The involvement of the Notch locus in *Drosophila* oogenesis. Development. **115:** 913–922.
10. HAYASHI, T., T. YAMADA, Y. KAGEYAMA, *et al.* 2004. Expression failure of the Notch signaling system is associated with the pathogenesis of maturation arrest in male infertility patients. Fertil. Steril. **81:** 697–699.
11. HAYASHI, T., Y. KAGEYAMA, K. ISHIZAKA, *et al.* 2001. Requirement of Notch 1 and its ligand jagged 2 expressions for spermatogenesis in rat and human testes. J. Androl. **22:** 999–1011.
12. DIRAMI, G., N. RAVINDRANATH, M.V. ACHI & M. DYM. 2001. Expression of Notch pathway components in spermatogonia and Sertoli cells of neonatal mice. J Androl. **22:** 944–952.
13. DHO, S.E., M.B. FRENCH, S.A. WOODS & C.J. MCGLADE. 1999. Characterization of four mammalian numb protein isoforms: identification of cytoplasmic and membrane-associated variants of the phosphotyrosine binding domain. J. Biol. Chem. **274:** 33097–33104.
14. MENG, X., D.G. DE ROOIJ, K. WESTERDAHL, *et al.* 2001. Promotion of seminomatous tumors by targeted overexpression of glial cell line-derived neurotrophic factor in mouse testis. Cancer Res. **61:** 3267–3271.
15. SARIOLA, H. & M. SAARMA. 2003. Novel functions and signalling pathways for GDNF. J. Cell Sci. **116:** 3855–3862.

P450 Oxidoreductase Deficiency

A New Disorder of Steroidogenesis

WALTER L. MILLER, NINGWU HUANG, AMIT V. PANDEY, CHRISTA E. FLÜCK, AND VISHAL AGRAWAL

Department of Pediatrics, University of California, San Francisco, San Francisco, California 94143-0978, USA

ABSTRACT: Microsomal P450 enzymes, which metabolize drugs and catalyze steroid biosynthesis. require electron donation from NADPH via P450 oxidoreductase (POR). POR knockout mice are embryonically lethal, but we found recessive human POR missense mutations causing disordered steroidogenesis and Antley-Bixler syndrome (ABS), a skeletal malformation syndrome featuring craniosynostosis. Dominant mutations in exons 8 and 10 of fibroblast growth factor receptor 2 (FGFR2) cause phenotypically related craniosynostosis syndromes and were reported in patients with ABS and normal steroidogenesis. Sequencing POR and FGFR2 exons in 32 patients with ABS and/or hormonal findings suggesting POR deficiency showed complete genetic segregation of POR and FGFR2 mutations. Fifteen patients carried POR mutations on both alleles, four carried POR mutations on 1 allele, nine carried FGFR2/3 mutations on one allele and no mutation was found in three patients. The 34 affected POR alleles included 10 with A287P, 7 with R457H, 9 other missense mutations and 7 frameshifts. These 11 missense mutations and 10 others identified by database mining were expressed in *E. coli*, purified to apparent homogeneity, and their catalytic capacities were measured in four assays: reduction of cytochrome *c*, oxidation of NADPH, and support of the 17α-hydroxylase and 17,20 lyase activities of human P450c17. As assessed by V_{max}/K_m, 17,20 lyase activity provided the best correlation with clinical findings. Modeling human POR on the X-ray crystal structure of rat POR shows that these mutant activities correlate well with their locations in the structure. POR deficiency is a new disease, distinct from the craniosynostosis syndromes caused by FGFR mutations.

KEYWORDS: adrenal hyperplasia; steroidogenesis; Antley-Bixler syndrome; P450; FGFR2; steroid hormone

INTRODUCTION

Cytochrome P450 enzymes, so-named because of their characteristic spectral shift at 450 nm, are heme-containing proteins that catalyze a broad array of oxidative reactions. Cytochrome P450 enzymes falls into two categories: Type I enzymes,

Address for correspondence: Professor Walter L. Miller, Department of Pediatrics, Bldg. MR-IV, Room 209, University of California, San Francisco, San Francisco 94143-0978, CA. Voice: 415-476-2598 ; fax: 415-476-6286.
wmlab@itsa.ucsf.edu

**Ann. N.Y. Acad. Sci. 1061: 100–108 (2005). © 2005 New York Academy of Sciences.
doi: 10.1196/annals.1336.012**

which are found in bacteria and in mitochondria, receive electrons from NADPH through a flavoprotein (ferredoxin reductase) and an iron/sulfur protein (ferredoxin); Type II enzymes, which are found in the endoplasmic reticulum, receive electrons from NADPH through a single protein, P450 oxidoreductase (POR).[1] The human genome contains 57 genes for P450s, of which 50 are Type II. Among these 50 Type II P450s, about 15 are involved in drug metabolism, 20 are involved in the biosynthesis of sterols, fatty acids, and eicosanoids, and 15 are "orphan" enzymes, whose activities remain unknown or ill-defined.[2] Steroidogenic Type II P450 enzymes include P450c17 (17α-hydroxylase/17,20 lyase), P450c21 (21-hydroxylase), and P450aro (aromatase). P450c17 is the predominant human testicular Type II P450, but the others are expressed in the testes of some other species. Disruption of POR should affect all of these enzymes, with predictably disastrous consequences, and disruption of the POR gene in knockout mice results in early embryonic lethality.[3,4] By contrast, liver-specific knockouts of POR result in a profound disruption of hepatic drug metabolism, but these mice are morphologically and reproductively normal.[5–7] Similarly, a hypomorphic POR mouse expressing only 5–25% of normal amounts of POR has limited embryonic lethality and decreased drug metabolism despite an apparent compensatory increase in total liver P450 content.[8] Thus one would not have predicted that severe mutations of human POR would yield a viable phenotype.

In 1985, a clinical report described a patient with genital ambiguity and an abnormal urinary steroid profile suggesting partial combined deficiencies of what was then thought to be three distinct steroidogenic enzymes: 17α-hydroxylase, 17,20 lyase, and 21-hydroxylase.[9] As these activities were known to be catalyzed by Type II cytochrome P450 enzymes, we suggested that this patient had a disorder in the electron donor for these enzymes, P450 oxidoreductase.[10] Although several similar patients were reported subsequently, this hypothesis lay dormant, and the human POR gene was never cloned. However, the description of the human POR gene though the Human Genome Project, plus the advent of rapid, inexpensive PCR-based sequencing, permitted us to examine the POR gene in such patients, resulting in the first four reported cases of POR deficiency.[11] In the subsequent year, we reported another 19 patients with this disorder[12] and three other groups reported 15 more patients.[13–15]

Although the initial patient described in 1985 did not have skeletal malformations,[9] the majority of patients described to date with POR deficiency also had skeletal deformities usually diagnosed as Antley–Bixler syndrome (ABS). This disorder is characterized by craniosynostosis (premature fusion of bones of the skull), radio-ulnar or radio-humeral synostosis, bowed femurs and other, more variable skeletal disorders.[16,17] Several other congenital malformation syndromes, including the Apert, Crouzon, Jackson-Weiss, Pfeiffer, and Beare-Stevenson syndromes are also characterized by craniosynostosis. Studies between 1994 and 1996 identified dominant, gain-of-function mutations in the gene for fibroblast growth factor receptor 2 (FGFR2) in these five syndromes;[18–22] including identical missense mutations in several of these,[23–25] indicating that these syndromes are variants of a single disease. Examination of several patients with ABS also identified mutations in the FGFR2 gene[26,27] but in a series of 16 patients with ABS, only 7 had FGFR2 mutations.[28] These 7 patients had normal genitalia and no evidence of disordered steroidogenesis, but the 9 patients without FGFR2 mutations had genital anomalies and/or abnormal steroid profiles.[28] This study provided genetic evidence that ABS was really two distinct genetic disorders: one due to dominant, gain-of-function FGFR2 mutations (and hence a variant of the other craniosynostosis syndromes caused by gain-of-function mutations in FGFR2), and the other due to an apparently recessive mutation in an unidentified gene. Our studies have now established that recessive POR mutations are responsible for the form of ABS that includes genital malformations (in both sexes) and disordered steroidogenesis, but we have also found relatively milder POR mutations in people with disordered steroidogenesis who do not have ABS, indicating that disorders of POR may, in fact, be rather common.[11,12,29]

P450 OXIDOREDUCTASE

POR is an 82-kDa membrane-bound protein containing 680 residues encoded by a 32-kb gene containing 15 exons on chromosome 7q11,2. The structure of POR is well-understood from the 2.6 Å-resolution X-ray crystal structure of a soluble, N-terminally deleted (N-56) form of rat POR.[30] This structure shows that POR contains two distinct domains, one containing the NADPH-binding site and the FAD-binding domain, the other containing the FMN domain that eventually interacts with the redox-partner binding site of the P450. Thus POR is the only factor required to transport electrons from NADPH to a microsomal P450 enzyme. The domains binding the FAD and FMN moieties are separated by a flexible hinge region. Binding of NADPH and receipt of electrons by the FAD moiety appears to elicit flexion of the hinge, aligning the isoalloxazine rings of the FAD and FMN moieties, so that electrons can pass from FAD to FMN. Upon doing so, the hinge flexes once more, permitting the FMN domain to become associated with the redox-partner binding site of the P450 (FIG. 1). The surface charge of the FMN domain of POR is negative, produced by acidic residues,[30–32] whereas the redox-partner binding sites of microsomal P450 enzymes have a positive surface charge produced by basic residues.[33–38] The redox-partner binding site of the P450 is on the opposite side of the P450 heme group from the substrate-binding site, and hence electrons from the FMN moiety of the POR must travel about 18 Å to reach the heme iron of the P450.[39] It appears likely that there are multiple pathways for this electron flow in various P450 enzymes.

Studies with P450c17, the steroidogenic enzyme that catalyzes both 17α-hydroxylase and 17,20 lyase activities[40] show that the levels of some P450 activities are determined, at least in part, by the stereochemistry of the interaction of POR with the P450. In the case of P450c17, the 17,20 lyase reaction, but not the 17α-hydroxylase reaction, is very sensitive to this stereochemistry, as shown by three lines of evidence. First, mutations of basic residues in the redox-partner binding site of P450c17 selectively reduce the 17,20 lyase activity.[35] Second, cytochrome b_5 acts as an allosteric factor to promote the interaction of P450c17 with POR, selectively increasing 17,20 lyase activity.[41–43] Third, phosphorylation of P450c17 selectively increases the 17,20 lyase activity of P450c17.[43–45] Thus it seems likely that some changes in POR will affect some P450 enzymes to a greater or lesser extent than some other P450 enzymes. As the great majority of drugs in current clinical use are metabolized by hepatic type II P450 enzymes, a thorough understanding of disease-causing mutations and polymorphic variants of POR could provide useful information about the genetic basis of variations in drug metabolism. Therefore we have sought POR mutations and sequence variants and characterized their enzymology.

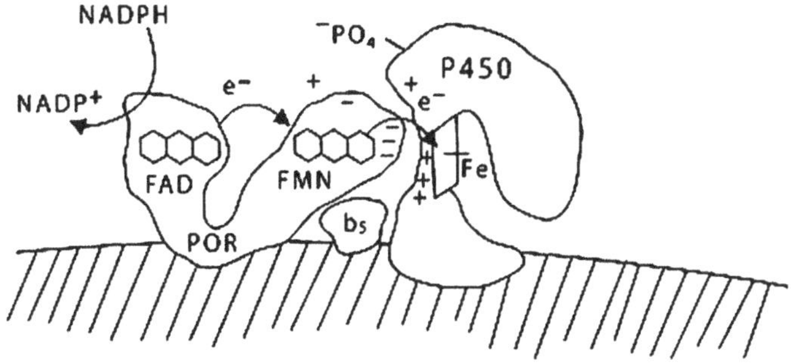

FIGURE 1. Diagram of electron transfer by microsomal (Type II) P450 enzymes. NADPH interacts with P450 oxidoreductase (POR), bound to the endoplasmic reticulum, and gives up a pair of electrons, which are received by the FAD moiety. Electron receipt elicits a conformational change, permitting the isoalloxazine rings of the FAD and FMN moieties to move close together, so that the electrons pass from the FAD to the FMN. After another conformational change that returns the POR to its original orientation, the FMN domain interacts with the redox-partner binding site of the P450, so that electrons reach the P450 heme iron to achieve catalysis. The interaction of POR and the P450 is coordinated by acidic residues on the surface of the FMN domain of POR, and basic residues in the redox-partner binding site of the P450. In the case of human P450c17, this interaction is facilitated by the allosteric action of cytochrome b5, and by serine phosphorylation of P450c17, which optimize the interaction of the two proteins.

POR MUTATIONS AND VARIANTS

The 38 patients with POR deficiency studied to date[1–15] represent only 68 distinct alleles as the studies included four pairs of siblings. This is a sufficiently large sample to draw the following conclusions: First, no patient described to date has carried two null alleles (e.g., deletion or frame-shift); all individuals carried a missense mutation on at least one allele and most individuals carried missense mutations on both alleles, suggesting that a low level of residual activity may be required for embryonic viability, even though several of the identified missense mutations lacked detectable activity when assayed in vitro. Second, the mutations A287P and R457H are quite common. A287P was found on 13 of 35 alleles (37%) in individuals of European ancestry, and this mutation was not found in any non-European individuals. R457H was found on 21 alleles and accounted for 57% (17 of 30) of alleles in individuals of Japanese ancestry, but was also seen in individuals of European and African ancestry. Thus, these two missense mutations account for half (34/68, or 50%) of affected alleles identified to date. To date, 21 missense mutations have been reported in POR; five from patients studied in our initial report,11 11 from patients in our second report, 12 four from our database mining,12 and one from another clinical report.[14] Thus experiments of nature have already provided excellent scanning mutagenesis of POR. However, assessing the activities of these mutants has not been straightforward.

ENZYMOLOGY OF THE AMINO ACID REPLACEMENT MUTATIONS

From the time of its initial identification and isolation, the standard assay for POR has been the reduction of cytochrome *c*. Although cytochrome *c* is a soluble protein and is not a physiologic substrate for membrane-bound POR, it is readily reduced by POR, and this reduction is easily assayed as a change in absorption at 550 nm. To measure reduction of cytochrome *c* and oxidation of NADPH, we expressed a form of POR lacking 27 amino-terminal residues (N-27 POR) in bacteria: this is analogous to rat N-24 POR, as human POR has three additional amino-terminal residues. This form of POR remains associated with membranes,[46] as does wild-type POR, and gives results that differ from those obtained with the soluble, cytoplasmic N-49 POR used by another group.[14] TABLE 1 shows the V_{max}/K_m (an estimate of catalytic efficiency) given as a percentage of the wild-type control, for the capacity of 21 POR missense mutants to catalyze the reduction of cytochrome *c* and the oxidation of NADPH. These data show a broad range of enzymatic impairment, and they show

TABLE 1. Enzymatic activities of POR mutants

	Cytochrome *c* Reduction	NADPH Oxidation	17-Hydroxylase	17,20 Lyase
Wild type	100	100	100	100
A115V	63	41	80	71
T142A	49	52	60	54
Q153R	9	11	31	27
Y181D	nil	nil	nil	nil
P228L	75	72	100	41
M263V	76	57	9	13
A287P	9	16	40	21
R316W	61	77	94	141
G413S	76	100	83	110
R457H	0.7	nil	3	nil
Y459H	0.4	nil	11	nil
V492E	0.3	nil	3	nil
A503V	69	86	60	56
G504R	53	47	91	103
G539R	9	0.2	46	8
L565P	14	1.4	32	19
C569Y	6	2	28	13
V608F	8	3	80	57
R616X	nil	nil	6	nil
V631I	74	23	51	40
F646del	36	94	97	46

NOTE: Values of K_m/V_{max} are given as percent of wild-type activity.

that there is generally good agreement between these two assays based on cytochrome *c*.

However, since soluble cytochrome *c* is not a physiologic substrate for membrane-bound POR, we also used a direct assay of the capacity of the POR mutants to support catalysis by a cytochrome P450 enzyme. For this purpose we used human P450c17 expressed in a genetically modified strain of yeast that lacks the endogenous yeast equivalent of POR, thus eliminating any background capacity to support P450-based catalysis.[41] The yeast membranes containing P450c17 were combined with *E. coli* membranes containing human N-27 POR in the presence of phosphatidyl choline to facilitate membrane fusion. As the 17,20 lyase activity of P450c17 is facilitated by the allosteric action of cytochrome b_5,[41] we also added bacterially expressed human cytochrome b_5 to support the lyase reaction. As defective androgen synthesis is a hallmark of the POR-deficient patients and as the 17,20 lyase activity of human P450c17 is exquisitely sensitive to the topology of its interactions with POR,[35,36,42] this assay provides an excellent indication of the physiologic impact of each POR mutation. The data show general agreement between the 17-hydroxylase and 17,20 lyase activities, and between the assays based on P450c17 and those based on cytochrome *c*, but there are some noteworthy exceptions (TABLE 1). For example, the V608F mutant, which had 3–8% of wild-type activity in the cytochrome *c* assays, had 57–80% of activity in the P450c17 assays, and the G539R mutant had 46% of 17-hydroxylase, but only 8% of wild-type 17,20 lyase activity. These and other differences showed that the P450c17 assays correlated much better with the clinical and hormonal findings in the patients than did the cytochrome *c* assays.[11,12]

CONCLUSIONS

Clinical observations led to the apparently improbable hypothesis that POR mutations caused an unusual syndrome of disordered steroidogenesis and skeletal malformations. Analysis of the POR gene in these patients yielded 21 mutations, providing a broad array of enzymatic dysfunctions assayed in four different fashions. The mechanism by which POR mutations cause skeletal malformations is unclear, but may relate to disorders in cholesterol synthesis (as in the Smith-Lemli-Opitz syndrome) and defective cholesterol derivitization of hedgehog proteins.[29] The available enzymologic data indicate that POR mutations and polymorphisms will be important in many areas of medicine and biology. The amino acid sequences of yeast, *Drosophila, Xenopus*, rat and human POR are well-conserved, especially in the FMN, FAD and NADPH binding sites, and computational modeling of human POR permits rational explanation of the degrees of catalytic impairment of each of the mutants.[12] These analyses also strongly suggest that the impact of specific POR mutants on individual P450 enzymes will differ substantially. Thus the degree to which any of the mutants identified will or will not affect the various hepatic drug-metabolizing P450 enzymes cannot be predicted, and hence each P450 will have to be studied with each POR variant. We expect that such studies will soon reveal that POR is an essential variable in the genetics of inter-individual variations in the metabolism of many commonly used drugs.

ACKNOWLEDGMENTS

This work was supported by NIH Grant GM 73020 to WLM. We thank all the patients and families who participated in this study.

REFERENCES

1. MILLER, W. L. 1988. Molecular biology of steroid hormone synthesis. Endocr. Rev. **9:** 295–318.
2. GUENGERICH, F. P. 2004. Cytochrome P450: what have we learned and what are the future issues? Drug Metab. Rev. **36:** 159–197.
3. SHEN, A. L., K. A. O'LEARY & C. B. KASPER. 2002. Association of multiple developmental defects and embryonic lethality with loss of microsomal NADPH-cytochrome P450 oxidoreductase. J. Biol. Chem. **277:** 6536–6541.
4. OTTO, D. M. *et al.* 2003. Identification of novel roles of the cytochrome P450 system in early embryogenesis: effects on vasculogenesis and retinoic acid homeostasis. Mol. Cell. Biol. **23**: 6103–6116.
5. HENDERSON, C. J. *et al.* 2003. Inactivation of the hepatic cytochrome P450 system by conditional deletion of hepatic cytochrome P450 reductase. J. Biol. Chem. **278:** 13480–13486.
6. WU, L. *et al.* 2003. Conditional knockout of the mouse NADPH-cytochrome P450 reductase gene. Genesis **36:** 177–181.
7. GU, J. *et al.* 2003. Liver-specific deletion of the NADPH-cytochrome P450 reductase gene: impact on plasma cholesterol homeostasis and the function and regulation of microsomal cytochrome P450 and heme oxygenase. J. Biol. Chem. **278:** 25895–25901.
8. WU, L. *et al.* 2005. Transgenic mice with a hypomorphic NADPH-cytochrome P450 reductase gene: effects on development, reproduction, and microsomal cytochrome P450. J. Pharmacol. Exp. Ther. **312:** 35–43.
9. PETERSON, R.E., *et al.* 1985. Male pseudohermaphroditism due to multiple defects in steroid-biosynthetic microsomal mixed-function oxidases: a new variant of congenital adrenal hyperplasia. N. Engl. J. Med. **313:** 1182–1191.
10. MILLER, W.L. 1986. Congenital adrenal hyperplasia. N. Engl. J. Med. **314:** 1321–1322.
11. FLUCK, C.E., *et al.* 2004. Mutant P450 oxidoreductase causes disordered steroidogenesis with and without Antley-Bixler syndrome. Nat. Genet. **36:** 228–230.
12. HUANG, N. *et al.* 2005. Diversity and function of mutations in P450 oxidoreductase in patients with Antley-Bixler syndrome and disordered steroidogenesis. Am. J. Hum. Genet. 76: 729–749.
13. ADACHI, M. *et al.* 2004. Compound heterozygous mutations of cytochrome P450 oxidoreductase gene (POR) in two patients with Antley-Bixler syndrome. Am. J. Med. Genet. A. **128:** 333–339.
14. ARLT, W. *et al.* 2004. Congenital adrenal hyperplasia caused by mutant P450 oxidoreductase and human androgen synthesis: analytical study. Lancet. **363:** 2128–2135.
15. FUKAMI, M. *et al.* 2005. Cytochrome P450 oxidoreductase gene mutations and Antley-Bixler syndrome with abnormal genitalia and/or impaired steroidogenesis: molecular and clinical studies in 10 patients. J. Clin. Endocrinol. Metab. **90:** 414–426.
16. ANTLEY, R. & D. BIXLER. 1975. Trapezoidocephaly, midfacial hypoplasia and cartilage abnormalities with multiple synostoses and skeletal fractures. Birth Defects Orig. Art. Ser. **11:** 397–401.
17. CRISPONI, G., C. PORCU & M.E. PIU. 1997. Antley-Bixler syndrome: case report and review of the literature. Clin. Dysmorphol. **6:** 61–68.
18. REARDON, W. *et al.* 1994. Mutations in the fibroblast growth factor receptor 2 gene cause Crouzon syndrome. Nat. Genet. **8:** 98–103.
19. MUENKE, M. *et al.* 1994. A common mutation in the fibroblast growth factor receptor 1 gene in Pfeiffer syndrome. Nat. Genet. **8:** 269–274.

20. JABS, E. W. *et al.* 1994. Jackson-Weiss and Crouzon syndromes are allelic with mutations in fibroblast growth factor receptor 2. Nat. Genet. 8: 275–279.
21. WILKIE, A.O. *et al.* 1995. Apert syndrome results from localized mutations of FGFR2 and is allelic with Crouzon syndrome. Nat. Genet. 9: 165–172.
22. PRZYLEPA, K.A. *et al.* 1996. Fibroblast growth factor receptor 2 mutations in Beare-Stevenson cutis gyrata syndrome. Nat. Genet. 13: 492–494.
23. RUTLAND, P. *et al.* 1995. Identical mutations in the FGFR2 gene cause both Pfeiffer and Crouzon syndrome phenotypes. Nat. Genet. 9: 173–176.
24. MEYERS, G. A., *et al.* 1996. FGFR2 exon IIIa and IIIc mutations in Crouzon, Jackson-Weiss, and Pfeiffer syndromes: evidence for missense changes, insertions, and a deletion due to alternative RNA splicing. Am. J. Hum. Genet. 58: 491–498.
25. SCHAEFER, F. *et al.* 1998. Novel mutation in the FGFR2 gene at the same codon as the Crouzon syndrome mutations in a severe Pfeiffer syndrome type 2 case. Am. J. Med. Genet. 75: 252–255.
26. CHUN, K. *et al.* 1998. FGFR2 mutation associated with clinical manifestations consistent with Antley-Bixler syndrome. Am. J. Med. Genet. **77:** 219–224.
27. TSAI, F.J. *et al.* 2001. Further evidence that fibroblast growth factor receptor 2 mutations cause Antley-Bixler syndrome. Acta Paediatr. **90:** 595–597.
28. REARDON, W. *et al.* 2000. Evidence for digenic inheritance in some cases of Antley-Bixler syndrome? J. Med. Genet. **37:** 26–32.
29. MILLER, W.L. 2004. P450 oxidoreductase deficiency: a new disorder of steroidogenesis with multiple clinical manifestations. Trends Endocrinol. Metab. 15: 311–315.
30. WANG, M.. *et al.* 1997. Three-dimensional structure of NADPH-cytochrome P450 reductase: prototype for FMN- and FAD-containing enzymes. Proc. Natl. Acad. Sci. USA. **94:** 8411–8416.
31. SHEN, A.L. & C.B. KASPER. 1995. Role of acidic residues in the interaction of NADPH-cytochrome P450 oxidoreductase with cytochrome P450 and cytochrome c. J. Biol. Chem. **270:** 27475–27480.
32. ESTABROOK, R.W. *et al.* 1996. The interaction of NADPH-P450 reductase with P450: an electrochemical study of the role of the flavin mononucleotide-binding domain. Arch. Biochem. Biophys. **333:** 308–315.
33. HASEMANN, C.A. *et al.* 1995. Structure and function of cytochromes P450: a comparative analysis of three crystal structures. Structure **3:** 41–62.
34. FISHER, C.W., M.S. SHET & R.W. ESTABROOK. 1996. Construction of plasmids and expression in *Escherichia coli* of enzymatically active fusion proteins containing the heme-domain of a P450 linked to NADPH-P450 reductase. Methods Enzymol. **272:** 15–25.
35. GELLER, D.H. *et al.* 1997. The genetic and functional basis of isolated 17,20-lyase deficiency. Nat. Genet. **17:** 201–205.
36. AUCHUS, R.J. & W.L. MILLER. 1999. Molecular modeling of human P450c17 (17α-hydroxylase/17,20-lyase): insights into reaction mechanisms and effects of mutations. Mol. Endocrinol. **13:** 1169–1182.
37. KONDO, S. *et al.* 1999. Electrostatic interaction between cytochrome P450 and NADPH-P450 reductase: comparison of mixed and fused systems consisting of rat cytochrome P450 1A1 and yeast NADPH-P450 reductase. Biochem. Biophys. Res. Commun. **257:** 273–278.
38. DAVYDOV, D.R. *et al.* 2000. Association of cytochromes P450 with their reductases: opposite sign of the electrostatic interactions in P450BM-3 as compared with the microsomal 2B4 system. Biochemistry **39:** 6489–6497.
39. SEVRIOUKOVA, I.F. *et al.* 1999. Structure of a cytochrome P450-redox partner electron-transfer complex. Proc. Natl. Acad. Sci. USA 96: 1863–1868.
40. MILLER, W.L., R.J. AUCHUS & D.H. GELLER. 1997. The regulation of 17,20 lyase activity. Steroids **62:** 133–142.
41. AUCHUS, R.J., T.C. LEE & W.L. MILLER. 1998. Cytochrome b5 augments the 17,20-lyase activity of human P450c17 without direct electron transfer. J. Biol. Chem. **273:** 3158–3165.
42. GELLER, D.H., R.J. AUCHUS & W.L. MILLER. 1999. P450c17 mutations R347H and R358Q selectively disrupt 17,20-lyase activity by disrupting interactions with P450 oxidoreductase and cytochrome b5. Mol. Endocrinol. **13:** 167–175.

43. Pandey, A.V. & W.L. Miller. 2005. Regulation of 17,20 lyase activity by cytochrome b5 and by serine phosphorylation of P450c17. J. Biol. Chem. **280:** 13265–13271.
44. Zhang, L.H. *et al.* 1995. Serine phosphorylation of human P450c17 increases 17,20-lyase activity: implications for adrenarche and the polycystic ovary syndrome. Proc. Natl. Acad. Sci. USA. **92:** 10619–10623.
45. Pandey, A.V., S.H. Mellon & W.L. Miller. 2003. Protein phosphatase 2A and phosphoprotein SET regulate androgen production by P450c17. J. Biol. Chem. **278:** 2837–2844.
46. Dierks, E.A., S.C. Davis & P.R. Ortiz de Montellano. 1998. Glu-320 and Asp-323 are determinants of the CYP4A1 hydroxylation regiospecificity and resistance to inactivation by 1-aminobenzotriazole. Biochemistry **37:** 1839–1847.

Neuroendocrine Regulation of Leydig Cell Development

P.J. O'SHAUGHNESSY, P.J. BAKER, AND H. JOHNSTON

Institute of Comparative Medicine, University of Glasgow Veterinary School, Glasgow, Scotland, United Kingdom

ABSTRACT: During development in the mouse, two populations of Leydig cells arise sequentially. The fetal Leydig cell population arises shortly after testicular differentiation and functions primarily to produce androgens that are essential for masculinization of the fetus. The origin of the fetal Leydig stem cells remains uncertain, but it has been suggested that adrenocortical cells and fetal Leydig cells may share a common origin in an adrenogenital primordium. The fetal Leydig cells require an intact pituitary for normal development and are sensitive to both luteinizing hormone (LH) and adrenocorticotrophic hormone (ACTH). Loss of either one of these hormones does not, however, affect fetal androgen production, suggesting that both LH and ACTH may act to maintain fetal Leydig cell function *in vivo* in a redundant fashion. The adult Leydig cell population starts to develop soon after birth in the mouse. Initial differentiation does not appear to require gonadotropin input, but subsequent development and function are completely dependent upon LH. The adult Leydig cells do not require circulating follicle-stimulating hormone, provided that LH is present, but androgen stimulation, through the androgen receptor, is required for normal Leydig cell development in the mouse. It is likely that the effects of androgen are mediated directly in the Leydig cells or indirectly through the peritubular cells.

KEYWORDS: Leydig cell; fetal; adult; development; testis; luteinizing hormone; adrenocorticotrophic hormone; gonadotropin; androgen

INTRODUCTION

In all mammalian species so far studied, two or three populations of Leydig cells arise sequentially during development. Two populations arise in the mouse and rat, and it is now generally accepted from functional and morphological studies that these represent two distinct cell lineages.[1–6] In humans it is reported that there are three phases of Leydig cell development,[7] although the extent to which these represent distinct populations remains to be determined. Studies using mouse models are now providing good insight into the development and regulation of the different Leydig cell populations, although our understanding of these processes in other species remains patchy, particularly with regard to the human.

Address for correspondence: P.J. O'Shaughnessy, Division of Cell Sciences, Institute of Comparative Medicine, University of Glasgow Veterinary School, Bearsden Rd., Glasgow G61 1QH Scotland, U.K.. Voice: 44-141-330-5793; fax: 44-141-330-5797.
p.j.oshaughnessy@vet.gla.ac.uk

**Ann. N.Y. Acad. Sci. 1061: 109–119 (2005). © 2005 New York Academy of Sciences.
doi: 10.1196/annals.1336.013**

LEYDIG CELL ORIGIN AND DIFFERENTIATION

The origins of the fetal Leydig cell population remain uncertain. The Leydig cells appear around 12.5 dpc (days postconception) in the mouse[8] and the stem cells could arise from either the coelomic epithelium or the mesonephros. The coelomic epithelium gives rise to the Sertoli cells and contributes significantly to the interstitial cells of the developing testis.[9] By tracking the fate of cells derived from the coe-lomic epithelium, Capel and colleagues suggest, however, that the interstitial cells derived from this source are unlikely to include a significant number of fetal Leydig cells.[10] During testis differentiation there is significant movement of cells from the mesonephros to the developing gonad about 11.5 dpc,[11] and it has been reported that some of these cells have the potential to differentiate into fetal Leydig cells.[12,13]

At 11.5 dpc in the rat (equivalent to about 10.5 dpc in the mouse), a population of cells positive for the orphan nuclear receptor steroidogenic factor 1 (Sf1) can be seen at the cranial end of the mesonephros.[14] These cells are reported to separate subsequently into two distinct cell populations, one in the region of the primordium of the adrenal cortex and the other near the coelomic epithelium where the gonad will arise.[14] Sf1 is essential for differentiation of steroidogenic tissues,[15] and Hatano *et al.* suggest that, on the basis of the distribution and development of *Sf1*-positive cells, Leydig cells and adrenal cells share a common origin. This is an attractive hypothesis, which is supported by the similarities between fetal Leydig cells and fetal adrenal cells as described below.

Differentiation of the fetal Leydig cells is likely to depend on Sertoli cell function and the Sertoli cell products Desert hedgehog (Dhh) and platelet-derived growth factor A(*Pdgf-A*) have both been implicated in fetal Leydig cell differentiation, possibly through expansion of the Leydig cell precursor population.[10,16] In addition to these signaling molecules, expression of the X-linked Aristaless-related homeobox gene (*Arx*) is required for normal Leydig cell development.[17] Expression of *Arx* is restricted to peritubular myoid cells, endothelial cells, and interstitial fibroblasts, indicating that these cells are also involved in fetal Leydig cell differentiation.[17]

The adult Leydig cell population starts to develop in the mouse about postnatal day 7 and increases in number thereafter to reach adult levels by about day 30.[18,19] The initial evidence that this population of Leydig cells was distinct from the fetal population came from morphological data.[1–3] Thus, for example, in the rat there are clear differences in the abundance of cytoplasmic lipid droplets between fetal and adult cells.[20] More recently, it has become clear that these cells are also functionally distinct, and a number of genes have been identified that are expressed specifically in fetal or adult cells, including 3β-hydroxysteroid dehydrogenase type VI (*3βHSD* VI), melanocortin type 2 receptor (*Mc2r*), estrogen sulfotransferase (*Est*), and prostaglandin D synthetase.[4,6,21–23]

The origin of the stem cells that give rise to the adult population of Leydig cells remains unknown. Surprisingly, it also remains uncertain which cells in the developing testis give rise to the adult Leydig cells. Most studies have indicated that the adult Leydig cells differentiate from peritubular mesenchymal cells.[24,25] Other studies, however, have identified vascular smooth muscle cells and pericytes as the adult Leydig cell progenitors.[26] It is possible that there may be multifocal development of the adult Leydig cell population with different locales of initial differentiation apparent during development[27] although further work is clearly required in this area.

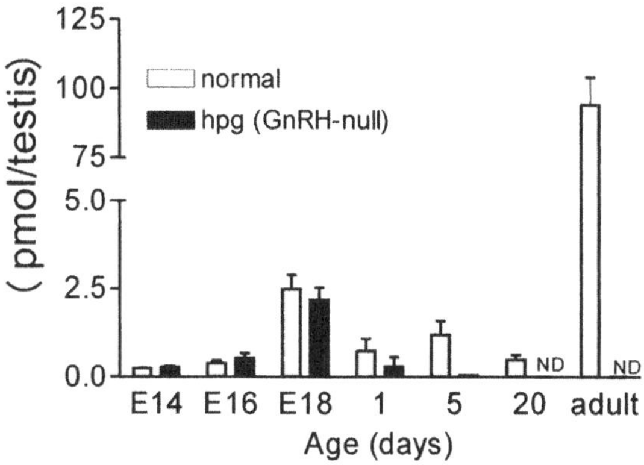

FIGURE 1. Testicular testosterone levels during development in the normal mouse and in the *hpg* mouse, which lacks GnRH and, therefore, circulating gonadotropins.

REGULATION OF FETAL LEYDIG CELL FUNCTION

The adult Leydig cell population is dependent upon LH for development and function (as discussed below). The fetal Leydig cells in the mouse are also sensitive to LH and will show a marked steroidogenic response to LH *in vitro,*[6,28,29] but masculinization of the fetus occurs normally in mice that lack LH or the LH receptor,[30–33] and androgen levels have been shown to be normal during fetal development in mice that lack circulating gonadotropins.[34] It is only after birth that testicular androgen levels decline in mice lacking circulating gonadotropins, becoming barely detectable by postnatal day 5 (FIG. 1).[34] Thus, LH appears not to be essential for fetal Leydig cell function.

To identify fetal-specific transcripts which may be involved in the regulation of fetal Leydig cell function, we carried out a virtual subtraction between adult and fetal testes using expressed sequence tag (EST) libraries available through the U.S. National Center for Biotechnology Information (NCBI).[6] The fetal-specific mRNA species identified from this subtraction included *Mc2r*, and real-time polymerase chain reaction (PCR) confirmed that *Mc2r* is differentially expressed in the fetal testis.[6] *Mc2r* is the specific receptor for adrenocorticotrophic hormone (ACTH) and subsequent studies showed that fetal androgen production is stimulated more than 10-fold by ACTH.[6] ACTH also stimulates androgen production by neonatal testes, but has no effect on testes from mice aged over 15 days when adult Leydig cells start to predominate in the testis.[4] Like LH, ACTH is not essential for fetal testicular androgen production; fetal testicular androgen levels are normal in mice lacking ACTH.[6] Interestingly, however, in mice lacking a pituitary gland because of a null-mutation in the *T/ebp* gene, testosterone levels are markedly reduced in late gestation,[35] indicating that fetal Leydig cell function is critically dependent upon pituitary hormone support. There is a strong possibility, therefore, that regulation of fetal Leydig cell function may show redundancy, and that both LH and ACTH are able to maintain cell activity. This is consistent with a series of experiments in the rat showing that fetal decapitation will significantly reduce testicular androgen levels.[36]

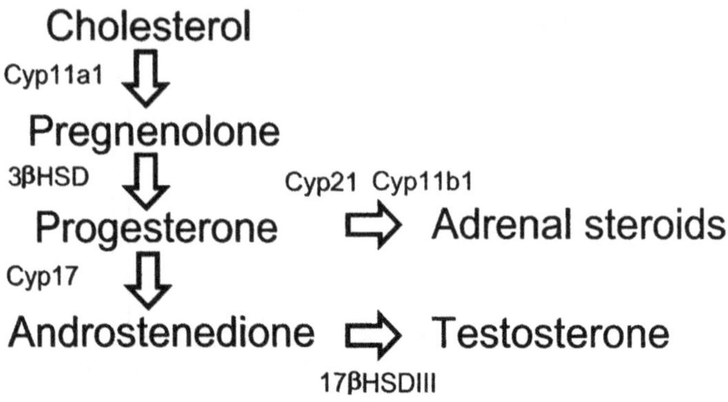

FIGURE 2. Major steroidogenic pathways and converting enzymes in the fetal adrenal cortex and Leydig cells. The major adrenal steroid produced in the mouse fetus is corticosterone.

THE ADRENAL CORTEX AND FETAL LEYDIG CELL ORIGIN

The cells of the adrenal cortex develop adjacent to the gonad on the urogenital ridge, forming a distinct cluster of cells at about 11 dpc in the mouse. The cells can be shown to be expressing the genes coding the key adrenal steroidogenic enzyme *Cyp21* (21-hydroxylase) at 11.5 dpc[37] with expression of *Cyp11b1* (11β-hydroxylase) widespread thereafter.[38] Normal corticosteroid production by the adrenal is dependent upon ACTH stimulation of cortical cells, primarily though Mc2r.[39]

Adrenocortical cells and Leydig cells are clearly similar in that they are both primarily steroidogenic (FIG. 2) and they are both largely under pituitary endocrine control. Differences in the major steroid secreted are due mainly to the expression of *Cyp21* and *Cyp11b1* in the adrenal cells and *Cyp17* (17α-hydroxylase) in the Leydig cells. This difference is only a matter of degree, however, in that Cyp11b1 has been shown to be expressed in the rat testis,[14] and we have now shown that it is expressed specifically in the fetal mouse testis (FIG 3). In addition, we have also shown that the fetal mouse testis expresses *Cyp21* (FIG. 3). It is known that adult mouse adrenals lack expression of *Cyp17*,[40] but this gene is expressed in the fetal mouse adrenal gland and, thus, fetal adrenal cells and fetal Leydig cells express essentially the same steroidogenic enzyme genes although, clearly, the relative levels of expression will be different. This, linked with the demonstration of functional Mc2r in the fetal Leydig cells, highlights further the similarity between the two steroidogenic cell types.

A clear implication of these studies would be that, in evolutionary terms, there is a link between the fetal Leydig cells and adrenocortical cells. It has been proposed that Leydig cells arrived relatively late within the vertebrate testis and that they evolved originally as a source of circulating steroids.[41] Before this time, Sertoli cells are likely to have been the major steroidogenic cell within the testis. In mammals, the Leydig cells came to dominate steroid production and Sertoli cells lost most of their steroidogenic potential,[42] although remnants of steroidogenic enzyme activity can be seen in the mammalian Sertoli cell.[43,44] Comparative studies show that adrenal glands (or their homologues) are significantly older in an evolutionary sense than

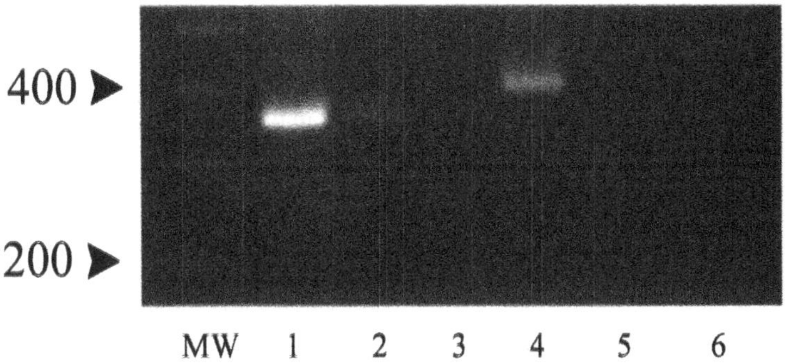

FIGURE 3. Expression of *Cyp21* and *Cyp11b1* in fetal and adult mouse testis. Adult and fetal testes were used to prepare cDNA, which was amplified by PCR using primers specific to *Cyp21* (*lanes* 1–3) or *Cyp11b1* (*lanes* 4–6). *Lanes* 1 and 4 show amplification from fetal testis cDNA, *lanes* 2 and 5 show amplification from adult testis cDNA. and *lanes* 3 and 6 are negative controls.

the Leydig cells. For example, species in which Leydig cells have not evolved, such as the elasmobranches, have functional adrenal homologues. Thus, the similarities between the fetal Leydig cells and the fetal adrenocortical cells suggest that the fetal Leydig cells may have evolved from slight modifications of the fetal adrenal cells.

REGULATION OF ADULT LEYDIG CELL FUNCTION

Luteinizing Hormone

The adult Leydig cell population is critically dependent upon LH for normal function. This can be seen clearly in mice that lack LH (*hpg* and LHKO) or the LH-receptor (LuRKO). In each case, androgen levels are very low or undetectable[31,34,45–47] and Leydig cell number is markedly reduced.[18] Treatment with LH,[45] in turn, stimulates steroidogenesis and induces Leydig cell hyperplasia. LH does not, however, appear to be required for initial adult Leydig cell differentiation. In the *hpg* mouse there is a small, but clear, increase in Leydig cell number around puberty, and these cells express *3βHSDVI*, a marker of the adult phenotype.[18,48] This enzyme can also be detected in adult testes of LuRKO mice.[30] The initial stimulus for adult Leydig cell differentiation remains unknown, although thyroid-stimulating hormone appears to be required[49,50] along with the Sertoli cell products Dhh and PDGF-A.[51,52] Whatever other factor or factors are involved, it appears likely that at least some will be of Sertoli cell origin, because these cells are closely involved in regulation of the Leydig cells and are necessary for Leydig cell survival and function.[53]

Follicle-Stimulating Hormone

A considerable body of evidence suggests that FSH can act to regulate Leydig cell development, function, and survival. Initial studies, dating back 30 years, showed that FSH acts to increase Leydig cell activity in hypophysectomized animals.[54–57] More recent studies using recombinant FSH and other models such as the *hpg* mouse have confirmed this earlier work.[58–60] FSH receptors are confined to the Sertoli cell population of the testis, providing further strong evidence that Leydig cell function can be regulated by the Sertoli cell in the adult animal.

One important caveat about these studies, however, is that LH is lacking in each of the models used, and the role of FSH in regulating Leydig cell function in the presence of normal LH was not addressed. To tackle this issue, we examined Leydig cell function in mice specifically lacking the FSH subunit (FSHβKO) or the FSH receptor (FSHRKO).[61–63] In FSHβKO mice, Leydig cell number and function appear largely normal, indicating that FSH is not required for Leydig cell development in the presence of normal circulating levels of LH.[64] In contrast, in the FSHR-KO mouse circulating and testicular androgen levels are significantly reduced,[64,65] and this is associated with a reduction in Leydig cell number.[64] The reason for this apparent discrepancy is not clear, but it may be related to constitutive activity in the FSH receptor.[64]

These studies suggest that FSH is not required for normal Leydig cell development in the presence of LH, provided FSH receptors are present. The lack of effect of the FSHβ-null mutation on Leydig cell number contrasts with the marked failure of postpubertal development of Leydig cell number in *hpg* mice, which lack LH and FSH.[18] Thus LH alone is sufficient for normal postnatal Leydig cell development (in the presence of the FSHR), but in the absence of LH, evidence from earlier studies suggests that FSH can induce stimulation of Leydig cell function.[58–60]

Androgen

The effects of androgens on Leydig cell function are complex. It has been shown that androgen can inhibit Leydig cell function directly in culture.[66] Other studies have indicated in contrast that androgens are essential for normal Leydig cell development, in the rodent at least. Initial studies supporting a developmental role for androgen showed that Leydig cell precursor activity was increased by androgen.[67] This was supported by studies using the testicular feminized (*Tfm*) mouse, which is characterized by total loss of functional androgen receptors through a frameshift mutation in the receptor gene.[68] In adult *Tfm* mice, the testes have Leydig cell numbers that are about 60% normal,[22] but androgen production is severely reduced.[69] This failure of androgen production is associated with loss of expression of several genes that are essential for normal Leydig cell function, including *Cyp17*, *3βHSD VI,* and *17β-hydroxysteroid dehydrogenase III.*[22] The developmental expression pattern of these genes suggests that adult Leydig cell differentiation occurs in the *Tfm* mouse but that there is a failure of subsequent Leydig cell development.[22]

In the adult animal, androgen receptors are found in Leydig cells, peritubular cells, and Sertoli cells. Studies in the *Tfm* mouse could not indicate through which cell type androgen acts to regulate Leydig cell development. The recent independent development of three separate mouse models which lack androgen receptors only in the Sertoli cells has provided a means to determine the role of the Sertoli cell in mediating androgen action on the Leydig cell.[70–72] A detailed study of Leydig cell function in one of these models suggests that the major effect of androgens on Leydig cell development is not mediated through the Sertoli cell but rather through direct action on the Leydig cells or, possibly, through the peritubular cells.[73]

In the human, loss of androgen receptor function has a less marked effect on the adult Leydig cell population. In contrast to the mouse, complete androgen insensitivity in the human is usually associated with elevated testosterone and LH levels at puberty.[74] It is likely, however, that Leydig cell number is also increased in these individuals and it is possible that androgen output per Leydig cell may be reduced despite elevated levels of LH.[75] Overall, however, it is clear that the human Leydig cell is less sensitive to the effects of androgen receptor dysfunction and that species differences exist in the importance that the androgen receptor is likely to play in the development and maintenance of adult Leydig cell function.

CONCLUSIONS

Normal development of fetal Leydig cell function is essential for masculinization of the fetus. In the mouse, the pituitary gland is essential for this process, and our studies indicate that both LH and ACTH may act, in a redundant fashion, to assure normal development of fetal Leydig cells. It is clear that LH is essential for postnatal Leydig cell development and function, although the cells can start the process of differentiation without LH stimulation. In the presence of LH there is no requirement for FSH to develop or maintain normal adult Leydig cell function. However, androgens, acting either directly on the Leydig cell or through the peritubular cells, are required for normal Leydig cell development in the mouse.

ACKNOWLEDGMENTS

This work was supported by awards from the Wellcome Trust and the Biotechnology and Biological Sciences Research Council, United Kingdom.

REFERENCES

1. Roosen-Runge, E.C. & D. Anderson. 1959. The development of the interstitial cells of the albino rat. Acta Anat. **37:** 125–137.
2. Lording, D.W. & D.M. De Kretser. 1972. Comparative ultrastructural and histochemical studies of the interstitial cells of the rat testis during fetal and postnatal development. J. Reprod. Fertil. **29:** 261–269.
3. Vergouwen, R.P.F.A., S.G.P.M. Jacobs, R. Huiskamp, *et al.* 1991. Proliferative activity of gonocytes, Sertoli cells and interstitial cells during testicular development in mice. J. Reprod. Fertil. **93:** 233–243.
4. Baker, P.J., J.A. Sha, M.W. McBride, *et al.* 1999. Expression of -hydroxysteriod dehydrogenase type I and VI isoforms in the mouse testis during development. Eur. J. Biochem. **260:** 911–916.
5. Nef, S. & L.F. Parada. 1999. Cryptorchidism in mice mutant for Insl3. Nat. Genet. **22:** 295–299.
6. O'Shaughnessy, P.J., L.M. Fleming, G. Jackson, *et al.* 2003. Adrenocoricotrophic hormone directly stimulates testosterone production by the fetal and neonatal mouse testis. Endocrinology **144:** 3279–3284.
7. Prince, F.P. 2001. The triphasic nature of Leydig cell development in humans, and comments on nomenclature. J. Endocrinol. 168: 213–216.
8. Byskov, A. 1986. Differentiation of mammalian embryonic gonad. Physiol. Rev. **66:** 77–112.

9. Karl, J. & B. Capel. 1998. Sertoli cells of the mouse testis originate from the coelomic epithelium. Dev. Biol. **203:** 323–333.
10. Brennan, J., C. Tilmann & B. Capel. 2003. Pdgfr-alpha mediates testis cord organization and fetal Leydig cell development in the XY gonad. Genes Dev. **17:** 800–810.
11. Buehr, M., S. Gu & A. McLaren. 1993. Mesonephric contribution to testis differentiation in the fetal mouse. Development **117:** 273–281.
12. Merchant-Larios, H. & N. Moreno-Mendoza. 1998. Mesonephric stromal cells differentiate into Leydig cells in the mouse fetal testis. Exp. Cell Res. **244:** 230–238.
13. Nishino, K., K. Yamanouchi, K. Naito & H. Tojo. 2001. Characterization of mesonephric cells that migrate into the XY gonad during testis differentiation. Exp. Cell Res. **267:** 225–232.
14. Hatano, O., A. Takakusu, M. Nomura & K. Morohashi. 1996. Identical origin of adrenal cortex and gonad revealed by expression profiles of Ad4BP/SF-1. Genes Cells **1:** 663–671.
15. Luo, X.R., Y.Y. Ikeda & K.L. Parker. 1994. A cell-specific nuclear receptor is essential for adrenal and gonadal development and sexual-differentiation. Cell **77:** 481–490.
16. Yao, H.H., W. Whoriskey & B. Capel. 2002. Desert Hedgehog/Patched 1 signaling specifies fetal Leydig cell fate in testis organogenesis .Genes Dev. **16:** 1433–1440.
17. Kitamura, K., M. Yanazawa, N. Sugiyama, *et al.* 2002. Mutation of ARX causes abnormal development of forebrain and testes in mice and X-linked lissencephaly with abnormal genitalia in humans. Nat. Genet. **32:** 359–369.
18. Baker, P.J. & P.J. O'Shaughnessy. 2001. Role of gonadotrophins in regulating numbers of Leydig and Sertoli cells during fetal and postnatal development in mice. Reproduction **122:** 227–234.
19. Vergouwen, R.P.F.A., R. Huiskamp, R.J. Bas, *et al.* 1993. Postnatal development of testicular populations in mice. J. Reprod. Fertil. **99:** 479–485.
20. Ariyaratne, H.B. & S. Chamindrani Mendis-Handagama. 2000. Changes in the testis interstitium of Sprague Dawley rats from birth to sexual maturity. Biol. Reprod. **62:** 680–690.
21. Baker, P.J. & P.J. O'Shaughnessy. 2001. Expression of prostaglandin D synthetase during development in the mouse testis Reproduction 122: 553–559.
22. O'Shaughnessy, P.J., H. Johnston, L. Willerton & P.J. Baker. 2002. Failure of normal adult Leydig cell development in androgen-receptor-deficient mice. J. Cell Sci. **115:** 3491–3496.
23. Song,W.C., Y. Qian, X. Sun & M. Negishi. 1997. Cellular localization and regulation of expression of testicular estrogen sulfotransferase Endocrinology **138:** 5006–5012.
24. Ariyaratne, H.B., S.M. Mendis-Handagama, D.B. Hales & J.I. Mason. 2000. Studies of the onset of Leydig precursor cell differentiation in the prepubertal rat testis Biol. Reprod. **63:** 165–171.
25. Hardy, M.P., B.R. Zirkin & L.L. Ewing. 1989. Kinetic studies on the development of the adult population of Leydig cells in testes of the pubertal rat. Endocrinology **124:** 762–770.
26. Davidoff, M.S., R. Middendorff, G. Enikolopov, *et al.* 2004. Progenitor cells of the testosterone-producing Leydig cells revealed. J. Cell Biol. **167:** 935–944.
27. Russell, L.D., L.R. De Franca, R. Hess & P. Cooke. 1995. Characteristics of mitotic cells in developing and adult testes with observations on cell lineages. Tissue Cell **27:** 105–128.
28. Feldman, S.C. & E. Bloch. 1978. Developmental pattern of testosterone synthesis by fetal rat testes in response to luteinizing hormone. Endocrinology **102:** 999–1007.
29. El Gehani, F., F.P. Zhang, P. Pakarinen, *et al.* 1998. Gonadotropin-independent regulation of steroidogenesis in the fetal rat testis. Biol. Reprod. **58**: 116–123.
30. Zhang, F.P., T. Pakarainen, F. Zhu, *et al.* 2004. Molecular characterization of postnatal development of testicular steroidogenesis in luteinizing hormone receptor knockout mice. Endocrinology **145:** 1453–1463.
31. Lei, Z.M., S. Mishra, W. Zou, *et al.* 2001. Targeted disruption of luteinizing hormone/human chorionic gonadotrophin receptor gene. Mol. Endocrinol. **15:** 184–200.

32. KENDALL, S.K., L.C. SAMUELSON, T.L. SAUNDERS, *et al.* 1995. Targeted disruption of the pituitary glycoprotein hormone β-subunit produces hypogonadal and hypothyroid mice. Genes Dev. **9:** 2007–2010.
33. CATTANACH, B.M., C.A. IDDON, H.M. CHARLTON *et al.* 1977. Gonadtrophin releasing hormone deficiency in a mutant mouse with hypogonadism. Nature **269:** 338–340.
34. O'SHAUGHNESSY, P.J., P. BAKER, U. SOHNIUS, *et al.* 1998. Fetal development of Leydig cell activity in the mouse is independent of pituitary gonadotroph function. Endocrinology **139:** 1141–1146.
35. PAKARINEN, P., S. KIMURA, F. EL GEHANI, *et al.* 2002. Pituitary hormones are not required for sexual differentiation of male mice: phenotype of the T/ebp/Nkx2.1 null mutant mice. Endocrinology **143:** 4477–4482.
36. MIGRENNE, S., C. PAIRAULT, C. RACINE, *et al.* 2001. Luteinizing hormone-dependent activity and luteinizing hormone-independent differentiation of rat fetal Leydig cells. Mol. Cell Endocrinol. **172:** 193–202.
37. HEIKKILA, M., H. PELTOKETO, J. LEPPALUOTO, *et al.* 2002. Wnt-4 deficiency alters mouse adrenal cortex function, reducing aldosterone production. Endocrinology **143:** 4358–4365.
38. MITANI, F., K. MUKAI, H. MIYAMOTO, *et al.* 1999. Development of functional zonation in the rat adrenal cortex. Endocrinology **140:** 3342–3353.
39. ABDEL-MALEK, Z.A. 2001. Melanocortin receptors: their functions and regulation by physiological agonists and antagonists. Cell Mol. Life Sci. **58:** 434–441.
40. PERKINS, L.M. & A.H. PAYNE. 1988. Quantification of P450scc, P450(17) alpha, and iron sulfur protein reductase in Leydig cells and adrenals of inbred strains of mice. Endocrinology **123:** 2675–2682.
41. CALLARD, G.V. 1996. Endocrinology of Leydig cells in nonmammalian vertebrates. *In* The Leydig Cell. A.H. Payne, M.P. Hardy & L.D. Russell, Eds.: 307–331. Cache River Press. Vienna, IL.
42. PUDNEY, J. 1996. Comparative cytology of the Leydig cell. *In* The Leydig Cell. A.H. Payne, M.P. Hardy & L.D. Russell, Eds.: 97–142. Cache River Press. Vienna, IL.
43. WELSH, M.J. & J.P. WIEBE. 1978. Sertoli cell capacity to metabolize C19 steroids: variation with age and the effect of follicle-stimulating hormone. Endocinology **103:** 838–844.
44. DORRINGTON, J.H. & I.B. FRITZ. 1975. Androgen synthesis and metabolism by preparations from the seminiferous tubule of the rat testis. *In* Hormonal Regulation of Spermatogenesis. F.S. French, V. Hansson, E.M. Ritzen & S.H. Nayfeh, Eds.: 37–52. Plenum Press. New York.
45. SCOTT, I.S., H.M. CHARLTON, B.S. COX, *et al.* 1990. Effect of LH injections on testicular steroidogenesis, cholesterol side-chain cleavage P450 messenger RNA content and Leydig cell morphology in hypogonadal mice. J. Endocrinol. **125:** 131–138.
46. ZHANG, F.-P., M. POUTANEN, J. WILBERTZ & I. HUHTANIEMI. 2001. Normal prenatal but arrested postnatal sexual development of luteinizing hormone receptor knockout (LuRKO) mice. Mol. Endocrinol. **15:** 172–183.
47. MA, X., Y. DONG, M.M. MATZUK & T.R. KUMAR. 2004. Targeted disruption of luteinizing hormone beta-subunit leads to hypogonadism, defects in gonadal steroidogenesis, and infertility. Proc. Natl. Acad. Sci. USA **101:** 17294–17299.
48. BAKER, P.J., H. JOHNSTON, M.H. ABEL, *et al.* 2003. Differentiation of adult-type Leydig cells occurs in gonadotrophin-deficient mice. Reprod. Biol. Endocrinol. **1:** 4.
49. MENDIS-HANDAGAMA, S.M.L.C., H.B.S. ARIYARATNE, K.R. TEUNISSEN VAN MANEN & R.L. HAUPT. 1998. Differentiation of adult Leydig cells in the neonatal rat testis is arrested by hypothyroidism. Biol. Reprod. **59:** 351–357.
50. TEERDS, K.J., D.G. DE ROOIJ, F.H. DE JONG & L.H. VAN HAASTER. 1998. Development of the adult-type Leydig cell population in the rat is affected by neonatal thyroid hormone levels. Biol. Reprod. **59:** 344–350.
51. CLARK, A.M., K.K. GARLAND & L.D. RUSSELL. 2000. Desert hedgehog (Dhh) gene is required in the mouse testis for formation of adult-type Leydig cells and normal development of peritubular cells and seminiferous tubules. Biol. Reprod. **63:** 1825–1838.

52. GNESSI, L., S. BASCIANI, S. MARIANI, *et al.* 2000. Leydig cell loss and spermatogenic arrest in platelet-derived growth factor (PDGF)-A-deficient mice. J. Cell Biol. **149:** 1019–1026.
53. RUSSELL, L.D., J. WARREN, L. DEBELJUK, *et al.* 2001. Spermatogenesis in Bclw-deficient mice. Biol. Reprod. **65:** 318–332.
54. ODELL, W.D., R.S. SWERDLOFF, H.S. JACOBS & M.A. HESCOX. 1973. FSH induction of sensitivity to LH: one cause of sexual maturation in the male rat. Endocrinology **92:** 160–165.
55. CHEN, Y.I., A.H. PAYNE & R.P. KELCH. 1976. FSH stimulation of Leydig cell function in the hypophysectomized immature rat. Proc. Soc. Exp. Biol. Med. **153:** 473–475.
56. VAN BEURDEN, W.M., B. ROODNAT, F.H. DE JONG,, *et al.* 1976. Hormonal regulation of LH stimulation of testosterone production in isolated Leydig cells of immature rats: the effect of hypophysectomy, FSH, and estradiol-17beta. Steroids **28:** 847–866.
57. SELIN, L.K. & W.H. MOGER. 1977. The effect of FSH on LH induced testosterone secretion in the immature hypophysectomized male rat. Endocr. Res. Commun. **4:** 171–182.
58. VIHKO, K.K., P.S. LAPOLT, K. NISHIMORI & A.J. HSUEH. 1991. Stimulatory effects of recombinant follicle-stimulating hormone on Leydig cell function and spermatogenesis in immature hypophysectomized rats. Endocrinology **129:** 1926–1932.
59. O'SHAUGHNESSY, P.J., M.K. BENNETT, I.S. SCOTT & H.M. CHARLTON. 1992. Effects of FSH on Leydig cell morphology and function in the hypogonadal mouse. J. Endocrinol. **135:** 517–525.
60. MATIKAINEN, T., J. TOPPARI, K.K. VIHKO & I. HUHTANIEMI. 1994. Effects of recombinant human FSH in immature hypophysectomized male rats: evidence for Leydig cell-mediated action on spermatogenesis. J. Endocrinol. **141:** 449–457.
61. KUMAR, T.R., Y. WANG, N. LU & M.M. MATZUK. 1997. Follicle stimulating hormone is required for ovarian follicle maturation but not male fertility Nat. Genet. **15:** 201–204.
62. ABEL, M.H., A.N. WOOTTON, V. WILKINS *et al.* 2000. The effect of a null mutation in the follicle-stimulating hormone receptor gene on mouse reproduction. Endocrinology **141:** 1795–1803.
63. DIERICH,.A., M.R. SAIRAM, L. MONACO, *et al.* 1998. Impairing follicle-stimulating hormone (FSH) signaling in vivo: targeted disruption of the FSH receptor leads to aberrant gametogenesis and hormonal imbalance. Proc. Natl. Acad. Sci. USA **95:** 13612–13617.
64. BAKER, P.J., P. PAKARINEN, I.T. HUHTANIEMI, *et al.* 2003. Failure of normal Leydig cell development in follicle-stimulating hormone (FSH) receptor-deficient mice, but not FSH beta-deficient mice: role for constitutive FSH receptor activity. Endocrinology **144:** 138–145.
65. KRISHNAMURTHY, H., R. KATS, N. DANILOVICH, *et al.* 2001. Intercellular communication between Sertoli cells and Leydig cells in the absence of follicle-stimulating hormone-receptor signaling. Biol. Reprod. **65:** 1201–1207.
66. HALES, D.B., L.L. SHA & A.H. PAYNE. 1987. Testosterone inhibits cAMP-induced de novo synthesis of Leydig cell cytochrome P-450 (17 alpha) by an androgen receptor-mediated mechanism. J. Biol. Chem. **262:** 11200–11206.
67. HARDY, M.P., W.R. KELCE, G.R. KLINEFELTER & L.L. EWING. 1990. Differentiation of Leydig cell precursors in vitro: a role for androgen. Endocrinology **127:** 488–490.
68. CHAREST, N.J., Z. ZHOU, D.B. LUBAHN, *et al.* 1991. A frameshift mutation destabilizes androgen receptor messenger RNA in the Tfm mouse. Mol. Endocrinol. **5:** 573–581.
69. MURPHY, L. & P.J. O'SHAUGHNESSY. 1991. Testicular steroidogenesis in the testicular feminized (Tfm) mouse: loss of 17β-hydroxylase activity. J. Endocrinol. **131:** 443–449.
70. DEGENDT, K., J.V. SWINNEN, P.T. SAUNDERS, *et al.* 2004. A Sertoli cell-selective knockout of the androgen receptor causes spermatogenic arrest in meiosis. Proc. Natl. Acad. Sci. USA 101: 1327–1332.
71. CHANG, C., Y.T. CHEN, S.D. YEH, *et al.* 2004. Infertility with defective spermatogenesis and hypotestosteronemia in male mice lacking the androgen receptor in Sertoli cells. Proc. Natl. Acad. Sci. USA **101:** 6876–6881.

72. HOLDCRAFT, R.W. & R.E. BRAUN. 2004. Androgen receptor function is required in Sertoli cells for the terminal differentiation of haploid spermatids. Development **131:** 459–467.
73. DE GENDT, K., N. ATANASSOVA, K.A. TAN, *et al.* 2005. Development and function of the adult generation of Leydig cells in mice with Sertoli cell-selective (SCARKO) or total (ARKO) ablation of the androgen receptor. Endocrinology **146:** 4117–4126.
74. BRINKMANN, A.O. 2001. Molecular basis of androgen insensitivity. Mol. Cell Endocrinol. **179:** 105–109.
75. BARDIN, C.W., L.P. BULLOCK, R.J. SHERINS, *et al.* 1973. Androgen metabolism and mechanism of action in male pseudohermaphroditism: a study of testicular feminization. Recent Prog. Horm. Res. **29:** 65–109.

Mitochondrial Function in Leydig Cell Steroidogenesis

DALE B. HALES,[a,b] JOHN A. ALLEN,[a] TRISTAN SHANKARA,[a] PAUL JANUS,[b] STEVE BUCK,[b] THORSTEN DIEMER,[c] AND KAREN HELD HALES[a]

[a]*Departments of Physiology and Biophysics, University of Illinois at Chicago, Chicago, Illinois 60612, USA*

[b]*Department of Bioengineering, University of Illinois at Chicago, Chicago, Illinois 60612, USA*

[c]*Department of Urology, University Hospital of the Justus-Liebig University, Giessen, Germany*

ABSTRACT: **The first and rate-limiting step in the biosynthesis of steroid hormones is the transfer of cholesterol into mitochondria, which is facilitated by the steroidogenic acute regulatory (StAR) protein. Recent studies of Leydig cell function have focused on the molecular events controlling steroidogenesis; however, few studies have examined the importance of the mitochondria. The purpose of this investigation was to determine which aspects of mitochondrial function are necessary for Leydig cell steroidogenesis. MA-10 tumor Leydig cells were treated with 8-bromo-cAMP (cAMP) and site-specific mitochondrial disrupters, pro-oxidants, and their effects on progesterone synthesis, StAR expression, mitochondrial membrane potential ($\Delta\Psi_m$) and ATP synthesis were determined. Dissipating $\Delta\Psi_m$ with CCCP inhibited progesterone synthesis, even in the presence of newly synthesized StAR protein. The electron transport inhibitor antimycin A significantly reduced cellular ATP, inhibited steroidogenesis, and reduced StAR protein expression. The F_0/F_1 ATPase inhibitor oligomycin reduced cellular ATP and inhibited progesterone synthesis and StAR protein expression, but had no effect on $\Delta\Psi_m$) Disruption of pH with nigericin significantly reduced progesterone production and StAR protein, but had minimal effects on $\Delta\Psi_m$. Sodium arsenite at low concentrations inhibited StAR protein but not mRNA expression and inhibited progesterone without disrupting $\Delta\Psi_m$. The mitochondrial Ca^{2+} inhibitor Ru360 also inhibited StAR protein expression. These results demonstrate that $\Delta\Psi_m$, ATP synthesis, ΔpH and $[Ca^{2+}]_{mt}$ are all required for steroid biosynthesis, and that mitochondria are sensitive to oxidative stress. These results suggest that mitochondria must be energized, polarized, and actively respiring to support Leydig cell steroidogenesis and alterations in the state of mitochondria may be involved in regulating steroid biosynthesis.**

ABBREVIATIONS: $\Delta\Psi_m$, mitochondrial membrane potential; pH, mitochondrial pH gradient; cAMP, 8-bromo-cyclic adenosine monophosphate; IOD, integrated optical density; LH. leutinizing hormone; P450scc, P450 side-chain cleavage enzyme; ROS, reactive oxygen species; SFM, serum-free medium; StAR, steroidogenic acute regulatory protein; TMRE, tetramethylrhodamine ethyl ester dye; CCCP, carbonyl cyanide *m*-chlorophenyl hydrazone; nd, not detectable.

Address for correspondence: Dale B. Hales, Department of Physiology and Biophysics, University of Illinois at Chicago, 835 S. Wolcott Avenue, Chicago, IL 60612-7342. Voice: 312-413-2891; fax: 312-996-0195.

dbhale@uic.edu

Ann. N.Y. Acad. Sci. 1061: 120–134 (2005). © 2005 New York Academy of Sciences.
doi: 10.1196/annals.1336.014

KEYWORDS: **Leydig cell; steroid hormone; mitochondria; StAR; reactive oxygen; calcium; membrane potential; ATP; arsenite; inhibitors**

INTRODUCTION

Mitochondria are a key control point for the regulation of steroid hormone biosynthesis. The first and rate-limiting step in steroidogenesis is the transfer of cholesterol across the inner membrane space from the outer mitochondrial membrane to the inner mitochondrial membrane, a process dependent on the action of StAR.[1,2] Testosterone production by Leydig cells is primarily under the control of LH, which acts via its intracellular second messenger cAMP to regulate testosterone production acutely at the level of cholesterol transport into the mitochondria via the action of StAR, and chronically at the level of steroidogenic enzyme gene transcription. The extra-mitochondrial 37 kDa precursor form of StAR is believed to be the form active in cholesterol transfer, whereas the intra-mitochondrial 30/32 kDa processed forms of StAR are no longer active.[1] The outer mitochondrial membrane protein peripheral benzodiazapene receptor (PBR) is also essential for cholesterol transfer across the inner mitochondrial space (IMS) into the steroidogenic pathway.[3] LH stimulation of the Leydig cell results in the activation of StAR transcription and translation, and StAR facilitates the transfer of cholesterol into the mitochondrial matrix to cholesterol side-chain cleavage cytochrome P450 (P450scc, CYP11A), which converts cholesterol to pregnenolone.[4] Pregnenolone diffuses to the smooth endoplasmic reticulum where it is further metabolized via the actions of 3β-hydroxysteroid dehydrogenase Δ5-Δ4-isomerase (3β-HSD), 17α-hydroxylase/C17-20 lyase (P450c17, CYP17). and 17β-hydroxysteroid dehydrogenase type III (17HSD3) to testosterone.[5]

During aerobic respiration, ATP is generated through oxidative phosphorylation within the mitochondrion, which involves the transport of electrons through four enzyme complexes located at the inner mitochondrial membrane. Transport of electrons through these complexes yields free energy that is used to pump protons from the matrix into the inner membrane space, thereby creating a proton gradient in the mitochondria. This proton gradient generates a pH differential (ΔpH) and the mitochondrial membrane potential ($\Delta\Psi_m$), which provides the proton motive force that drives ATP synthesis by the F_0/F_1 ATPase (see FIGURE 1 for schematic illustration).

Pharmacological mitochondrial disruptors can directly affect mitochondrial activity. Ionophores such as carbonyl cyanide *m*-chlorophenyl-hydrazone (CCCP) perturbs $\Delta\Psi_m$ and ΔpH by disrupting the hydrogen ion gradient;[6-8] nigericin disrupts ΔpH by exchanging K^+ for H^+, effectively equilibrating pH without affecting $\Delta\Psi_m$.[7] CCCP and nigericin perturb the mitochondria and inhibit StAR-mediated cholesterol transfer.[9-16] Previous studies have shown that during oxidative stress, reactive oxygen species and reactive nitrogen species (ROS/RNS) perturb Leydig cell mitochondria, causing an abrupt cessation to cholesterol transfer and steroid hormone production.[17-19] In addition, acute inflammation results in oxidative stress of Leydig cells, which perturbs mitochondria and inhibits steroidogenesis.[17,18] Antimycin A is a complex III inhibitor which prevents electron transport, and oligomycin inhibits the F_0/F_1 ATPase, preventing ATP synthesis. Ru360 is a cell-permeable derivative of ruthenium red that inhibits the mitochondrial calcium uniporter, and sodium arsenite is a pro-oxidant which affects mitochondria at multiple sites (for schematic of mitochondrial inhibitors see FIG. 1).

The purpose of the present study was to determine which aspects of mitochondrial function are necessary for Leydig cell steroidogenesis. Results of these studies demonstrate that $\Delta\Psi_m$, ΔpH, mitochondrial ATP synthesis, and $[Ca^{2+}]_{mt}$ are all essential for mitochondrial steroidogenesis. This study indicates that disrupting mitochondria results in post-transcriptional inhibition of StAR, and that mitochondria must be energized, polarized, and actively respiring to support Leydig cell steroidogenesis.

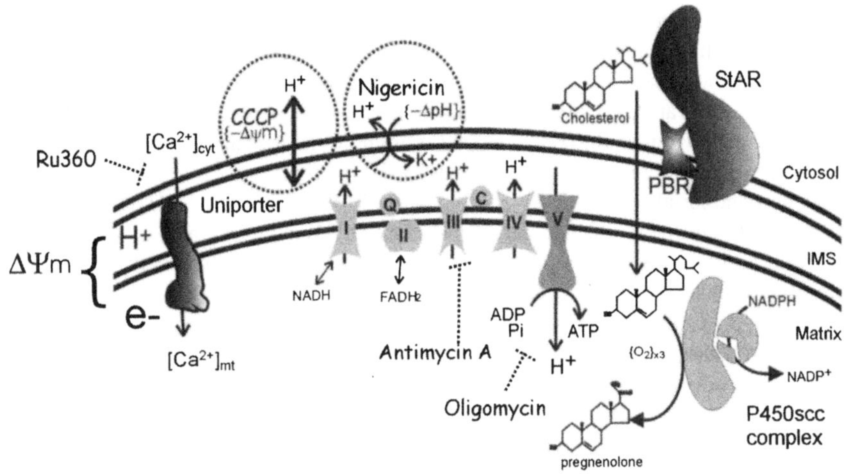

FIGURE 1. Composite drawing showing the mitochondrial electron transport systems and mitochondrial components of cholesterol transfer and metabolism. Cholesterol is transferred from the cytosol to the mitochondrial outer membrane, across the inner membrane space (IMS) to the matrix via the action of steroidogenic acute regulatory protein (StAR) and the peripheral benozodiazapene receptor (PBR). Once cholesterol enters the matrix, it is converted to pregnenolone via the action of the cytochrome P450scc complex. The action of StAR requires an intact $\Delta\Psi_m$, ΔpH, and ongoing ATP synthesis. The electron transport chain is composed of five protein complexes: I (complex I, NADH-ubiquinol oxidoreductase); II (complex II, succinate-ubiquinol oxidoreductase); III (complex III, ubiquinol cytochrome *c* oxidoreductase); IV (complex IV, cytochrome *c* oxidase); and V (complex V, F_0/F_1 ATP-ase). The electrons are transferred from complex I and complex II via reduction of ubiquin-one (Q) to complex III, and from complex III via reduction of cytochrome *c* (c) to complex IV. As protons (H^+) are pumped from the matrix into the inner membrane space (IMS), a proton gradient develops which provides the proton motive force for complex V, which converts ADP to ATP. The membrane potential ($\Delta\Psi_m$) is created by the pumping of protons from the matrix into the IMS via the action of the electron transport complexes. The calcium uniporter transports calcium from the cytosol $[Ca^{2+}]_{cyt}$ into the matrix $[Ca^{2+}]_{mt.}$ The mitochondria-specific inhibitors used in this study were Ru360, a permeable derivative of ruthenium red, which inhibits the calcium uniporter; CCCP, a protonophore that dissipates $\Delta\Psi_m$ and ΔpH; nigericin, a potassium/hydrogen exchanger that dissipates ΔpH without affecting $\Delta\Psi_m$; antimycin A, which inhibits complex III; and oligomycin, which inhibits complex V.

MATERIALS AND METHODS

Materials

All materials used were of the highest quality and obtained from standard sources as previously described.[18,19]

Cell Culture

MA-10 cells were a generous gift of Dr. Mario Ascoli and were cultured essentially as originally described.[20] MA-10 cells were cultured in Waymouth's complete medium MB 752/1 (Gibco BRL Life Technologies) containing 15% heat-inactivated donor herd horse serum (Gibco BRL Life Technologies). Pretreatment of confluent cell layers in culture dishes was performed in serum-free Waymouth's medium (SFM) for 1 hour before the onset of experimental treatments. Cells were incubated at 37°C in 5% CO_2 in a humidified incubator.

Western Blotting

Western blotting was performed as described previously.[21–23] The preparation of the polyclonal antiserum to StAR and P450scc has been described previously.[24,25] Detection and visualization of StAR and P450scc proteins on the blot was performed as described.[18] Detected bands were quantitated, after densitometry using Imagequant software (Molecular Dynamics, Sunnyvale, CA). Expression data for protein are represented as Integrated Optical Density (IOD).

Northern Blotting

Total cellular RNA was extracted by the acid guanidinium thiocyanate–phenol–chloroform method[26] as described previously.[21] The RNA was analyzed by Northern blotting as described.[23,27] Northern hybridization was performed by use of a $\alpha[^{32}P]$-dGTP-labeled random primed cDNA probe specific for StAR. Radioactivity was visualized after exposure to a phosphor screen for 24 hours. Hybridization signals were quantified and documented by use of a PhosphorImager (Storm 860TM, Molecular Dynamics).

Analysis of $\Delta\Psi_m$ and Reactive Oxygen Species using Fluorimetry

Mitochondrial membrane potential ($\Delta\Psi_m$) and the presence of reactive oxygen species were assessed using fluorescence dye indicators. $\Delta\Psi_m$ was assessed by TMRE fluorescence, while the presence of reactive oxygen species was determined by H_2DCFDA fluorescence (both obtained from Molecular Probes, Eugene, OR). MA-10 cells were grown in culture to 2.5×10^5 cells per cm^2 (approximately 75% confluency) in a 96-well fluorescence assay plate (Costar 3603, Fisher Scientific). Cells were treated for 3 hours with agents (as indicated in legends), then incubated in 200 µL of a 50-nM solution of TMRE or 100 µL of a 5-µM solution of H_2DCFDA in SFM for 20 minutes at 37°C. Cells were analyzed for overall fluorescence using the Synergy HT microplate fluorescence reader (Bio-Tek, Winooski, VT) and the accompanying computer software (KC-4, version 3.3). TMRE fluorescence values are proportional to the magnitude of $\Delta\Psi_m$, and provided a quantitative assessment of the mitochondrial membrane potential. H_2DCFDA fluorescence is proportional to the concentration of reactive oxygen species.

ATP Assay

Cellular ATP levels were assessed using the Promega Luminescent Cell Viability Assay (Promega G7570, Fisher Scientific). 2.5×10^5 MA-10 cells were seeded and grown to 75% confluency in a 96-well luminescence assay plate (Costar, Fisher Scientific). Cells were treated with agents as described in legends to FIGURES 1–5, and ATP luminescence was determined by assay using the suggested procedures of the manufacturer (Promega). Luminescent values are proportional to total cellular ATP and provided a quantitative assessment of cellular ATP levels. Measurement of cellular ATP after treatments with H_2O_2 and xanthine oxidase was determined using an ATP bioluminescent assay kit (Sigma Chemical Inc.) as described previously.[19]

Radioimmunoassay

After treatment, culture media were removed and boiled for 5 min and centrifuged at 2,000 × *g* for 20 min at 4°C. The supernatant was stored at –20°C until assayed for progesterone using Coat-A-Count RIA kits (DPC, Los Angeles, CA) as previously described.[19]

Statistical Analysis

Data are presented as means ± SEM of three or more independent experiments. For group comparisons, one-way analysis of variance was performed followed by a Student-Newman-Keuls' multiple range test, both using GraphPad InStat, version 3.0, statistical software package (GraphPad Software, San Diego, CA). Differences were considered as significant at $P < 0.05$.

RESULTS

Effects of Site-Specific Mitochondrial Disruptors and Reactive Oxygen on StAR Expression, Steroidogenesis, Cellular ATP, and $\Delta\Psi_m$

To determine which properties of mitochondria are important for Leydig cell steroidogenesis, MA-10 cells were treated with different mitochondria-disrupting agents or reactive oxygen species and the effects on steroidogenesis, StAR expression, cellular ATP, and $\Delta\Psi_m$ were determined. In initial experiments, expression of P450scc and StAR protein in response to alterations in mitochondria was investigated. Treatment of MA-10 cells with 8-Br-cAMP (cAMP) results in the induction of StAR protein expression; however, this expression is significantly altered by disrupting mitochondria (FIG. 1). Treatment of cells with the complex III inhibitor antimycin A, the $\Delta\Psi_m$ uncoupler CCCP, the reactive oxygen species H_2O_2, or the complex V inhibitor oligomycin all significantly reduced StAR protein expression, while P450scc protein was unchanged (FIG. 2 and TABLE 1). These data indicate that StAR protein expression is highly sensitive to alterations in mitochondria and that $\Delta\Psi_m$ and ATP synthesis are required for StAR expression. Of note is that the 37-kDa

TABLE 1. Effects of mitochondrial disruptors and reactive oxygen on MA-10 cell mitochondria and steroidogenesis[a]

	Control	cAMP	CCCP	H_2O_2	Xanthine oxidase	Anti-mycin A	Oligo-mycin	Nigericin
Progesterone	1 ± 0.5	100[c]	2 ± 0.7[c]	42 ± 11[b,c]	71 ± 4[b,c]	16 ± 7[b]	64 ± 11[b,c]	12 ± 5[b]
StAR mRNA	nd	100[c]	81 ± 10[c]	85 ± 5[c]	×	×	94 ± 9[c]	×
StAR protein	nd	100[c]	nd	24 ± 10[a]	10 ± 3[b]	4 ± 2[b]	66 ± 1[b,c]	nd
ATP	101 ± 2	100	86 ± 2	144 ± 51[b,c]	78 ± 11	76 ± 2	52 ± 2[b,c]	94 ± 5
$\Delta\Psi_m$	72 ± 3	100[c]	52 ± 3[b,c]	|50[f]	|50[f]	84 ± 6	103 ± 8	69 ± 8[b]

[a]Table represents data from multiple experiments which tested the effect of 1 mM 8-Br-cAMP (cAMP) vs. control or cAMP plus 5 μM CCCP, 250 μM H_2O_2, 10 mU xanthine oxidase with 100 μM xanthine; 1 μM antimycin A; 1 μM oligomycin; or 1 μM nigericin on progesterone (ng/mg protein), StAR mRNA and protein levels, total cellular ATP and $\Delta\Psi_m$. Cells were treated for 3 h, media were collected for progesterone radioimmunoassay, and cells were lysed for RNA and protein isolation. ATP was determined in whole-cell lysates in parallel, and $\Delta\Psi_m$ was measured by TMRE fluorescence as described in MATERIAL AND METHODS. Data are presented as mean ± SEM and are expressed as percent cAMP. Data were pooled from three or more independent experiments (n = 3). ×, experiment was not performed; nd, could not be detected; f, $\Delta\Psi_m$ was decreased approximately 50% versus control as assessed by fluorescence microscopy.[19]

[b]P <0.05 versus cAMP.

[c]P <0.05 versus control.

form of StAR remains constant while the 30- and 32-kDa processed forms of StAR are decreased after CCCP treatments. This lack of 30-kDa StAR expression due to CCCP suggests that StAR import into the mitochondrial matrix and its subsequent processing is dependent on an intact $\Delta\Psi_m$. The import and processing of many matrix targeted proteins are dependent on $\Delta\Psi_m$ and ATP hydrolysis,[28] and results of these CCCP experiments indicate that import and processing of StAR also require $\Delta\Psi_m$. StAR is thought to act on the outside of the mitochondria,[14,29] suggesting that StAR import and processing are not necessary for cholesterol transfer. In contrast, other reports have indicated that StAR import and processing are central to cholesterol transfer.[30] One interpretation of the data in this report is that preventing StAR import and processing with CCCP inhibits steroidogenesis.

The effects of the mitochondrial disrupting agents on steroidogenesis, StAR expression, cellular ATP, and $\Delta\Psi_m$ were determined and are presented in TABLE 1. Data were pooled from several experiments and are expressed as percent of cAMP. CCCP, antimycin A, oligomycin, and nigericin all significantly reduced progesterone synthesis versus cAMP alone. Similarly, reactive oxygen in the form of H_2O_2 or generated by xanthine oxidase also significantly decreased steroidogenesis. All of these agents significantly inhibited StAR protein expression versus cAMP, and this appears to occur post-transcriptionally, given that StAR mRNA was unchanged. Of the many agents tested, only antimycin A and oligomycin significantly decreased cellular ATP versus control or cAMP. CCCP, H_2O_2, xanthine oxidase, and nigericin all significantly decreased $\Delta\Psi_m$. These results indicate that Leydig cell steroidogenesis and StAR protein are highly sensitive to oxidative stress and to alterations in mitochondria. These results suggest that ongoing Leydig cell steroidogenesis depends upon the state of the mitochondrion.

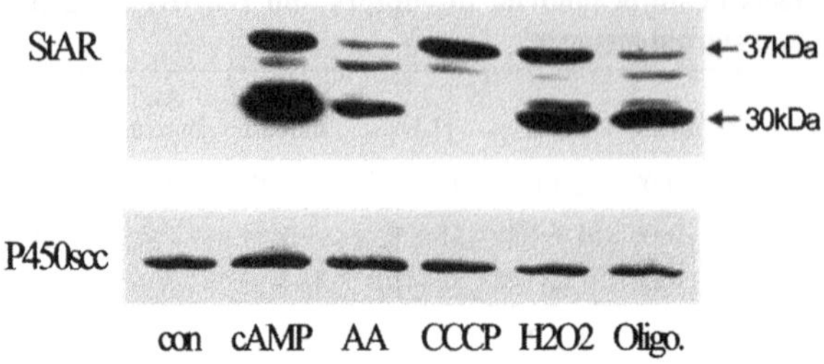

FIGURE 2. Effects of mitochondrial disrupting agents and reactive oxygen on StAR and P450scc protein expression. MA-10 cells were grown in culture and treated for 3 h with 1mM 8-Br-cAMP (cAMP) or cAMP plus 1 μM antimycin A (AA), 5 μM CCCP, 250 μM H_2O_2 or 1 μM oligomycin (Oligo). After treatment, cells were lysed and subjected to Western blot analysis for StAR and P450scc protein as described in the MATERIALS AND METHODS section. Data are representative of results from three or more independent experiments ($n = 3$).

Effects of Na^+Arsenite on Steroidogenesis, StAR Expression, and $\Delta\Psi_m$

Na^+ arsenite was investigated to characterize further the sensitivity of Leydig cell steroidogenesis to pro-oxidant–induced inhibition of Leydig cell function. Arsenic is an essential trace element required for growth and metabolism, but is toxic at higher concentrations.[31] Arsenic produces a variety of stress responses in mammalian cells, including metabolic abnormalities accompanied by growth inhibition and eventually apoptosis.[32] However, at concentrations below a certain threshold (50 μM), Na^+ arsenite significantly inhibits cAMP-stimulated progesterone production (FIG. 3A), and post-transcriptionally inhibits StAR protein (FIG. 3C), without affecting StAR mRNA (FIG. 3B). Higher concentrations of arsenite abrogated StAR expression. Treatment with cAMP plus Na^+ arsenite reduced $\Delta\Psi_m$ to control levels, but caused no further reduction in TMRE fluorescence (FIG. 3D). In addition, Na^+ arsenite treatments did not alter total cellular ATP levels, indicating that its inhibitory effects were not due to decreasing ATP (control = 4550 ± 160; 1 mM cAMP = 4288 ± 169, 1 mM cAMP plus 50 μM Na^+ arsenite = 4374 ± 147; 1 mM cAMP plus 100 μM Na^+ arsenite = 4348 ± 205; 1 mM cAMP plus 200 μM Na^+ arsenite = 3926 ± 279, $n = 3$). These data demonstrate that Na^+ arsenite significantly inhibits steroidogenesis in an ATP-independent manner via the post-transcriptional inhibition of StAR.

Effects of Mitochondrial Disruptors on the Production of Reactive Oxygen Species

Mitochondria are highly sensitive to oxidative stress and they are also a major source of intracellular reactive oxygen, which is generated during oxidative phosphorylation. To determine whether mitochondria-disrupting agents increase the level of intracellular reactive oxygen, MA-10 cells were treated with the disrupting drugs using the minimal concentrations previously shown to inhibit steroidogenesis (as in

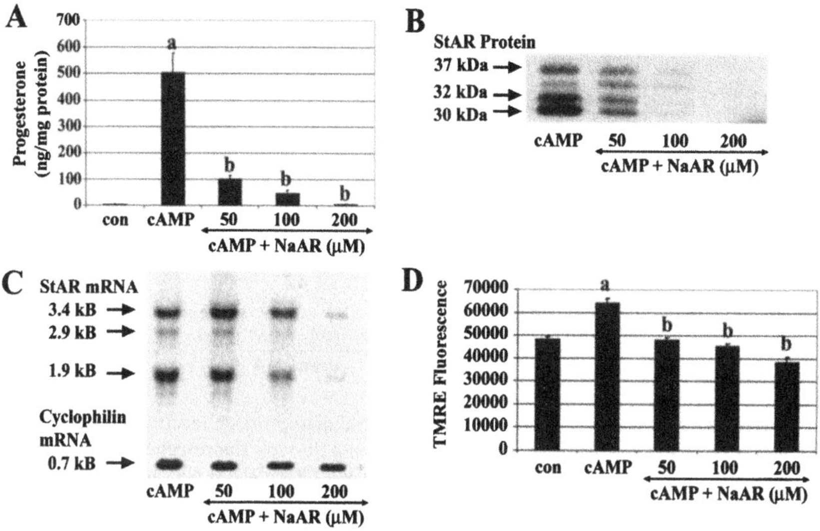

FIGURE 3. Effect of Na^+ arsenite on progesterone production, StAR expression and $\Delta\Psi_m$ in MA-10 cells. MA-10 cells were grown in culture and treated for 3 h with 1 mM 8-Br-cAMP (cAMP) or cAMP plus increasing concentrations of Na^+ arsenite (50–200 μM). After treatment, media were collected and subjected to a progesterone RIA, and protein lysates or total RNA obtained for Western or Northern blot analysis. (**A**) Progesterone concentrations were determined by RIA, were normalized to protein concentrations for each sample, and are represented as means ± SEMs. ($n = 3$) **a**: $P < 0.05$ vs. control; **b**: $P < 0.05$ vs. cAMP. (**B**) Representative Western blot for StAR protein after cAMP and Na^+ arsenite treatments ($n = 3$). (**C**) Representative Northern blot analysis for StAR mRNA after cAMP and Na+ arsenite treatments. Expression of the gene cyclophilin was also determined, and StAR expression was normalized to that of cyclophilin ($n = 3$). (**D**) MA-10 cells were cultured in a 96-well fluorescence assay plate and treated as in **A**. Cells were subsequently incubated with 50 nM TMRE for 10 min, and $\Delta\Psi_m$ was quantitated by measuring TMRE fluorescence as described in MATERIALS AND METHODS. Data are represented as means ± SEMs for 16 independent experiments ($n = 16$); **a**: $P < 0.05$ versus control; **b**: $P < 0.05$ versus cAMP.

TABLE 1), and the level of intracellular reactive oxygen was assessed by measuring fluorescence of H_2DCFDA, a dye indicator that fluoresces upon oxidation. Treatment of MA-10 cells with either cAMP or the different mitochondrial disrupting drugs did not significantly increase DCF fluorescence versus untreated cells (FIG. 4). However, data indicate that treatment of cells with cAMP, 10 μM antimycin A, or 1 μM oligomycin increased DCF fluorescence versus control (FIG. 4).

Dependence of StAR Protein Expression on $[Ca^{2+}]_{mt}$

Mitochondria are storage organelles for free calcium and calcium uptake, and the retention of calcium in the matrix is dependent upon the mitochondrial membrane potential. The calcium uniporter, which is located on the inner mitochondrial membrane, transfers cytosolic calcium into the matrix in a $\Delta\Psi_m$-dependent manner.[33–35] To examine whether $[Ca^{2+}]_{mt}$ is required for StAR expression, MA-10 cells were

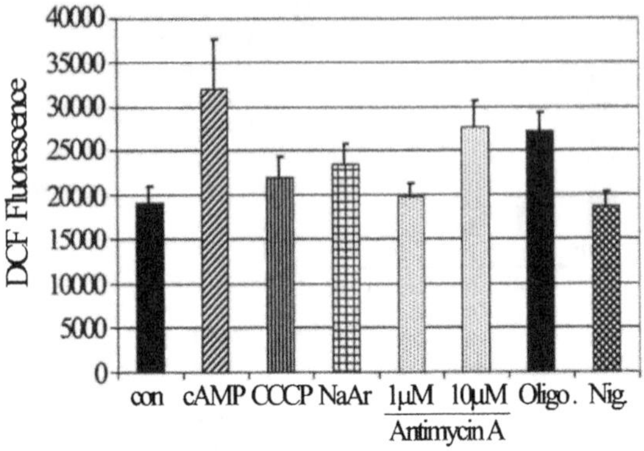

FIGURE 4. Effects of cAMP and mitochondrial disruptors on reactive oxygen production in MA-10 cells. MA-10 cells were cultured in a 96-well fluorescence assay plate and treated for 3 h with 1mM 8-Br-cAMP (cAMP), 5 μM CCCP, 50 μM Na^+ arsenite (Na-arsenite, 1 μM or 10 μM antimycin A, 1 μM oligomycin (Oligo), or 1 μM nigericin (Nig.). After treatment, cells were incubated with the reactive oxygen species dye indicator H_2DCFDA and dichlorofluorescien (DCF) fluorescence was quantitated as described in MATERIALS AND METHODS. Data are represented as means ± SEMs for three independent experiments ($n = 3$).

treated with the cell-permeable derivative of ruthenium red, Ru360, which is a specific inhibitor of the calcium uniporter.[36] Treatment of cells with cAMP plus Ru360 did not alter the expression of StAR protein (FIG. 5). In parallel, cells were pretreated with CCCP for 30 min to disrupt $\Delta\Psi_m$, which leads to a depletion of inner mitochondrial calcium;[37] then cells were treated with cAMP or cAMP plus Ru360. In CCCP-pretreated cells, Ru360 significantly reduced StAR protein expression (by approximately 50%) versus cAMP alone (FIG. 5B). These data suggest that cAMP-induced StAR protein expression is dependent upon $[Ca^{2+}]_{mt}$.

DISCUSSION

This study indicates that disrupting mitochondria results in post-transcriptional inhibition of StAR, and that mitochondria must be energized, polarized and actively respiring to support Leydig cell steroidogenesis. By using a mechanistic model and pharmacological approaches, this investigation demonstrates that several intrinsic properties of mitochondria are important for steroidogenesis. As shown in TABLE 1, CCCP, antimycin A, oligomycin, nigericin, H_2O_2, and xanthine oxidase all cause a significant inhibition in progesterone synthesis and StAR protein levels, without affecting StAR mRNA. Disrupting the mitochondrial electrochemical gradient with CCCP results in a significant inhibition in progesterone synthesis, even in the presence of newly synthesized StAR protein, indicating that $\Delta\Psi_m$ is essential for steroid biosynthesis. Preventing electron transport in mitochondria with antimycin A significantly reduced cellular ATP, potently inhibited steroidogenesis, reduced StAR protein expression, and dissipated $\Delta\Psi_m$. Similarly, preventing ATP synthesis with oligomycin reduced cellular ATP and inhibited progesterone synthesis and StAR

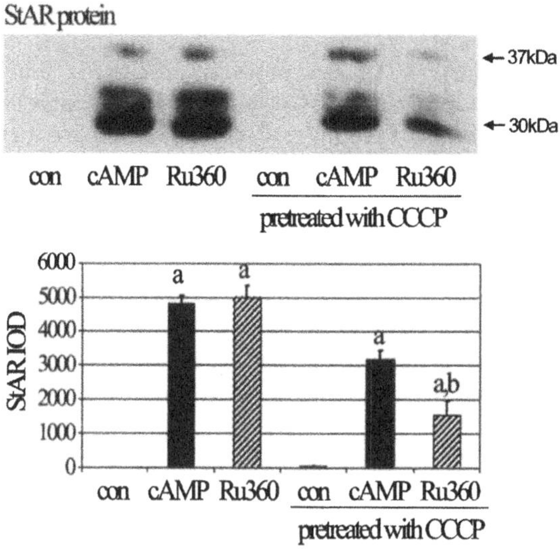

FIGURE 5. Effects of altering $[Ca^{2+}]_{mt}$ on StAR protein expression in MA-10 cells. MA-10 cells were treated for 3 h with 1mM 8-Br-cAMP (cAMP) or cAMP plus 5 μM Ru360. In parallel, cells were pretreated with 5 μM CCCP for 30 min (pretreated with CCCP) to disrupt $\Delta\Psi_m$ and reduce inner mitochondrial calcium, and cells were subsequently treated for 3 h with cAMP or cAMP plus Ru360. (**A**) Representative Western blot for StAR protein after the respective treatments. (**B**) Quantitation of StAR protein by scanning densitometry presented as means ± SEMs for three independent experiments ($n = 3$); **a**: $P < 0.05$ versus control; **b**: $P < 0.05$ versus cAMP. StAR IOD, integrated optical density of 30 + 37 kDa StAR protein bands.

protein expression, but had no effect on $\Delta\Psi_m$. These data indicate that preventing mitochondrial ATP synthesis with antimycin A or oligomycin inhibits steroidogenesis. Disruption of ΔpH with nigericin inhibited progesterone production and StAR protein but had minimal effects on $\Delta\Psi_m$. The oxidants H_2O_2, xanthine oxidase, and Na^+ arsenite caused a significant inhibition of progesterone and StAR protein, but not of StAR mRNA, and did not affect P450scc protein levels. Depleting $[Ca^{2+}]_{mt}$ with Ru360 resulted in a significant inhibition of StAR protein, demonstrating the importance of mitochondrial calcium to steroidogenesis. These results indicate that $\Delta\Psi_m$, mitochondrial ATP synthesis, ΔpH, and $[Ca^{2+}]_{mt}$ are all essential for the support of Leydig cell steroidogenesis.

Arsenic produces a variety of stress responses in mammalian cells, including metabolic abnormalities accompanied by growth inhibition and eventually apoptosis, and has been equated to the heat0shock response.[32] Considerable interest in arsenite as a pro-apoptotic agent with potential for chemotherapeutic treatment of cancer has focused on mechanisms of cytotoxicity. Arsenite has been shown to antagonize the anti-apoptotic Bcl-2 proteins and to promote the mitochondrial permeability transition, resulting in activation of caspases and culminating in cell death.[38–40] At high concentrations, arsenite inhibits respiration and ATP synthesis.[41,42]

In contrast to the cytotoxic effects observed at higher concentrations of arsenite, trace amounts are known to be essential for cell growth and metabolism, similar to selenium.[31,43] Results from this study demonstrate that at concentrations below a certain threshold (50 μM), Na^+ arsenite significantly inhibits cAMP-stimulated progesterone production (FIG. 3A), and post-transcriptionally inhibits StAR protein (FIG. 3C) without affecting StAR mRNA (FIG. 3B). Higher concentrations of arsenite abrogated StAR mRNA and protein expression after 3 hours of exposure, without causing a significant decrease in $\Delta\Psi_m$. It is likely that prolonged exposure to these higher concentrations of Na^+ arsenite would cause cell death. The results with low concentrations of Na^+ arsenite are strikingly similar to those observed under heat-shock conditions, which cause a post-transcriptional inhibition of StAR, presumably due to heat-shock protein–induced inhibition of StAR protein synthesis,[44] or to increased degradation of 37 kDa StAR. The effect of heat shock on mitochondrial function, however, was not examined. The results from the present study further demonstrate that perturbation of the mitochondria is deleterious to steroidogenesis without any evidence of frank cytotoxicity.

Leydig cell mitochondrial steroidogenesis is sensitive to oxidative stress (FIGS. 2 and 3), as has been shown previously.[18,19,45,46] Reactive oxygen is a byproduct of cellular respiration, and inhibition of the electron transport chain results in increased production of ROS.[47] Steroidogenic enzyme reactions catalyzed by cytochrome P450 enzymes are also a source of ROS.[48–51] To assess the endogenous production of ROS, DCF fluorescence was measured. The results shown in FIGURE 4 demonstrate that treatment with cAMP, antimycin A, and oligomycin cause an increase in ROS production in MA-10 cells. These results suggest that Leydig cells are particularly sensitive to inhibition by oxidative stress owing to the high levels of reactive oxygen produced during cAMP-stimulated steroidogenesis.

The importance of intracellular calcium to steroidogenesis has been demonstrated by several investigators. In particular, work from Capponi and co-workers has established that the maintenance of $[Ca^{2+}]_{mt}$ is required for adrenal corticosteroid biosynthesis.[52–54] The results of the present study confirm and extend these findings. The cell-permeable derivative of ruthenium red, Ru360, a specific inhibitor of the mitochondrial calcium uniporter, was used to block re-uptake of calcium into the mitochondria after depletion with CCCP, resulting in a marked decrease in cAMP induction of StAR protein (FIG. 5). In contrast, pretreatment with CCCP and subsequent treatment with cAMP in the absence of Ru360 did not significantly inhibit StAR protein compared to cells that were not pre-treated with CCCP. These results indicate that maintenance of $[Ca^{2+}]_{mt}$, which depends on $\Delta\Psi_m$, is required for StAR protein expression. It is likely that one mechanism through which Ru360 exerted its effect was via inhibition of ATP synthesis. Several enzymes in the Krebs-TCA cycle are calcium-dependent, and the results of blocking calcium influx with RU360 parallel the effects of antimycin A and oligomycin.

A consistent result with all mitochondrial disrupting agents tested is that StAR protein expression is highly sensitive to alterations in mitochondria. This suggests that perturbing the mitochondria results in either degradation of StAR protein or in a post-transcriptional inhibition of StAR. The reduction in ATP due to oligomycin or antimycin A likely inhibits steroidogenesis owing to reduced StAR translation, which is dependent on ATP. However, dissipating $\Delta\Psi_m$ with CCCP is ATP-independent, and CCCP did not prevent synthesis of StAR, indicating that other processes are involved, such as activation of proteases resulting in StAR degradation. CCCP appears to decrease the half-life of the 30- and 32-kDa forms of StAR and activate proteases, [15,16] supporting the notion that degradation of intramitochondrial StAR may occur due to disruption of $\Delta\Psi_m$. In addition, ATP stores are important for numerous cellular processes, including phosphorylation. StAR phosphorylation enhances its activity,[13] and newly synthesized, phosphorylated StAR is necessary for cholesterol transfer.[30] Phosphorylation of StAR has also been suggested to increase the protein's stability.[55] Thus, reduced ATP due to mitochondrial disruption may reduce phosphorylation of StAR, thereby contributing to the inhibition of steroidogenesis.

The data in this report clearly demonstrate that disrupting mitochondria profoundly inhibits steroidogenesis, which is likely due to multiple effects that are in addition to alterations in StAR. The current unifying model for cholesterol transfer into mitochondria involves the peripheral-type benzodiazepine receptor (PBR),[56] which is thought to organize cholesterol into cholesterol-rich domains in the outer mitochondrial membrane. It is proposed that StAR acts in conjunction with PBR to form the cholesterol-transfer complex.[57] The effect of mitochondrial-specific disrupting agents on PBR also needs to be examined. It is likely that perturbation of mitochondria with these agents also affects PBR function.

Steroidogenic cells have evolved a strategy to exploit the unique membrane compartmentation of the mitochondria in order to regulate steroid synthesis. The hydrophilic barrier of the mitochondrial inner membrane space separates P450scc from cholesterol, and bridging this gap is a key control point for enabling steroidogenesis. Results from this study suggest that the state of the mitochondrion itself is an important determinant in steroidogenesis. We suggest that mitochondria must be energized, polarized, and actively respiring to support Leydig cell steroidogenesis, and that alterations in the state of mitochondria may be involved in regulating steroid biosynthesis.

ACKNOWLEDGMENTS

We wish to thank Cassie Mahon for expert technical assistance. This work was previously presented in preliminary form at the 37th Society for the Study of Reproduction meeting in July 2004 and the XVIII North American Testes Workshop in April 2005. This work was supported by National Institutes of Health Research Training Grant T32 HL07692 (J.A.A), a postdoctoral fellowship grant (Deutschen Forschungsgemeinschaft Di 723/1-1) (to T.D.), National Institutes of Health Grants HD-27571 and HD-35544, and American Cancer Society Grant ACS 04010 (to D.B.H.).

REFERENCES

1. STOCCO, D.M. 2001. StAR protein and the regulation of steroid hormone biosynthesis. Annu. Rev. Physiol. **63:** 193–213.
2. CHRISTENSON, L.K. & J.F. STRAUSS. 2001. Steroidogenic acute regulatory protein; an update on its regulation and mechanism of action. Arch. Med. Res. **32:** 576–586.
3. HAUET, T. *et al.* 2005. Peripheral-type benzodiazepine receptor-mediated action of steroidogenic acute regulatory protein on cholesterol entry into Leydig cell mitochondria. Mol. Endocrinol. **19:** 540–554.
4. STOCCO, D.M. 1999. Steroidogenic acute regulatory (StAR) protein: what's new? Bioessays **21:** 768–775.
5. PAYNE, A.H. & D.B. HALES. 2004. Overview of steroidogenic enzymes in the pathway from cholesterol to active steroid hormones. Endocr. Rev. **25:** 947–970.

6. KASIANOWICZ, J., R. BENZ & S. MCLAUGHLIN. 1984. The kinetic mechanism by which CCCP (carbonyl cyanide m-chlorophenylhydrazone) transports protons across membranes. J. Membr. Biol. **82:** 179–90.
7. PRESSMAN, B.C. 1976. Biological applications of ionophores. Annu. Rev. Biochem. **45:** 501–530.
8. SLATER, E.C. 1967. Application of inhibitors and uncouplers for a study of ozidative phosphorylation. Methods Enzymol. **10:** 48–57.
9. KING, S.R. & D.M. STOCCO. 1996. ATP and a mitochondrial electrochemical gradient are required for functional activity of the steroidogenic acute regulatory (StAR) protein in isolated mitochondria. Endocr. Res. **22:** 505–514.
10. KING, S.R. *et al.* 1999. Effects of disruption of the mitochondrial electrochemical gradient on steroidogenesis and the steroidogenic acute regulatory (StAR) protein. J. Steroid Biochem. Mol. Biol. **69:** 143–154.
11. KING, S.R., L.P. WALSH & D.M. STOCCO. 2000. Nigericin inhibits accumulation of the steroidogenic acute regulatory protein but not steroidogenesis. Mol. Cell. Endocrinol. **166:** 147–153.
12. STRAUSS, J.F., 3RD *et al.* 1999. The steroidogenic acute regulatory protein (StAR): a window into the complexities of intracellular cholesterol trafficking. Recent Prog. Horm. Res. **54:** 369–394.
13. ARAKANE, F. *et al.* 1997. Phosphorylation of steroidogenic acute regulatory protein (StAR) modulates its steroidogenic activity. J. Biol. Chem. **272:** 32656–32662.
14. WANG, X. *et al.* 1998. Effect of truncated forms of the steroidogenic acute regulatory protein on intramitochondrial cholesterol transfer. Endocrinology **139:** 3903–3912.
15. GRANOT, Z. *et al.* 2002. The life cycle of the steroidogenic acute regulatory (StAR) protein: from transcription through proteolysis. Endocr. Res. **28:** 375–386.
16. GRANOT, Z. *et al.* 2003. Proteolysis of normal and mutated steroidogenic acute regulatory proteins in the mitochondria: the fate of unwanted proteins. Mol. Endocrinol. **17:** 2461–2476.
17. HALES, D.B. 2002. Testicular macrophage modulation of Leydig cell steroidogenesis. J. Reprod. Immunol. **57:** 3–18.
18. ALLEN, J.A. *et al.* 2004. Bacterial endotoxin lipopolysaccharide and reactive oxygen species inhibit Leydig cell steroidogenesis via perturbation of mitochondria. Endocrine **25:** 265–276.
19. DIEMER, T. *et al.* 2003. Reactive oxygen disrupts mitochondria in MA-10 tumor Leydig cells and inhibits steroidogenic acute regulatory (StAR) protein and steroidogenesis. Endocrinology **144:** 2882–2891.
20. ASCOLI, M. 1981. Characterization of several clonal lines of cultured Leydig tumor cells: gonadotropin receptors and steroidogenic responses. Endocrinology **108**: 88–95.
21. HALES, D.B. 1992. Interleukin-1 inhibits Leydig cell steroidogenesis primarily by decreasing 17 alpha-hydroxylase/C17-20 lyase cytochrome P450 expression. Endocrinology **131:** 2165–2172.
22. LAEMMLI, U.K. 1970. Cleavage of structural proteins during the assembly of the head of bacteriophage T4. Nature **227:** 680–685.
23. XIONG, Y. & D.B. HALES. 1993. The role of tumor necrosis factor-alpha in the regulation of mouse Leydig cell steroidogenesis. Endocrinology **132:** 2438–2444.
24. ANAKWE, O.O. & A.H. PAYNE. 1987. Noncoordinate regulation of de novo synthesis of cytochrome P450 cholesterol side-chain cleavage and cytochrome P450 17α-hydroxylase/$C_{17\text{-}20}$ lyase in mouse Leydig cells: relation to steroid production. Molec. Endocrinol. **1:** 595–603.
25. HELD HALES, K. *et al.* 2000. Diametric effects of bacterial endotoxin lipopolysaccharide on adrenal and Leydig cell steroidogenic acute regulatory protein. Endocrinology **141:** 4000–4012.
26. CHOMCZYNSKI, P. & N. SACCHI. 1987. Single-step method of RNA isolation by acid guanidinium thiocyanate-phenol-chloroform extraction. Anal. Biochem. **162:** 156–159.
27. HALES, D.B., Y. XIONG & I. TUR-KASPA. 1992. The role of cytokines in the regulation of Leydig cell P450c17 gene expression. J. Ster. Biochem. Mol. Biol. **43:** 907–914.

28. NEUPERT, W. 1997. Protein import into mitochondria. Annu. Rev. Biochem. **66:** 863–917.
29. ARAKANE, F. *et al.* 1998. Steroidogenic acute regulatory protein (StAR) acts on the outside of mitochondria to stimulate steroidogenesis. Endocr. Res. **24:** 463–468.
30. ARTEMENKO, I.P. *et al.* 2001. Mitochondrial processing of newly synthesized steroidogenic acute regulatory protein (StAR), but not total StAR, mediates cholesterol transfer to cytochrome P450 side chain cleavage enzyme in adrenal cells. J. Biol. Chem. **276:** 46583–46596.
31. STOLZ, J.F., P. BASU & R.S. OREMLAND. 2002. Microbial transformation of elements: the case of arsenic and selenium. Int. Microbiol. **5:** 201–207.
32. BERNSTAM, L. & J. NRIAGU. 2000. Molecular aspects of arsenic stress. J. Toxicol. Environ. Health B Crit. Rev. **3:** 293–322.
33. GUNTER, T.E. & D.R. PFEIFFER. 1990. Mechanisms by which mitochondria transport calcium. Am. J. Physiol. **258:** C755–786.
34. GUNTER, T.E. *et al.* 1994. Mitochondrial calcium transport: physiological and pathological relevance. Am. J. Physiol. **267:** C313–339.
35. GUNTER, T.E. *et al.* 2000. Mitochondrial calcium transport: mechanisms and functions. Cell. Calcium **28:** 285–296.
36. MATLIB, M.A. *et al.* 1998. Oxygen-bridged dinuclear ruthenium amine complex specifically inhibits $Ca2^+$ uptake into mitochondria in vitro and in situ in single cardiac myocytes. J. Biol. Chem. **273:** 10223–10231.
37. GUNTER, T.E. *et al.* 1978. Efflux of Ca2+ and Mn2+ from rat liver mitochondria. Biochemistry **17:** 339–345.
38. CHEN, Y.C., S.Y. LIN-SHIAU & J.K. LIN. 1998. Involvement of reactive oxygen species and caspase 3 activation in arsenite-induced apoptosis. J. Cell Physiol. **177:** 324–333.
39. LAROCHETTE, N. *et al.* 1999. Arsenite induces apoptosis via a direct effect on the mitochondrial permeability transition pore. Exp. Cell Res. **249:** 413–421.
40. COSTANTINI, P. *et al.* 2000. Mitochondrion as a novel target of anticancer chemotherapy. J. Natl. Cancer Inst. **92:** 1042–1053.
41. CSILLAG, A. 1971. The effect of arsenite on rat-liver mitochondria. FEBS Lett. **17:** 342–344.
42. YIH, L.H. *et al.* 1991. Sodium arsenite induces ATP depletion and mitochondrial damage in HeLa cells. Cell Biol. Int. Rep. **15:** 253–264.
43. HEALY, S.M. *et al.* 1998. Enzymatic methylation of arsenic compounds. V. Arsenite methyltransferase activity in tissues of mice. Toxicol. Appl. Pharmacol. **148:** 65–70.
44. LIU, Z. & D.M. STOCCO. 1997. Heat shock-induced inhibition of acute steroidogenesis in MA-10 cells is associated with inhibition of the synthesis of the steroidogenic acute regulatory protein. Endocrinology **138:** 2722–2728.
45. HALES, K.H. *et al.* 2000. Diametric effects of bacterial endotoxin lipopolysaccharide on adrenal and Leydig cell steroidogenic acute regulatory protein. Endocrinology **141:** 4000–4012.
46. STOCCO, D.M., J. WELLS & B.J. CLARK. 1993. The effects of hydrogen peroxide on steroidogenesis in mouse Leydig tumor cells. Endocrinology **133:** 2827–2832.
47. DEGLI ESPOSTI, M. & H. MCLENNAN. 1998. Mitochondria and cells produce reactive oxygen species in virtual anaerobiosis: relevance to ceramide-induced apoptosis. FEBS Lett. **430:** 338–342.
48. QUINN, P.G. & A.H. PAYNE. 1985. Steroid product-induced, oxygen-mediated damage of microsomal cytochrome P-450 enzymes in Leydig cell cultures. Relationship to desensitization. J. Biol. Chem. **260:** 2092–2099.
49. QUINN, P.G. & A.H. PAYNE. 1984. Oxygen-mediated damage of microsomal cytochrome P-450 enzymes in cultured leydig cells: role in steroidogenic desensitization. J. Biol. Chem. **259:** 4130–4135.
50. QUINN, P.G. & A.H. PAYNE. 1984. Microsomal cytochrome P-450 enzyme damage in cultured Leydig cells: relation to steroidogenic desensitization. Ann. N.Y. Acad. Sci. **438:** 649–651.
51. HORNSBY, P.J. 1989. Steroid and xenobiotic effects on the adrenal cortex: mediation by oxidative and other mechanisms. Free Radic. Biol. Med. **6:** 103–115.

52. LIN, T. *et al.* 1980. The effects of calcium ionophore A23187 on interstitial cell steroidogenesis. Biochim. Biophys. Acta **627:** 157–164.
53. CAPPONI, A.M. *et al.* 1988. Calcium stimulates steroidogenesis in permeabilized bovine adrenal cortical cells. J. Biol. Chem. **263:** 16113–16117.
54. CHERRADI, N. *et al.* 1997. Submitochondrial distribution of three key steroidogenic proteins (steroidogenic acute regulatory protein, cytochrome P450scc and 3β-hydroxysteroid dehydrogrenase isomerase enzymes) upon stimulation by intracellular calcium in adrenal glomerulosa cells. J. Biol. Chem. **272**: 7899–7907.
55. CLARK, B.J., V.V. RANGANATHAN & R. COMBS. 2001. Steroidogenic acute regulatory protein expression is dependent upon post-translational effects of cAMP-dependent protein kinase A. Mol. Cell Endocrinol. **173:** 183–192.
56. LACAPERE, J.J. & V. PAPADOPOULOS. 2003. Peripheral-type benzodiazepine receptor: structure and function of a cholesterol-binding protein in steroid and bile acid biosynthesis. Steroids **68:** 569–585.
57. HAUET, T. *et al.* 2005. Peripheral-type benzodiazepine receptor-mediated action of steroidogenic acute regulatory protein on cholesterol entry into Leydig cell mitochondria. Mol. Endocrinol. **19:** 540–554.

Determining Structure and Function of Steroid Dehydrogenase Enzymes by Sequence Analysis, Homology Modeling, and Rational Mutational Analysis

WILLIAM L. DUAX,[a,e] JAMES THOMAS,[b] VLADIMIR PLETNEV,[c] ANTHONY ADDLAGATTA,[d] ROBERT HUETHER,[a] LUKAS HABEGGER,[a] AND CHARLES M. WEEKS[a]

[a]*Hauptman-Woodward Medical Research Institute Buffalo, Buffalo, New York 14203, USA*

[b]*Mercer University School of Medicine, Macon, Georgia 31207, USA*

[c]*Institute of Bioorganic Chemistry, RAS, Moscow, Russia*

[d]*University of Oregon, Eugene, Oregon 97403, USA*

[e]*SUNY at Buffalo, Buffalo, New York, USA*

ABSTRACT: The short-chain oxidoreductase (SCOR) family of enzymes includes over 6,000 members identified in sequenced genomes. Of these enzymes, ~300 have been characterized functionally, and the three-dimensional crystal structures of ~40 have been reported. Since some SCOR enzymes are steroid dehydrogenases involved in hypertension, diabetes, breast cancer, and polycystic kidney disease, it is important to characterize the other members of the family for which the biological functions are currently unknown and to determine their three-dimensional structure and mechanism of action. Although the SCOR family appears to have only a single fully conserved residue, it was possible, using bioinformatics methods, to determine characteristic fingerprints composed of 30–40 residues that are conserved at the 70% or greater level in SCOR subgroups. These fingerprints permit reliable prediction of several important structure-function features including cofactor preference, catalytic residues, and substrate specificity. Human type 1 3β-hydroxysteroid dehydrogenase isomerase (3β-HSDI) has 30% sequence identity with a human UDP galactose 4-epimerase (UDPGE), a SCOR family enzyme for which an X-ray structure has been reported. Both UDPGE and 3-HSDI appear to trace their origins back to bacterial 3α,20β-HSD. Combining three-dimensional structural information and sequence data on the 3α,20β-HSD, UDPGE, and 3β-HSDI subfamilies with mutational analysis, we were able to identify the residues critical to the dehydrogenase function of 3-HSDI. We also identified the residues most probably responsible for the isomerase activity of 3β-HSDI. We test our predictions by specific mutations based on sequence analysis and our structure-based model.

Address for correspondence: William L. Duax, Hauptman-Woodward Medical Research Institute, 73 High St., Buffalo, NY 14203. Voice: 716-856-9600; fax: 716-852-6086.
duax@hwi.buffalo.edu

Ann. N.Y. Acad. Sci. 1061: 135–148 (2005).
doi: 10.1196/annals.1336.015

KEYWORDS: rational mutational analysis; steroid dehydrogenase; homology modeling; prostate cancer; breast cancer; prediction of ligand specificity; testosterone 3-HSDII

INTRODUCTION

The short-chain oxidoreductase (SCOR) family of enzymes is an ancient family that includes many dehydrogenases and reductases vital to growth and development.[1] The features that have come to define the SCOR family are: (1) an approximate length of 250–350 amino acids that form a single-domain Rossmann fold; (2) NAD or NADP cofactor dependence; (3) a glycine-rich signature sequence (GxxxxxG) located near the N-terminus and involved in cofactor binding; and (4) a catalytic triad usually composed of a Ser residue and Tyr and Lys residues in a YxxxK sequence.[1–3] Over 6,000 probable SCOR genes have been identified by sequence similarity, but only 300 of these have been biochemically characterized.[5] We have identified structural subgroups of the SCOR family, determined the sequence fingerprints of these subgroups, and demonstrated that the cofactor specificity of 90% of all hypothetical members of the family can be reliably predicted.[2,4]

Although the SCOR enzymes have a common Rossmann fold, bind NAD(P) cofactors, and have common catalytic residues and a postulated mechanism of action, there is at most one fully conserved residue in the family, and the substrates of the SCOR family include a wide range of structure in terms of size and hydrophobicity. Because the known substrates for SCOR enzymes include steroids, prostaglandins, sugars, dyes, porphyrins, acids, and alcohols, the substrate-binding pocket is expected to be highly variable. We have used covariance analysis methods to identify the substrates of 25% of the SCOR subfamily that has a TGxxxGIG signature in β1α2 turn (FIG. 5).[1]

3α,20β-HYDROXYSTEROID DEHYDROGENASE

The first member of the SCOR family for which the crystal structure was determined was 3α,20β-hydroxysteroid dehydrogenase (3α,20β-HSD) from *Streptomyces hydrogenans,* an enzyme involved in the reversible oxidation of the 3α-group of androstane derivatives and the 20β-group of pregnane derivatives. At least two models had been proposed to explain the dual activity of the enzyme.[7] One model invoked a single stereospecific steroid-binding pocket with cofactor binding sites at either end, accounting for the 3α and 20β activity. A second model proposed a single cofactor-binding site and a substrate-binding pocket that would permit steroids to bind in two different orientations. The X-ray structure of the complex of the tetrameric enzyme and cofactor[8] revealed that each subunit of the tetramer contains a cofactor binding site and a putative steroid-binding site. The 245-amino acid monomer has essentially a single domain. The first 145 residues have the characteristic Rossmann fold,[9] composed of a five-stranded parallel β-sheet with two helices on either side (FIG. 1). The rest of the single-domain structure consists of two additional β-strands added to the β-sheet and two more helices. The cofactor resides on one side of the β-sheet in an extended conformation. The adenine-ribose end of the

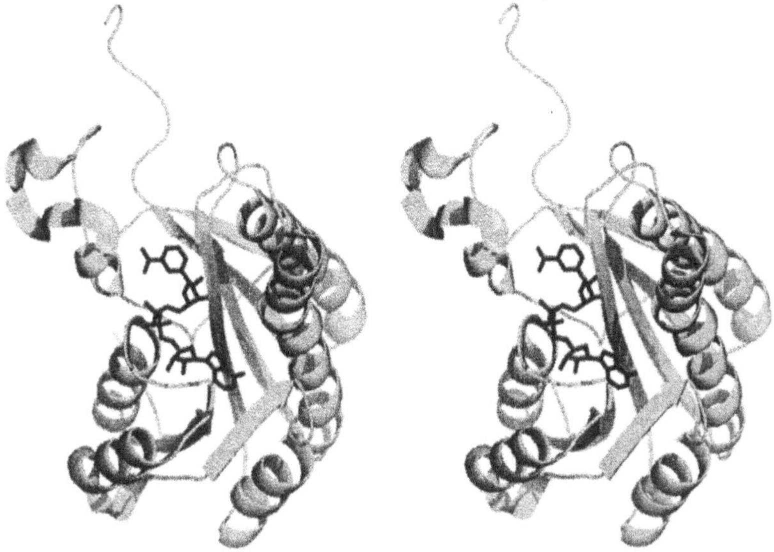

FIGURE 1. Stereo illustration of the crystallographically observed complex of 3α,20β-HSD and the NAD cofactor.

cofactor lies in a cleft surrounded by five peptide segments from one monomer of the protein. Hydroxy groups of the adenine-ribose ring form hydrogen bonds with the Asp37 side chain, and the *bis*-phosphate moiety interacts with the Arg16 side chain.

On the basis of the observed structure, we proposed a mechanism of action involving highly conserved Ser(139) and Tyr(152) xxx Lys(156) residues (FIG. 2). Numerous mutation studies have demonstrated that these residues are critical to oxidoreductase activity. Subsequent structure determinations and mutation studies suggested that as many as four additional residues [Asp(111) and Pro(184)-GlyxxxThr(187)] that are closely linked to the cofactor or between the cofactor and the substrate-binding pocket may have important roles in the mechanism of action. Without altering the structure of the enzyme it was possible to model a cortisol molecule in the substrate-binding cleft with either the 3β- or 20β-carbonyl oriented near the cofactor and catalytic tyrosine, suggesting that the dual function of 3α,20β-HSD is due to the fact that the binding site accepts steroids in two orientations.[8]

LICORICE EFFECTS ON STEROID METABOLISM

For over 50 years, physicians have warned people with hypertension not to eat licorice. Although a correlation between licorice consumption and elevation of blood pressure was well established, the physiological basis for the effect was not understood.[10] When glycyrrhizic acid, the active ingredient responsible for the hypertensive response, was isolated, its ability to compete for the mineralocorticoid receptor was tested, but no evidence of binding could be detected. In the 1970s, some

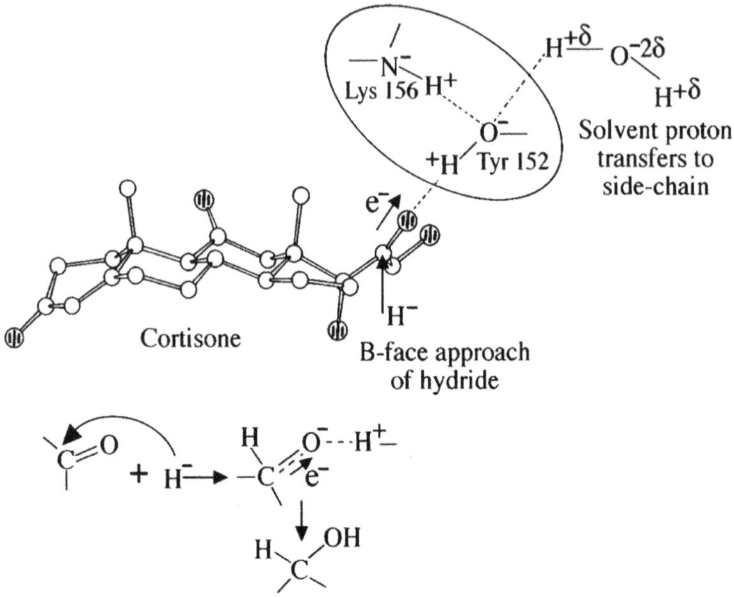

FIGURE 2. Proposed mechanism of action of 3α,20β-HSD.

newborn infants were found to have clinical signs of aldosteronism, no detectable circulating aldosterone, and an inability to oxidize cortisol to cortisone. Urinary 11β-hydroxycortisol metabolites in these infants were greatly increased, and the corresponding 11β-oxocortisone metabolites were diminished. This condition was named apparent mineralocorticoid excess (AME).[11,12] The altered steroid metabolism in AME was traced to a defective 11β-hydroxysteroid dehydrogenase (11β-HSD). The defect leads to a persistently active mineralocorticoid receptor saturated by excess accumulated cortisol. The discovery of the critical role of 11β-HSD in mineralocorticoid receptor binding and activity led investigators to question whether glycyrrhizic acid might inhibit 11β-HSD, producing an increased concentration of cortisol and associated hypertension. As predicted, inhibitor studies revealed that glycyrrhizic acid is a potent competitor for binding to the active site of 11β-HSD. The interactions between 11β-HSD, cortisol, cortisone, aldosterone, glycyrrhizic acid, and the mineralocorticoid receptor are illustrated in FIGURE 3.

Because 11β-HSD is a member of the SCOR family, we tested glycyrrhizic acid and its metabolite, carbenoxolone, and found them to be very potent inhibitors of 3α,20β-HSD.[13] Single crystals of the complex of carbenoxolone and 3α,20β-HSD were prepared, and the structure determination revealed the presence of the inhibitor in the previously postulated binding site.[14] The carbenoxolone molecule is anchored in the catalytic cleft by hydrogen bonding (FIG. 4) at both ends of the molecule. The hemisuccinate terminal oxygen forms a hydrogen bond (2.8Å) to Tyr152, one of the two strictly conserved residues of the catalytic triad. The cofactor is either prevented from binding or displaced by the inhibitor molecule. The fact that carbenoxolone

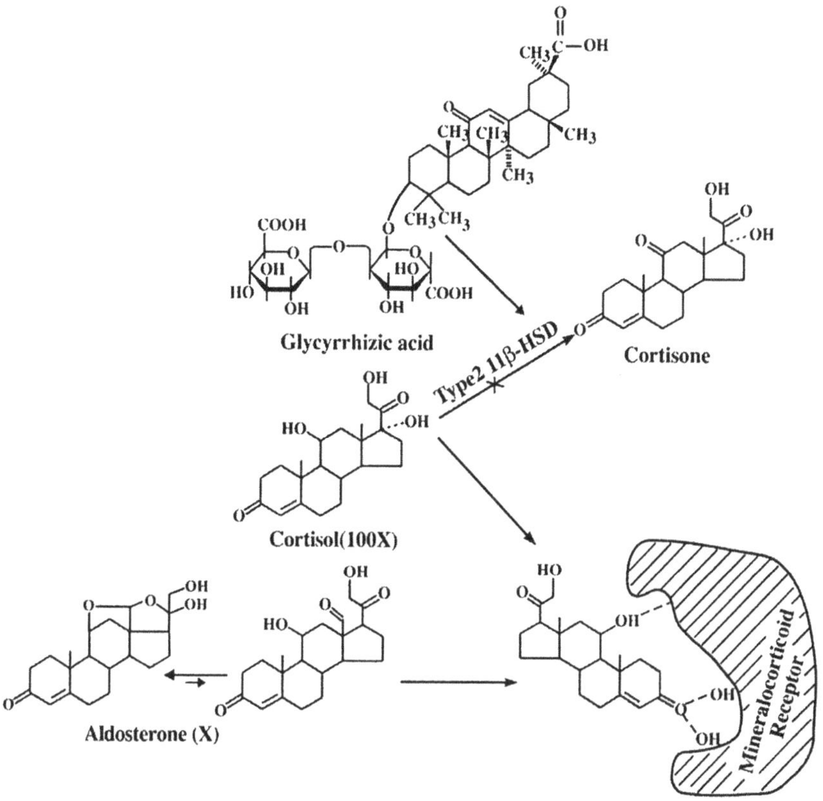

FIGURE 3. Schematic representation of interactions between aldosterone, the mineralocorticoid receptor, cortisol, cortisone, NADH-dependent 11β-hydroxysteroid dehydrogenase, and glycyrrhizic acid related to proper modulation of steroid balance in the corticoid-collecting duct of the kidney.

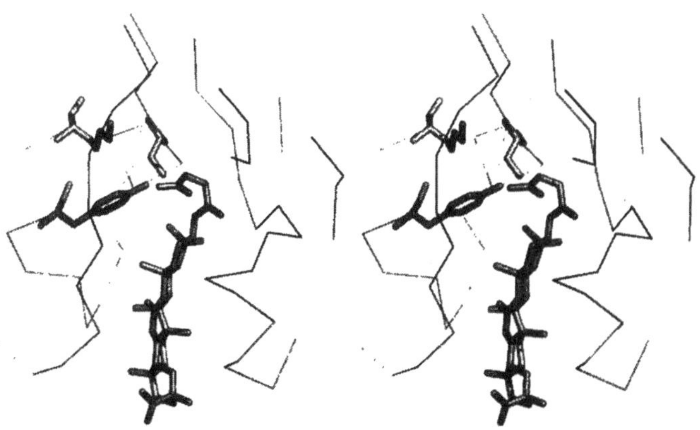

FIGURE 4. Stereo illustration of the monomer of 3α,20β-HSD in complex with carbenoxolone.

occupies the catalytic cleft, displaces the cofactor, and hydrogen bonds directly to a member of the catalytic triad makes it a very potent high-affinity inhibitor and a good model upon which to base the design of selective inhibitors for other enzymes, particularly other members of the SCOR family.

Licorice has also been linked to decreases in serum testosterone levels in women. This effect may be due to inhibition of 17β-hydroxysteroid dehydrogenase and 17–20 lyase in the adrenal glands of healthy young women.[15]

SHORT-CHAIN OXIDOREDUCTASE SUPERFAMILY

The three-dimensional crystal structures of over 40 unique SCOR enzymes had been reported and archived in the Protein Data Bank (PDB).[16] Although at most one residue in the SCOR enzyme is conserved in all 6,000 members of the superfamily, we have identified a group of ~40 sequence positions exhibiting 70% or greater conservation that define the fold fingerprint of the largest subgroup of the SCOR family (FIG. 5). Of these 40 fingerprint residues, 12–15 make hydrogen-bond contacts with the cofactor (FIG. 6). In current databases, the majority of proteins with SCOR fingerprint sequence patterns are identified as putative, hypothetical, or probable oxidoreductases/dehydrogenases without a defined function or simply as open reading frame products of unknown function.

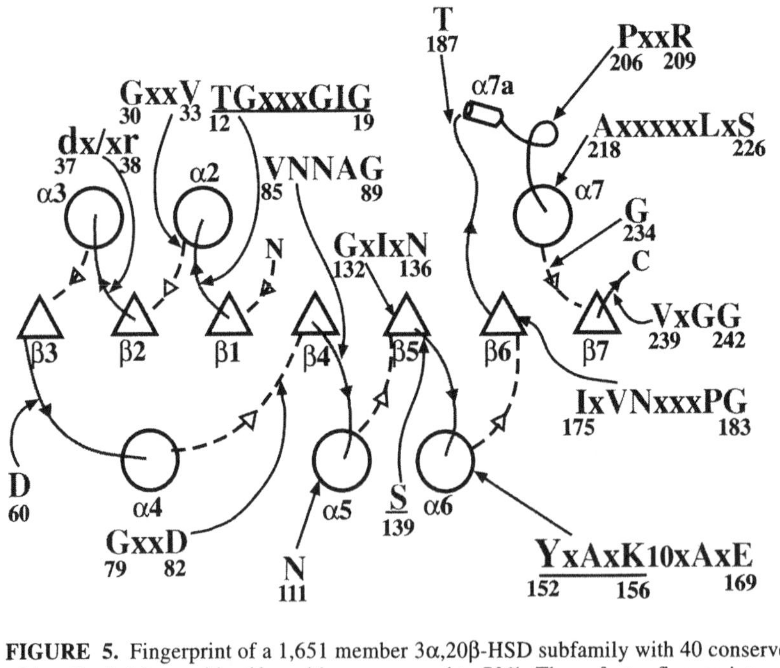

FIGURE 5. Fingerprint of a 1,651 member 3α,20β-HSD subfamily with 40 conserved identities. Capital letters identify residues conserved to 70%. The cofactor fingerprint residues (GxxxGxG) and catalytic residues Serine (S) and Tyrosine (Y) xxx Lysine (K) are underlined.

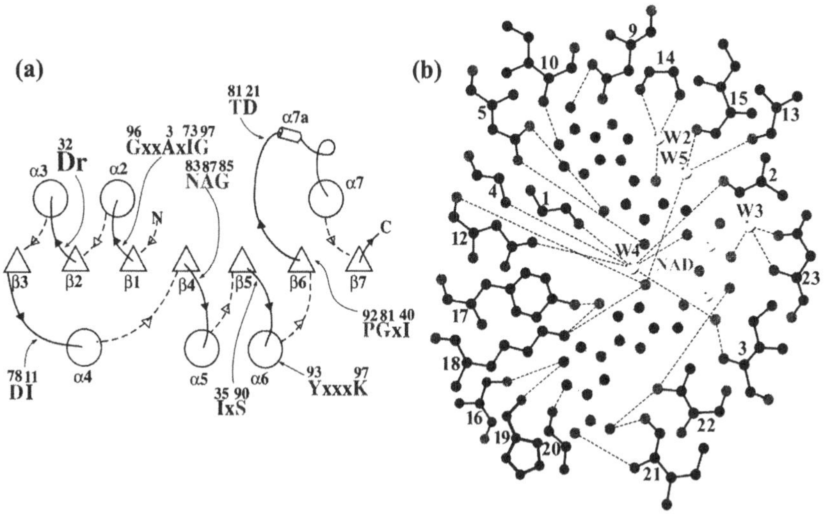

FIGURE 6. (**a**) A schematic diagram and (**b**) an atomic project of residues contacting the NAD cofactor either directly or through water molecules included 13 of the 34 fingerprint residues conserved at 65% identity in 1,651 structures that align with 3α,20β-HSD (1HDC). The numbers over the capital letters are the % conserved identity of that residue in the 1,651 gene products found to have 25% identity with 3α,20β-HSD.

We have shown that NAD or NADP cofactor preferences can be reliably predicted for ~70% of the SCORs by the presence of Asp or Arg recognition residues, respectively, in specific adjacent sequence positions in the β2α3 turn near the N-terminus (FIG. 5). The residues in the β1α2 turn of the Rossmann fold distinguish different subfamilies. The 1,651 structures having sequence homology with 3α,20β-HSD are characterized by the TGxxxGIG sequence in the β1α2 turn.[2] Because the known substrates for SCOR enzymes include steroids, prostaglandins, sugars, dyes, porphyrins, acids, and alcohols, the substrate-binding pocket is highly variable. In many crystal structures, competitive inhibitors have been cocrystallized with the enzyme steroid, permitting the determination of the substrate-binding site. Three flexible loops of hypervariable composition create a pocket on the surface of the enzyme that controls access to the catalytic residues and nicotinamide end of the cofactor. We have used covariance analysis[6] to determine groups of amino acids that line this pocket and covary with the type of substrate (FIG. 7). In contrast to the fingerprint of highly conserved residues that determine the Rossmann fold, interacting with the cofactor and including the catalytic residues, the substrate-defining residues are not highly conserved.

A search for covariance among residues predicted to line the substrate-binding pocket led to identification of a conserved set of 10 residues that define this pocket for the β-ketoacyl-[acyl carrier protein] (β-k-ACPR) subfamily[6] (FIG. 7). An identical pattern of 10 residues in three similarly spaced segments was found in 118 protein sequences in the SWISS-PROT/TrEMBL database. The majority of these structures had previously been identified as known or hypothetical β-k-ACPRs. These structures come from 95 different species ranging from bacteria to humans.

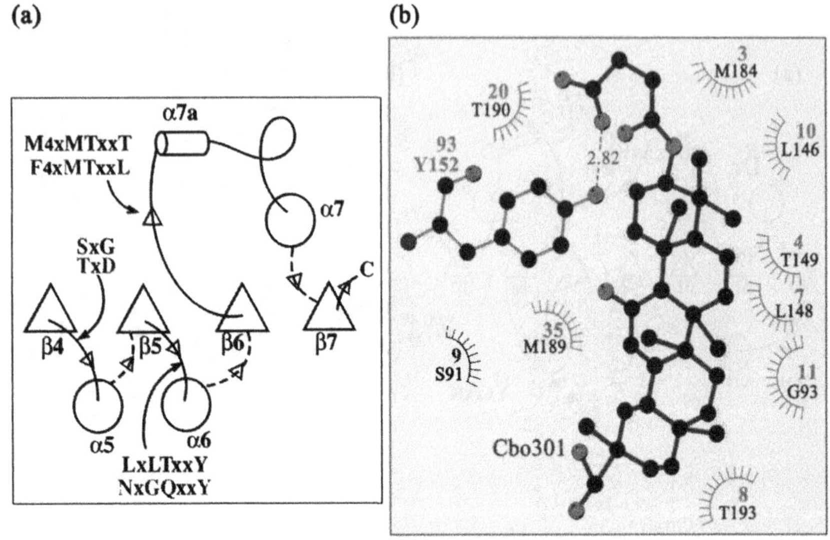

FIGURE 7. Illustrations of the residue on three flexible hypervariable loops of members of the SCOR family of enzymes that make up the substrate-binding pocket. (**A**) Schematic illustratation of the sequences of residues in the three flexible loops including the identities of residues in the sequence of the 3α,20β-HSD (*dark upper fingerprint*) and the 118 member β-keto acyl carrier reductase subfamily (*light lower fingerprint*) and (**B**) a projection of the residues in the flexible loop that are in contact with the carbonoxolone in the crystalline complex of 3α,20β-HSD and the inhibitor. With the exception of a hydrogen bond involving the catalytic Tyr152, all interactions are van der Waals contacts.

Fourteen gene products previously assigned ambiguous or erroneous identifications can now be reliably predicted to be β-k-ACPRs based on this fingerprint. For example, one of these proteins (Ke 6) is a gene product previously identified as a type 8 17β-hydroxysteroid dehydrogenase that is downregulated in a murine model for polycystic kidney disease.[6] This enzyme is likely to be involved in fattyacid processing rather than sex steroid metabolism. We subsequently discovered that different combinations of amino acid residues in the same 9 sequence position on the 3 flexible loops define the substrate specificity of 11 different SCOR enzyme subfamilies.

RATIONAL MUTATIONAL ANALYSIS BASED ON PROTEOMIC FINGERPRINTS

In humans, there are two isoforms of 3β-hydroxysteroid dehydrogenase encoded by separate genes and expressed in a tissue-specific manner. Human type 1 3β-hydroxysteroid dehydrogenase/isomerase (3β-HSDI) catalyzes the two sequential enzyme reactions on a single protein that converts dehydroepiandrosterone or pregnenolone to androstenedione or progesterone, respectively, in placenta, mammary gland, breast tumors, prostate, prostate tumors, and other peripheral tissues. Human type 2 3β-hydroxysteroid dehydrogenase/isomerase (3β-HSDII) is expressed in testis, ovary, and the adrenal gland and has 90% homology with human 3β-HSDI.[17,18]

```
                      G xx G xx G
3βHSD1       MTGWSCLVTGAGGFLGQRIIRLLVKEKELKEIRVLDKAFGPELREEFSKLQNKTKLTVLE
1NAH         ---MRVLVTGGSGYIGSHTCVQLLQNGHDVIILDNLCNSKRSVLPVIERLGGKHP-TFVE

3βHSD1       GDILDEPFLKRACQD--VSVIIHTACIIDVFG-VTHRESIMNVNVKGTQLLLEACVQASV
1NAH         GDIRNEALMTEILHDHAIDTVIHFAGLKAVGESVQKPLEYYDNNVNGTLRLISAMRAANV

                    SS                                   Y xx SK
3βHSD1       PVFIYTSSIEVAGPNSYKEIIQNGHEEEPLENTWPAPYPHSKKLAEKAVLAANGWNLKNG
1NAH         KNFIFSSSATVYGDN-PK--I-PYVESFP-TGTPQSPYGKSK-L---MV-EQILTDLQKA

3βHSD1       GTLYTCALRPMYIYGEGSRFLSASINEALNNNGILSSVGKFSTVNPVYVGNVAWAHILAL
1NAH         QPDWSIAL--LRYFNPVGAHPSGDMGEDPQ--GIPN------NLMP-YIAQVAVGRRDSL

                                             Y xx SK
3βHSD1       RALQDPKKAPSIRGQFYYISDDTPHQSYDNLNYTLSKEFGLRLDSRWSFPLSLMYWIGFL
1NAH         AIFGNDYPTEDGTGVRDYI-----HVMDLADGHVVAMEK---LANKPGVHI-YNLGAGVG

3βHSD1       LEIVSFLLRPIYTYRPPFNRHIVTLSNSVFTFSYK---KAQRDLAYKPLYSWEEAKQKTV
1NAH         NSVLDVVNAFSKACGKPVNYHFAPRREGDLPAYWADASKADRELNWRVTRTLDEMAQDTW

3βHSD1       EWVGSLVDRHKETLKSKTQ
1NAH         HWQS----RHPQGYPD---

                                                              CLUSTAL W(1.60)
```

FIGURE 8. Sequence alignment of UDP galactase epimerase (1NAH) and 3β-HSDI. The positions of the Rossmann fold signature (TGxxGxxG) and the catalytic residues (SS and YxxxK) are highlighted. A second YxxxK sequence found in 3β-HSDI is also identified.

Amino acid sequence analysis suggests that 3β-HSDI is a member of the SCOR family. Similarities include the presence of a YxxxK sequence, a Serine doublet, and a Gly-rich sequence near the N-terminus. However, 3β-HSDI is significantly larger than the typical SCOR enzyme (372 verses 250 amino acid residues). 3β-HSDI is an isomerase as well as a dehydrogenase, and the Gly-rich sequence in the β1α2 turn is different in detail. When the sequence of 3β-HSDI was compared with the sequence of all SCOR enzymes for which crystal structures have been reported, the structure found to have the greatest percentage identity was the UDP galactose-4-epimerase (UDPGE) structure from *Escherichia coli* (PDB code 1NAH).[19]

Although the percent conservation of identities between 1NAH and 3β-HSDI is only 20% and the alignment incorporates 11 insertions and 6 deletions (FIG. 8), there is no doubt about the fit because of the conservation of:

(1) The catalytic YxxxK and Ser.
(2) The presence in the 3β-HSDI sequence of 35 of the 95 fingerprint residues of UDPGE.
(3) The presence in the 3β-HSDI of the **TGxxGxxG** signature sequence in the β1α2 turn of the UDPGE family.[2]
(4) The conservation of many of the conserved residues in the UDPGE structural family that contact the NAD/NADP cofactor.

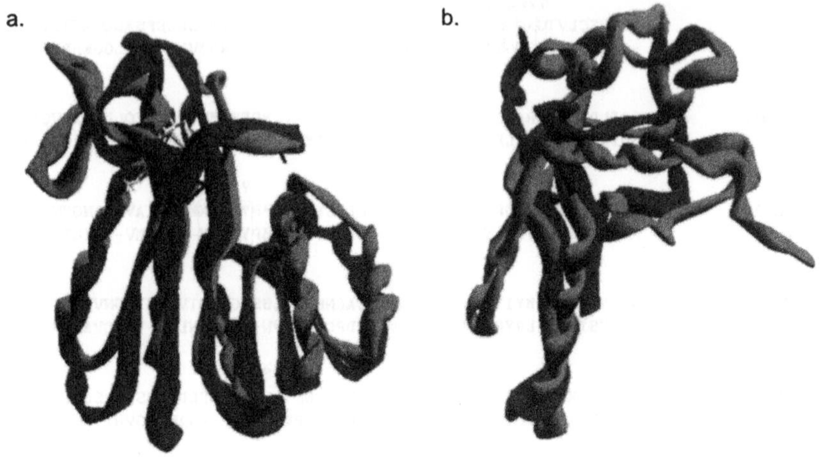

FIGURE 9. Overlap of the three-dimensional structure of 3α,20β-HSD (1HDC) and our model of 3β-HSD based on the UDP galactose epimerase (1NAH), illustrating (**a**) similarity in Rossmann fold cofactor-binding domain and catalytic S,YxxK,PGxxxT residues and (**b**) the significant difference in the C-terminal of the two structures associated with an additional 125 residues common to 1NAH and 3β-HSDI that are presumed to be associated with the isomerase activity of the latter.

(5) The presence of an aspartate (D) residue in theβ2α3 turn of 3β-HSDI isomerase that predicts NAD preference in cofactor binding.[2]

The validity of using the three-dimensional structure of UDPGE as a model for 3β-HSDI was tested biochemically by mutation studies. The superposition of the active site residues and cofactor positions in 3α,20β-HSDI and UDPGE (FIG. 9) together with the sequence alignment between UDPGE and 3β-HSDI allowed us to identify the probable catalytic residues in the 3β-HSDI. There are two YxxxK sequences in 3β-HSD [(Y(154)xxxK(158) and Y(269)xxxK(273)]. Mutation studies proved that the Y(154) and K(158) were the catalytic residues, and the superposition of numerous SCOR enzymes including UDPGE is consistent with this result and further validates UDPGE as a suitable model for 3β-HSDI.[20] We were able to change the cofactor dependence of 3β-HSDI from NAD to NADP by the double mutation D36A K37R.[21] This demonstrated that the model correctly identified the residues that distinguish between NAD and NADP binding. There are over a dozen serine residues in 3β-HSDI (FIG. 10), but the model indicated that the second serine in the doublet S123 S124 was the most probable candidate to be the catalytic serine. Mutation studies revealed that the S124 was the catalytic serine (FIG. 10).[22]

Sequence similarity between 3β-HSDI and 3α,20β-HSD is greatly reduced beyond the catalytic YxxxK. This is probably due to the fact that 3β-HSDI is a dehydrogenase and an isomerase. Our data indicated that the NAD binding associated with the D36 residue is critically important to both reactions. We postulate that the dual action is associated with the 120 additional residues on the carbonyl end of the enzyme compared to the other members of the SCOR family.

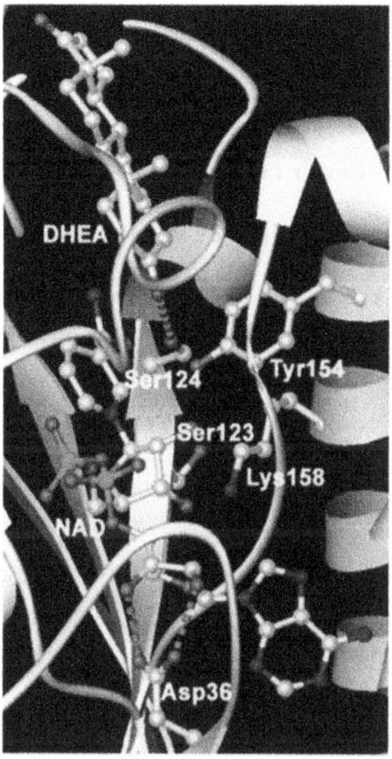

FIGURE 10. Our model suggests that the second serine in the S(123) S(124) doublet is the catalytic serine. In the model, S123 and S124 are on opposite sides of a β-strand with S124 proximal to the catalytic Tyr154, the nicotinamide ring, and the probable site of the substrate carbonyl. Mutational analysis confirmed the prediction that S124 is catalytic, but it also revealed that S123 influences the kinetics.[25] (Reprinted with permission.)

There is a strict tissue-specific distribution of the two human 3β-HSD isoforms. 3β-HSDI is expressed in mammary gland, breast tumors, prostate, prostate tumors, and placenta, and 3β-HSDII is expressed in the adrenals, testes, and ovaries. Comparison of the sequences of type 1 and type 2 3β-HSD (FIG. 11) reveals only 23 differences. Some of these variations in sequence are responsible for the profound differences measured for the kinetics of substrate utilization and inhibitor steroid binding by 3β-HSDI relative to 3β-HSDII. Tissue specificity is probably due to factors controlling expression of isomorphs in different tissues. Our initial efforts to probe the impact of amino acid differences between the isoforms have focused on the Tyr versus the His residue in the x_2 position in the helical turn containing the catalytic $Yx_1x_2x_3K$ sequences. We find that His156 contributes to the 14- to 17-fold higher affinity that 3β-HSDI exhibits for substrate and inhibitor steroids compared to the 3β-HSDII containing Tyr156 in the catalytic domain. Our structural model of 3β-HSDI correctly predicted that Glu105 on one enzyme subunit interacts differently with the Tyr156 and H156 residue and that the difference in dehydrogenase substrate kinetics has to do with dimerization of the enzyme.[23] Additional studies are in progress to further characterize the structural reasons for the substantially higher affinity for steroids of 3β-HSDI compared with 3β-HSDII. Because human 3β-HSDI is selectively expressed in breast and prostate tumors, the development of highly specific inhibitors will block the production of estradiol or testosterone to slow tumor growth without compromising the production of cortisol and aldosterone by 3β-HSDII in the adrenals.[23]

```
             1                       13          19        24                33          39
Human Type 1 T G W S C L V T G A G G F L G Q R I I R L L V K E K E L K E I R V L D K A F G P E L R E E
Human Type 2 . . . . . . . . . . . . L . . . . . V . . . . E . . . . . . . . A . . . . . R . . . . . .
                         52                                                78
             F S K L Q N K T K L T V L E G D I L D E P F L K R A C Q D V S V I I H T A C I I D V F G V
             . . . . . . R . . . . . . . . . . . . . . . . . . . . . . . . . V . . . . . . . . . . . .
                                                                               124
             T H R E S I M N V N V K G T Q L L L E A C V Q A S V P V F I Y T S S I E V A G P N S Y K E I
             . . . . . . . . . . . . . . . . . . . . . . . . . . . . . . . . . . . . . . . . . . . . . .
                                           152     156                                       177
             I Q N G H E E E P L E N T W P A P Y P H S K K L A E K A V L A A N G W N L K N G G T L Y T C
             . . . . . . . . . . . . . . . T . . . Y . . . . . . . . . . . . . . . . . . . . D . . . . .
                     187           194 195
             A L R P M Y I Y G E G S R F L S A S I N E A L N N N G I L S S V G K F S T V N P V Y V G N V
             . . . . T . . . . . . G P . . . . . . . . . . . . . . . . . . . . . . . . . . . . . . . . .
                                   240             248                                         270
             A W A H I L A L R A L Q D P K K A P S I R G N F Y Y I S D D T P H Q S Y D N L N Y T L S K
             . . . . . . . . . . . R . . . . . . . V . . . . . . . . . . . . . . . . . . . . . I . . .
                                   285   288                   299         305     309 311         318
             E F G L R L D S R W S F P L S L M Y W I G F L L E I V S F L L R P I Y T Y R P P F N R H I V
             . . . . . . . . . . . L . . T . . . . . . . . . . V . . . . . S . . . S . Q . . . . . . T .

             T L S N S V F T F S Y K K A Q R D L A Y K P L Y S W E E A K Q K T V E W V G S L V D R H K
             . . . . . . . . . . . . . . . . . . . . . . . . . . . . . . . . . . . . . . . . . . . . .
                           372
             E T L K S K T Q
             . . . . . . . .
```

FIGURE 11. Sequence alignment of 3β-HSDI Types 1 and 2 with the 23 differences that influence substrate specificity.

USING RATIONAL MUTATIONAL ANALYSIS TO PROBE FUNCTION AND SPECIFICITY

1. Identify a homologous sequence having a PDB structure.
2. Determine fingerprints for fold and cofactor.
3. Use LIGPOT[24] to determine the amino acids that directly contact the cofactor substrate and inhibitor in any crystal complexes.
4. Use the three-dimensional structure to predict the role of highly conserved fingerprint residues.
5. Test the predictions by mutations.
6. Use results to refine models for mechanism of action and to design inhibitors.

The technique described in this paper and summarized in the list above identifies the amino acid residues in an enzyme-binding protein or receptor that are critical to various aspects of function including cofactor and substrate specificity and can be applied to any protein for which a homologous sequence exists in the protein database.[16] A rational mutational analysis mechanism of protein function and the design of potent agonist, antagonist, or inhibitors involves the preceding sequence of steps.

ACKNOWLEDGMENTS

This work was supported by National Institutes of Health Grants DK26546 (W.L.D.) and CA114717 (J.L.T.).

REFERENCES

1. DUAX, W.L., J.F. GRIFFIN & D. GHOSH. 1996. The fascinating complexities of steroid binding enzymes. Curr. Opin. Struct. Biol. **6:** 813–823.
2. DUAX, W.L., V. PLETNEV, A. ADDLAGATTA, *et al.* 2003. Rational proteomics I. Fingerprint identification and cofactor specificity in the short-chain oxidoreductase (SCOR) enzyme family. Proteins Struct. Funct. Genet. **53:** 931–943.
3. JÖRNVALL, H., B. PERSSON, M. KROOK, *et al.* 1995. Short-chain dehydrogenase/reductases (SDR). Biochemistry **34:** 6003–6013.
4. PLETNEV, V.Z. & W.L. DUAX. 2004. Rational proteomics II. Electrostatic nature of cofactor preference in the short-chain oxidoreductase (SCOR) enzyme family. Protein Struct. Funct. Genet. **57:** 294–301.
5. BATEMAN, A., L. COIN, R. DURBIN, *et al.* 2004. The Pfam protein families database. Nucleic Acids Res. Database Issue **32:** D138–141.
6. PLETNEV, V. & W.L. DUAX. 2005. Rational proteomics IV. Modeling the primary function of the mammalian 17β-hydroxysteroid dehydrogenase type 8. J. Steroid Biochem. Molec. Biol. **94:** 327–335.
7. SWEET, F. & B.R. SAMANT. 1980. Bifunctional enzyme activity at the same active site: study of 3 and 20 activity by affinity alkylation of 3α,20β-hydroxysteroid dehydrogenase with 17-(Bromoacetoxy) steroids. Biochemistry **19:** 978–986.
8. GHOSH, D., C.M. WEEKS, P. GROCHULSKI, *et al.* 1991. Three-dimensional structure of holo 3α,20β hydroxysteroid dehydrogenase: a member of a short-chain dehydrogenase family. Proc. Natl. Acad. Sci. USA **88:** 10064–10068.
9. BIRKTTOFT, J.J. & L.J. BANASZAK. 1984. Structure-function relationships among nicotinamide-adenine dinucleotide-dependent oxidoreductases. Peptide and Protein. Rev. Vol. 4: 1–47. Edited by Milton and T.W. Hearn. Marcel Dekker, Inc. New York.
10. GIRERO, R J., L. RASSAERT, G. DIPASQUALE & L. KROC. 1958. Endocrine Involvement in licorice hypertension. Am. J. Physiol. **194:** 241–245.
11. ULICK, S., L.S. LEVINE, P. GUNCZLER, *et al.* 1979. A syndrome of apparent mineralocorticoid excess associated with defects in the peripheral metabolism of cortisol. J. Clin. Endocrinol. Metab. **49:** 757.
12. NEW, M.I., S.E. OBERFIELD, R. CAREY, *et al.* 1982. A genetic defect in cortisol metabolism as the basis for the syndrome of apparent mineralocorticoid excess. *In* Serono Symposium No. 50, Endocrinology of Hypertension. F. Mantero, E.G. Biglieri & C.R.W. Edwards, Eds. :85–101. Academic Press. London.
13. GHOSH, D., M. ERMAN, W. PANGBORN, *et al.* 1992. Inhibition of *Streptomyces hydrogenans* 3α,20β-hydroxysteroid dehydrogenase by licorice-derived compounds and crystallization of an enzyme-cofactor-inhibitor complex. J. Steroid Biochem. Molec. Biol. **42:** 849–853.
14. GHOSH, D., M. ERMAN, Z. WAWRZAK, *et al.* 1994. Mechanism of inhibition of 3α,20β-hydroxysteroid dehydrogenase by a licorice-derived steroidal inhibitor. Structure **2:** 973–980.
15. ARMANINI, D., M.J. MATTARELLO, C. FIORE, *et al.* 2004. Licorice reduces serum testosterone in healthy women. Steroids **69:** 763–766.
16. BERNSTEIN, F.C., T.F. KOETZLE, G.J. WILLIAMS, *et al.* 1977. The protein data bank: a computer-based archival File for Macromolecular Structures. J. Mol. Biol. **112:** 535–542.
17. RHEAUME, E., Y. LACHANCE, H.-F. ZHAO, *et al.* 1991. Structure and expression of a new complementary DNA encoding the almost exclusive 3 beta-hydroxysteroid dehydrogenase/delta 5-delta 4-isomerase in human adrenals and gonads. Mol. Endocrinol. **5:** 1147–1157.
18. THOMAS, J.L., R.P. MYERS & R.C. STRICKLER. 1989. Human placental 3β-hydroxy-5-ene-steroid dehydrogenase and steroid 5-4-ene-isomerase: purification from mitochondria and kinetic profiles, biophysical characterization of the purified mitochondrial and microsomal enzymes. J. Steroid Biochem. **33:** 209–217.
19. THODEN, J.B., P.A. FREY & H.M. HOLDEN. 1996. Crystal structures of the oxidized and reduced forms of UDP-galactose 4-epimerase isolated from *Escherichia coli*. Biochemistry **35:** 2557.

20. THOMAS, J.L., I.I. MASON, ST. BRANDT, B.R. SPENCER, JR. & W. NORRIS. 2002. Structure/function relationships responsible for the kinetic differences between human Type 1 and Type 2 3β-hydroxysteroid dehydrogenase and for the catalysis of the Type 1 activity. J. Biol. Chem. **277:** 42795–42801.
21. THOMAS, J.L., W.L. DUAX, A. ADDLAGATTA, *et al.* 2003. Structure/function relationships responsible for coenzyme specificity and the isomerase activity of human Type 1 3β-hydroxysteroid dehydrogenase/isomerase. J. Biol. Chem. **278:** 35483–35490.
22. THOMAS, J.L., W.L. DUAX, W.A. ADDLAGATTA, *et al.* 2004. Serine 124 completes the Tyr, Lys and Ser triad responsible for catalysis of human Type 1 3β-hydroxysteroid dehydrogenase. J. Mol. Endocrinol. **33:** 253–261.
23. THOMAS J.L., E.L. BOSWELL, L.A. SCACCIA, *et al.* 2005. Identification of key amino acids responsible for the substantially higher affinities of human type 1 3-hydroxysteroid dehydrogenase/isomerase (3-HSD1) for substrates, coenzymes and inhibitors relative to human 3-HSD2. J. Biol. Chem. **280:** 21321–21328.
24. WALLACE, A.C., R.A. LASKOWSKI & J.M. THORNTON. 1995. A program to generate schematic diagrams of protein-ligand interactions. Protein Eng. **8:** 127–134.
25. TOMAS, J.L. *et al.* 1994. J. Mol. Endocrinol. **33:** 253.

Postnatal and Pubertal Development of the Rhesus Monkey (*Macaca mulatta*) Testis

TONY M. PLANT,[a] SURESH RAMASWAMY,[a] DAVID SIMORANGKIR,[a] AND GARY R. MARSHALL[b]

[a]*Department of Cell Biology and Physiology, University of Pittsburgh School of Medicine, Pittsburgh, Pennsylvania 15261, USA*

[b]*Department of Medicine, University of Pittsburgh School of Medicine, Pittsburgh, Pennsylvania 15261, USA*

ABSTRACT: This review examines the neurobiology, endocrinology, and cell biology underlying the development of the testis from birth until puberty in the rhesus monkey, a representative higher primate.

KEYWORDS: Sertoli cell; kisspeptin; undifferentiated spermatogonia; stem cells; testosterone

INTRODUCTION

In man, initiation of spermatogenesis generally occurs in the second decade of life after a prolonged period of childhood and juvenile development. Although in the rhesus monkey this critical developmental event takes place during the third and fourth years of life, the neural, endocrine, and cellular mechanisms that delay the onset of fertility in these two higher primate species appear remarkably similar. The monkey, therefore, is an ideal paradigm for the study of testicular development in man.

The combined testis weight of the newborn male rhesus monkey is approximately 200 mg, which compares with a corresponding value in the adult of around 25,000 mg. Growth of the primate testis from birth until the onset of puberty is due primarily to increasing seminiferous tubule length which, in turn, reflects proliferation of Sertoli cells and undifferentiated spermatogonia. With the initiation of puberty, proliferation of Sertoli cells is initially accelerated and then terminated; proliferation of undifferentiated spermatogonia continues, and for the first time large numbers of differentiated spermatogonia are produced. In contrast to the tubular component of the primate testis, in which germ cell development is arrested prior to the onset of puberty at a pre-meiotic level, the steroidogenic interstitial component is activated temporarily but robustly during infancy, when secretion of testicular testosterone

Address for correspondence: Tony M. Plant, University of Pittsburgh School of Medicine, Department of Cell Biology and Physiology, 3550 Terrace Street, Rm. 828 Scaife Hall, Pittsburgh, PA 15261. Voice: 412-648-9395; fax: 412-383-7159.
Plant1@pitt.edu

Ann. N.Y. Acad. Sci. 1061: 149–162 (2005). © 2005 New York Academy of Sciences.
doi: 10.1196/annals.1336.016

produces circulating levels of this steroid that are in the range typically seen in adulthood.

The purpose of this review is to examine the neurobiology, endocrinology, and cell biology underlying this characteristic pattern of postnatal testicular development in the rhesus monkey, a representative higher primate. In the monkey, the prolonged period of development from birth to adulthood may be divided into four life history stages: infantile, juvenile, adolescent (pubertal), and adult.[1] Infancy is defined as the period when nourishment of the offspring is obtained via suckling; it extends from birth until approximately 6 months of age. The juvenile stage of development, during which time the young monkey is no longer dependent upon parental support for survival, follows infancy and is terminated by the onset of puberty, which in the male monkey is initiated at approximately 3½ years of age.[2]

THE ADULT TESTIS: ESSENTIAL FEATURES

The testis of the adult monkey contains approximately 2,000 million Sertoli cells,[3] which support the production of approximately 1.1×10^9 sperm/gm testis/day.[4] Spermatogenesis may be thought of as a process comprising three phases: stem cell renewal and differentiation, germ cell proliferation and differentiation, and spermiogenesis. Although novel imaging approaches are beginning to be applied to the study of primate spermatogenesis,[5,6] our understanding of male stem cell renewal and differentiation in the rhesus monkey is based primarily on the classic investigations of Clermont and his colleagues.[7,8] These early investigators recognized two types of "undifferentiated" spermatogonia based on their histochemical staining properties: dark type A (Ad) and pale type A (Ap) spermatogonia. Somewhat later, a third type of A spermatogonia, termed transitional A spermatogonia (At), was described.[9,10]

The classic, as well as contemporary, view of Ap spermatogonia in the rhesus monkey states that these cells divide once or twice during the 10.5-day cycle of the seminiferous epithelium both to renew themselves and to produce the first of four generations of "differentiated" B spermatogonia (FIG. 1).[6,11] Therefore, Ap spermatogonia have been viewed as the "renewing stem" cell of the macaque testis. Since there are approximately 300 million Ap spermatogonia in the adult testis, equivalent to 1 in every 50 germ cells,[12,13] and since it is generally recognized that stem cells are few in number and rarely divide,[14] it is unlikely that this spermatogonial cell type represents a stem cell, as currently defined. In analogy to epidermis, Ap spermatogonia may be viewed perhaps as a transit-amplifying cell.[15] If this is the case, the primate stem cell is most likely to reside within the population of Ad spermatogonia. While Ad spermatogonia divide rarely in the adult testis and therefore, classically, have been considered as a "reserve stem" cell, they are as numerous as Ap spermatogonia. Therefore, if Ad spermatogonia are indeed primate stem cells, only a very small fraction of these cells are expressing their potential for stemness under physiological conditions, and the majority must be viewed as resting stem cells.

The final resolution of the identity of stem cells in the monkey testis must await studies that correlate both the molecular and functional properties of the various sub-

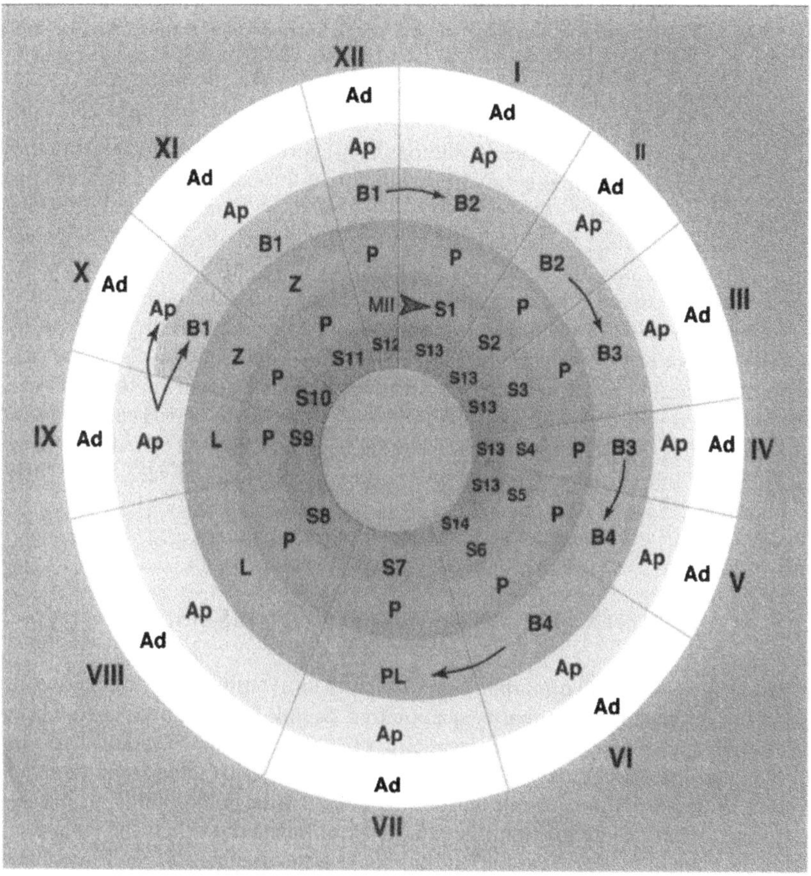

FIGURE 1. The cycle of the seminiferous epithelium of the rhesus monkey depicted in a highly schematic manner to emphasize its recurring nature.[11] The cellular associations that define the stages of the cycle are indicated with Roman numerals I-XII. The circumference represents time, and the width of each segment is proportional to the duration of each stage. The increasing density represents maturation of the cell types as the process of spermatogenesis unfolds. Spermatogenesis begins in stage IX when half of the population of Ap spermatogonia divide and produce the first generation of type B spermatogonia (B1). Coincidentally, the remaining Ap spermatogonia also divide to produce Ap spermatogonia thereby renewing the population of these stem cells. Spermatogenesis terminates in stage VI when S14 spermatids are released into the lumen of the seminiferous tubule. Ad = dark type A spermatogonia; Ap = pale type A spermatogonia; B1-B4 = four generations of type B spermatogonia; PL = preleptotene primary spermatocytes; L = leptotene spermatocytes; Z = zygotene spermatocytes; P = pachytene spermatocytes; MII = completion of meiosis; S1-S14 = spermatids at each of the 14 steps of spermiogenesis. The arrows and arrowhead indicate mitosis and the completion of meiosis, respectively. Reproduced with permission.

classes of undifferentiated spermatogonia classified according to histochemical phenotype. Parenthetically, in rodents, the cell type considered to be the stem cell (type A-single spermatogonia) number only a few thousand,[14] which is consistent with the results of functional analysis of cells for the property of stemness using the so called transplantation assay.[16]

In the monkey, the production of the first generation of B spermatogonia from undifferentiated Ap spermatogonia (FIG. 1) represents a critical step in the initiation and maintenance of spermatogenesis. Although Ap spermatogonia proliferate in the absence of LH and FSH stimulation, differentiation of this cell type is dependent on gonadotropin stimulation.[17] The somewhat hackneyed question of the relative importance of LH and FSH in the initial and subsequent steps of the spermatogenic process remains an area of current investigation,[11] but will not be considered here. LH is secreted in an episodic mode that is modulated diurnally and this leads to a corresponding pattern of testicular testosterone secretion producing marked moment-to- moment changes and nighttime elevations in circulating levels of the steroid.[18] LH secretion is regulated by a negative feedback action of testosterone, which is exerted indirectly to retard the frequency of pulsatile GnRH release from the hypothalamus.[19] FSH secretion, on the other hand, is regulated by a direct feedback action of inhibin B to suppress FSHβ gene expression and secretion of the intact hormone.[11]

POSTNATAL ONTOGENY OF THE TESTIS

The changes in testicular morphology from birth until termination of the juvenile phase of development are primarily quantitative. Sertoli cell number per testis increases from approximately 40 million at birth to 2,000 million in the adult,[4] and undifferentiated spermatogonia, including Ad spermatogonia that divide only rarely in the adult,[11] increase from one million at birth to 600 million in the adult.[13,20] In contrast to the rodent, where the Sertoli cell number of the adult testis is the major determinant of spermatogenic ceiling, in the monkey, the number of undifferentiated spermatogonia may also contribute to this important parameter of fertility.[11] The endocrine control and cell biology underlying the profound postnatal expansion of these critical cell types from birth until the completion of puberty therefore merits discussion. Since the impact of LH on the developing testis is mediated primarily by testosterone production from the Leydig cell, the ontogeny of this interstitial cell will also be considered.

In monkey and man, gonadotropin secretion during postnatal development exhibits a characteristic "primate" pattern, with robust LH and FSH release during infancy and during puberty, interrupted throughout juvenile development by markedly attenuated or absent LH and FSH release.[2] The hypogonadotropic juvenile phase of development in primates is directly responsible for maintaining the testis in a prepubertal state, which is characterized by low testosterone production and the failure to produce differentiated B spermatogonia (see below). The prepubertal testis, however, should not be viewed as a quiescent organ since Sertoli cells and undifferentiated spermatogonia are proliferating, and the spermatogenic ceiling of the adult testis is, therefore, being established.

SERTOLI CELLS

The time course of expansion of Sertoli cell number from birth to adulthood is shown in FIGURE 2. During infancy, when gonadotropin secretion is elevated, Sertoli cell number increases approximately four-fold. During this phase of development, the mitotic index of this cell type (as reflected by *in vivo* S-phase labeling with 5-bromo-2′-deoxyuridine [BrdU] coupled with a standard avidin-horseradish peroxidase-diaminobenzidine [DAB] reaction to visualize the anti-BrdU antibody, followed by counter-staining with PAS-hematoxylin) is 1.3%.[4] However, during juvenile development, when gonadotropin secretion is arrested, the rate of increase in Sertoli cell number is markedly reduced, corresponding with a labeling index of approximately 0.3%. At the time of puberty, the phase of postnatal development associated with a resurgence in gonadotropin secretion, the population of Sertoli cells is further amplified. The precise dynamics of this burst of Sertoli cell proliferation, however, has not been described in any primate species.

Since stimulation of the juvenile testis with gonadotropin elicits precocious pubertal proliferation of the Sertoli cell,[12] it is reasonable to conclude that Sertoli cell division is highly gonadotropin- dependent, and that the dynamics of Sertoli cell proliferation from birth to adulthood in the primate is a reflection of this dependence (FIG. 2). The relative role of FSH and LH in this regard is unclear, although initial pubertal activation of Sertoli cell proliferation may be induced precociously by stimulation of the juvenile testis with either FSH or LH/testosterone.[21–23] It is interesting to note that the classical view of the endocrine regulation of Sertoli cell proliferation in the rodent is now being modified to include a role for LH/testosterone in addition to that for FSH.[24-26] The signals responsible for the terminal differentiation of the primate Sertoli cell during puberty are unclear, but, as in the rodent, thyroid hormone appears to play a permissive role in this regard.[27]

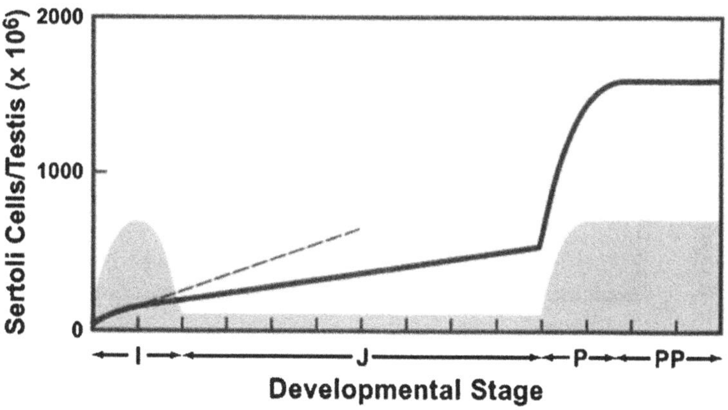

FIGURE 2. Schematic of Sertoli cell proliferation during postnatal development in the rhesus monkey.[3] Note that the rate of increase in Sertoli cell number during infancy (extended broken line), a stage of development characterized by elevated circulating gonadotropin levels (shaded ghost), is greater than that during juvenile development. A second gonadotropin- dependent burst in Sertoli cell proliferation occurs during puberty. Qualitatively, the model is applicable to the human situation. I, infant; J, juvenile; P, pubertal; PP, post-pubertal. Reprinted with permission.

UNDIFFERENTIATED SPERMATOGONIA

The postnatal ontogeny of Ad spermatogonia is of particular interest because the number of this cell type increases several hundred fold from birth to adulthood (FIG. 3), yet, in the postpubertal testis, this undifferentiated spermatogonia divides only rarely.[11] Although results were equivocal from our initial attempts to identify S-phase labeled Ad spermatogonia in the prepubertal testis using the procedure we had earlier applied to examine S-phase labeled Sertoli cells, when sections were "counter stained" before immunocytochemistry and an Alexafluor 488-labeled secondary antibody was employed instead of DAB, robust S-phase labeling of Ad spermatogonia was observed (FIG. 4).[20] Thus, it seems reasonable to conclude that the adult complement of Ad spermatogonia is generated by mitotic proliferation during prepubertal and pubertal development. The cell biology that subsequently drives the majority of these cells into G_0 in the adult testis remains to be established. Interestingly, the labeling index of Ad spermatogonia in the infantile and in the juvenile testis was similar, indicating that, in contrast to the Sertoli cell, proliferation of Ad spermatogonia is relatively independent of gonadotropin.[20]

The overall pattern of expansion of Ap spermatogonia during infancy and juvenile development and the labeling indices of this class of undifferentiated spermatogonia at these developmental stages were qualitatively similar to those for Ad spermatogonia.[20] The finding that prepubertal S-phase labeling of this cell type was robust was expected because Ap spermatogonia are mitotically active in the adult. Also, the maintenance of Ap proliferation during the hypogonadotropic juvenile phase[20] of development is consistent with recent observations that Ap spermato-

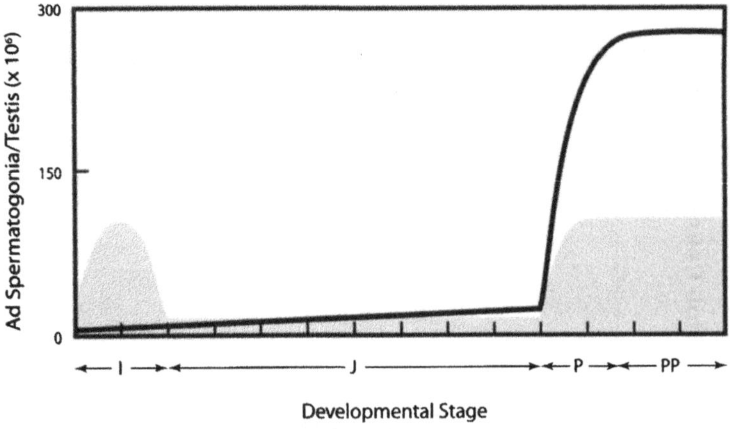

FIGURE 3. Schematic of proliferation of type Ad spermatogonia during postnatal development in the rhesus monkey. Proliferation of Ad spermatogonia during the juvenile period of development, when gonadotropin secretion is undetectable, results in a predicted 10-fold increase in the number of this undifferentiated spermatogonial type. A further 10-fold increase is posited to occur during puberty, when the gonadotropin- dependent burst in Sertoli-cell proliferation occurs (See FIG. 1). Gonadotropin levels were indicated by shaded ghost. I, infant; J, juvenile; P, pubertal; PP, postpubertal. Based on data reported previously.[12,20]

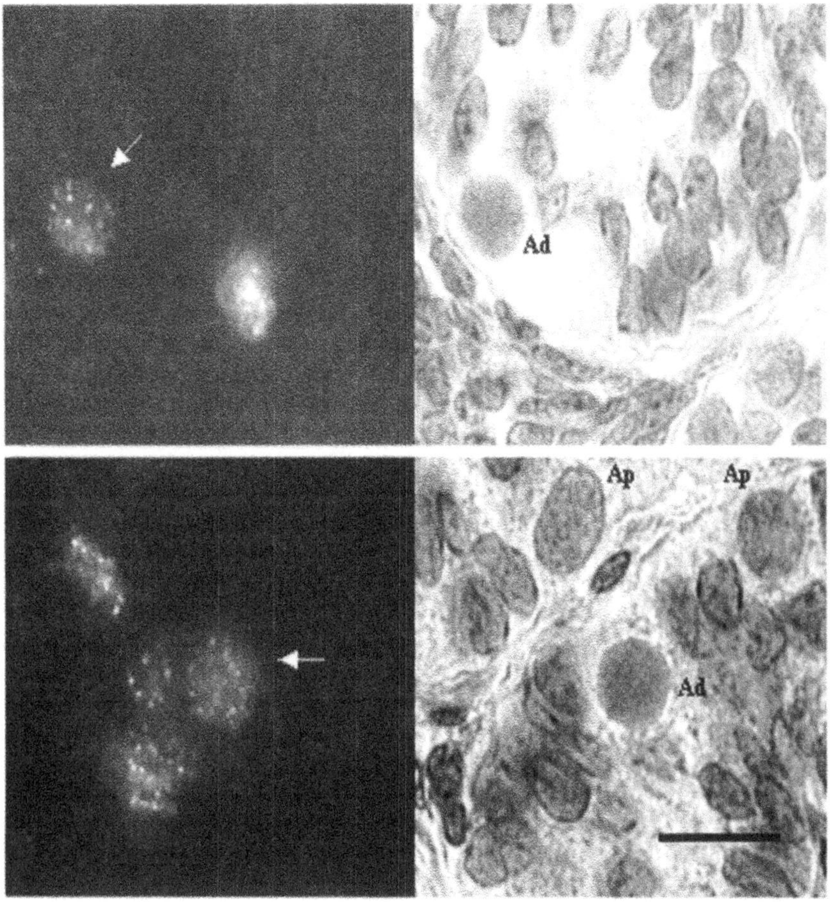

FIGURE 4. Photomicrograph of S-phase labeled Ad spermatogonia in testis of infant (4 months of age, top panel) and juvenile (15 months of age, lower panel) rhesus monkeys. The photomicrographs show fluorescent (left) and bright-field (right) views. S-phase labeled Ad spermatogonia (Ad) are indicated by the white arrow in the fluorescent fields. Bar: 10 μm.

gonia show marked S-phase labeling in adults rendered chronically hypogonadotropic by sustained treatment with a GnRH receptor antagonist.[17]

We had expected that proliferation of undifferentiated spermatogonia would follow in the wake of an expanding population of Sertoli cells. This, however, does not appear to be the case in the juvenile, as the ratio of both Ad spermatogonia and Ap spermatogonia to Sertoli cells was greatest at this stage of development when, as a result of the hypogonadotropic state, proliferation of the somatic cell is reduced. The "constitutive" proliferation of A spermatogonia in the juvenile testis appears to be amplified by increased gonadotropin secretion at the initiation of puberty (FIG. 2), probably as an indirect effect of a greatly expanded number of niches created by the gonadotropin-dependent increase in Sertoli cell number at this stage of development. Precocious stimulation of the juvenile testis with gonadotropin and/or testosterone results in an increase in number of A spermatogonia that is most striking in the case of Ap spermatogonia.[12,21,22] In the postpubertal testes, the number of Ap spermatogonia is reduced following surgical removal of the pituitary or after treatment with either a GnRH receptor antagonist or a contraceptive dose of testosterone,[28–31] perhaps also as a result of a decreased number of niches.

DIFFERENTIATED SPERMATOGONIA

Although the occasional type B or B-like spermatogonia has been reported in the prepubertal testis of man and macaques,[10,20,32] significant numbers of differentiated spermatogonia are not observed until the onset of puberty. Thus, in the monkey, the differentiation of Ap to B1 spermatogonia, which represents a late event in testicular development, may be viewed as the crucial juncture in the initiation of spermatogenesis. While the latter mitotic step is driven by gonadotropin secretion, the nature of the intratesticular signal responsible for both differentiation of Ap spermatogonia and survival of the B1 spermatogonia progeny has not been established. Such factors, however, are presumably generated by proliferating Sertoli cells (a function that is preserved following terminal differentiation of this somatic cell), via expression of growth factors, similar to those which support proliferation and survival of differentiated spermatogonia in other species.[33] Moreover, the intracellular pathways mediating FSH receptor and androgen receptor signaling in Sertoli cells on the one hand and growth-factor receptor signaling in spermatogonia on the other appear to be expressed in the juvenile testis. This conclusion is based on the kinetic consideration that leptotene-zygotene spermatocytes are found in the juvenile testis after combined treatment with FSH and LH for as little as 11 days, the time it takes in the adult testes for an Ap spermatogonia to progress to a primary spermatocyte.[11] It might be expected, therefore, that gonadotropin treatment of the juvenile would initiate full spermatogenesis after stimulation for approximately 44 days, the time it takes a spermatogonia to produce a testicular sperm in the adult testis.[11] The earliest we have observed mature spermatids in the prematurely stimulated juvenile testes, however, is after 16 weeks of gonadotropin exposure.[34] A longer delay has been reported when testicular fragments from juvenile monkeys are exposed to premature stimulation with heterologous gonadotropin after being grafted into nude mice.[35]

It is also of note that during infancy the testis of the monkey is exposed to an endocrine milieu similar to that of the adult gonad, but differentiation of Ap to B spermatogonia fails to occur. The most parsimonious explanation for this early block on spermatogenesis is that the infantile Sertoli cell is "immature". This immaturity may be related to the failure of the Sertoli cell to express androgen receptor during the first few months of postnatal life.[36,37]

TESTOSTERONE SECRETION

It is well established that, in man and other higher primates, testicular testosterone secretion during infancy is robust, and in the rhesus monkey both the magnitude and diurnal pattern of circulating levels of testosterone during this phase of development are similar to that in the fully adult male.[2] Following infancy, plasma testosterone concentrations fall to levels comparable to those in castrated adults, where they remain until the onset of puberty. This developmental pattern in testicular testosterone secretion, which is directly related to the timecourse of gonadotropin secretion described above, is associated with corresponding changes in the number and morphology of Leydig cells.[38,39] In the cynomolgus monkey, the number of Leydig cells increases from approximately 5 million per testis at birth to 30 million per testis in late infancy[40] and, as may be anticipated from the endocrine status, the cellular characteristics of Leydig cells during this period are similar to those seen in adulthood.[38]

Following infancy, the dramatic decline in testicular testosterone production resulting from the protracted hiatus in LH secretion during juvenile development is associated with regression and involution of Leydig cells, coupled with loss of their steroidogenic characteristics.[40,41] Although cells are present in the interstitium of the juvenile monkey testis, immunoactive 3β-hydroxysteroid dehydrogenase (3β-HSD) has not been detected (Ramaswamy, unpublished results), thereby supporting the view that very few steroidogenically active Leydig cells are present during this stage of development. The resurgence in gonadotropin secretion at the critical stage of puberty heralds the second phase of Leydig cell proliferation and development in primates. In the cynomolgus monkey, the pubertal proliferation of Leydig cells results in an adult population of approximately 100 million per testis.[40]

Leydig cell development and steroidogenic function during postnatal development is tightly and causally coupled to the circulating levels of gonadotropin, partic-

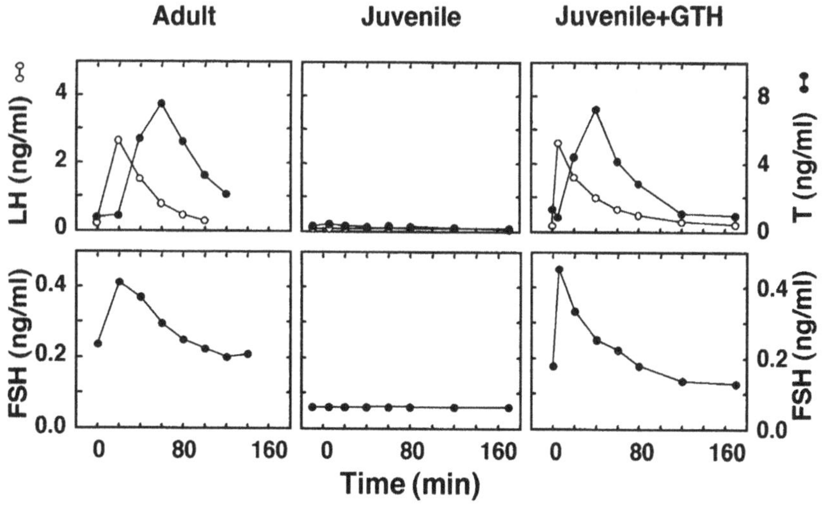

FIGURE 5. Circulating concentrations of LH and testosterone (top panels) and FSH (bottom panels) in an adult (left panels) and a juvenile monkey (middle panels), and in a juvenile monkey receiving an intermittent infusion of recombinant macaque LH and FSH (right panels). To achieve adult levels of gonadotropins in the juvenile animal, 600 ng/kg of LH and 30 ng/kg of FSH were injected *iv* once every 3 h (right panels). For the right panels, 0 min indicates time of gonadotropin (GTH) infusion. For other panels, time 0 was arbitrarily assigned.

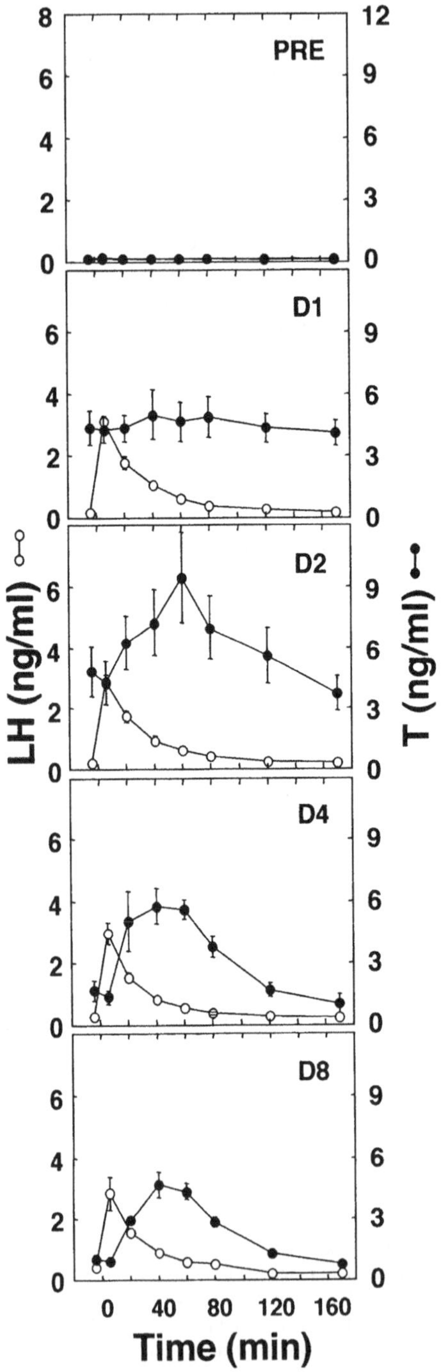

FIGURE 6. Time course of changes in circulating concentrations (mean ± SEM) of monkey LH (O) and testosterone (T, 0) before (PRE) and on days 1, 2, 4, and 8 after initiation of an intermittent i.v. infusion with rmLH (in combination with rmFSH). Note the gradual switch between days 2 and days 8 from an apulsatile to a pulsatile mode of testosterone release in association with a decrease in the mean concentration of testosterone in the face of an invariant LH stimulation.[42] Reprinted with permission.

ularly those of LH. In the juvenile monkey, pulsatile intravenous infusion (one brief infusion every 3 hours) of recombinant monkey LH, which mimics in the immature animal an adult pattern of circulating LH, elicits discharges of testosterone from the testis that are indistinguishable to those observed spontaneously in adulthood (FIG. 5).[42,43] Interestingly, such precocious pulsatile stimulation of the juvenile monkey testes results initially (first 24 hours of stimulation) in a continuous, apulsatile pattern of testosterone secretion (FIG. 6).[42] However, a switch from the apulsatile to a "mature" pulsatile mode of testosterone secretion emerges following 24-72 hours of intermittent LH stimulation, and a fully mature pulsatile profile of testicular testosterone is observed after a week of stimulation. At the present time, the cell biology underlying this curious switch in the juvenile testis from a continuous mode of testosterone production to one that is episodic is unclear, as is the question of whether the initiation of testicular secretion during spontaneous puberty involves an initial continuous mode of steroid release.[42] In any event, the number of Leydig cells following sustained (approximately 13 week), intermittent gonadotropin stimulation of the juvenile testis is approximately 90 million,[43] similar to that noted during spontaneous puberty in the closely related cynomolgus macaque.[40]

CONTROL OF THE TIMING OF INITIATION OF SPERMATOGENESIS

As noted above, by 15 months of age the testis of the juvenile monkey is able to respond to premature gonadotropin stimulation with the precocious initiation of spermatogenesis. The precise age at which the seminiferous cord of the monkey testis acquires the capacity to respond to gonadotropin stimulation allowing spermatogenesis to be initiated has not been established, but must lie between the termination of infancy, during which time spermatogonial differentiation is blocked, and 15 months of age. Moreover, the testis is able to be driven into an adult mode of operation well before the onset of puberty, and therefore the initiation of increased production of testosterone and differentiation of Ap spermatogonia to B spermatogonia during the onset of normal puberty is absolutely dependent on the increase in gonadotropin secretion at this critical stage of development. The pubertal reactivation in LH and FSH secretion, in turn, is dependent on the resurgence of pulsatile hypothalamic GnRH release that has been held in check since infancy.[2] Although the neurobiological trigger responsible for the pubertal resurgence in pulsatile GnRH release is incompletely understood, recent genetic studies of human subjects with hypogonadotropic hypogonadism and absent puberty[44,45] and experiments with the rhesus monkey,[46] suggest the involvement of a novel hypothalamic signaling system comprised of kisspeptin, a product of *KiSS-1,* and GPR54, a G-protein coupled receptor that is expressed by GnRH neurons.[47-49] *KiSS-1* expression in the hypothalamus of the monkey increases in association with the pubertal resurgence in GnRH release, and central or peripheral bolus administration of kisspeptin to the juvenile male monkey elicits precocious release of GnRH.[46] Whether *KiSS-1* expression and other molecular and structural changes in the hypothalamus at the end of juvenile development are activated by a "pubertal" clock or by circulating cues that relay information on somatic development to the hypothalamus remains a fundamental question in primate development.

REFERENCES

1. Bogin, B. 1995. Growth and development: recent evolutionary and biocultural research. *In* Biological Anthropology: The State of the Science. 49–70. International Institute for Human Evolutionary Research.
2. Plant, T.M. 1994. Puberty in Primates. *In* The Physiology of Reproduction, 2nd Edition. E. Knobil & J.D. Neill, Eds.: **2:** 453–485. Raven Press. New York.
3. Simorangkir, D.R., G.R. Marshall & T.M. Plant. 2003. Sertoli cell proliferation during prepubertal development in the rhesus monkey (*Macaca mulatta*) is maximal during infancy when gonadotropin secretion is robust. J. Clin. Endocrinol. Metab. **88:** 4984-4989.
4. Sharpe, R.M. 1994. Regulation of Spermatogenesis. *In* The Physiology of Reproduction, 2nd Edition. E. Knobil & J.D. Neill, Eds.: **1:** 1363–1434. Raven Press. New York.
5. Ehmcke, J., C.M. Luetjens & S. Schlatt. 2005. Clonal organization of proliferating spermatogonial stem cells in adult males of two species of non-human primates, *Macaca mulatta* and *Callithrix jacchus*. Biol. Reprod. **72:** 293–300.
6. Ehmcke, J., D.R. Simorangkir & S. Schlatt. 2005. Identification of the starting point for spermatogenesis and characterization of the testicular stem cell in adult male rhesus monkeys. Human Reprod. **20:** 1185–1193.
7. Clermont, Y. 1972. Kinetics of spermatogenesis in mammals: seminiferous epithelium cycle and spermatogonial renewal. Physiol. Rev. **52:** 298–236.
8. Clermont, Y. & M. Antar. 1973. Duration of the cycle of the seminiferous epithelium and the spermatogonial renewal in the monkey Macaca arctoides. Am. J. Anat. **136:** 153–165.
9. Fouquet, J.P. & J.P. Dadoune. 1986. Renewal of spermatogoina in the monkey (*Macaca fascicularis*). Biol. Reprod. **35:** 199–207.
10. Cavicchia, J.C. & M. Dym. 1978. Ultrastructural characteristics of monkey spermatogonia and preleptotene spermatocytes. Biol. Reprod. **18:** 219–228.
11. Plant, T.M. & G.R. Marshall. 2001. The functional significance of follicle-stimulating hormone in spermatogenesis and the control of its secretion in male primates. Endocrine Rev. **22:** 764–786.
12. Marshall, G.R. & T.M. Plant. 1996. Puberty occurring either spontaneously or induced precociously in rhesus monkey (*Macaca mulatta*) is associated with a marked proliferation of Sertoli cells. Biol. Reprod. **54:** 1192–1199.
13. Ramaswamy, S, G.R. Marshall, A.S. McNeilly *et al.* 2000. Dynamics of the follicle-stimulating hormone (FSH)-inhibin B feedback loop and its role in regulating spermatogenesis in the adult male rhesus monkey (*Macaca mulatta*) as revealed by unilateral orchidectomy. Endocrinology **141:** 18–27.
14. de Rooij, D.G. & L.D. Russell. 2000. All you wanted to know about spermatogonia but were afraid to ask. J. Androl. **21:** 776–798.
15. Watt, F.M. 2001. Stem cell fate and patterning in mammalian epidermis. Curr. Opinion Gen. Dev. **11:** 410–417.
16. Brinster, R.L. & J. W. Zimmerman. 1994. Spermatogenesis following male germ-cell transplantation. Proc. Natl. Acad. Sci. USA **91:** 11298–11302.
17. Marshall, G.R., S. Ramaswamy & T.M. Plant. 2005 Gonadotropin independent proliferation of the pale type A spermatogonia in the adult rhesus monkey (*Macaca mulatta*). Biol. Reprod. **73:** 222-229.
18. Plant, T.M. 1981. Time courses of concentrations of circulating gonadotropin, prolactin, testosterone, and cortisol in adult male rhesus monkeys (*Macaca mulatta*) throughout the 24 h light-dark cycle. Biol. Reprod. **25:** 244–252.
19. Plant, T.M. 1986. Gonadal regulation of hypothalamic gonadotropin-releasing hormone release in primates. Endocrine Rev. **7:** 75–88.
20. Simorangkir, D.R., G.R. Marshall, J. Ehmcke *et al.* 2005. Pre-pubertal expansion of dark pale type A spermatogonia in the rhesus monkey (*Macaca mulatta*) results from proliferation during infantile and juvenile development in a relatively gonadotropin independent manner. Biol. Reprod. **73:** 1109–1115.

21. ARSLAN, M. G.F. WEINBAUER, S. SCHLATT *et al.* 2000. FSH and testosterone, alone or in combination, initiate testicular growth and increase the number of spermatogonia and Sertoli cells in a juvenile non-human primate (*Macaca mulatta*). J. Endocrinol. **136:** 235–243.
22. SCHLATT, S., M. ARSLAN, F.G. WEINBAUER *et al.* 1995. Endocrine control of testicular somatic and premeiotic germ cell development in the immature testis of the primate *Macaca mulatta*. Eur. J. Endocrinol. **133:** 235–247.
23. RAMASWAMY, S., T.M. PLANT & G.R. MARSHALL. 2000. Pulsatile stimulation with recombinant single chain human luteinizing hormone elicits precocious Sertoli cell proliferation in the juvenile male rhesus monkey (*Macaca mulatta*). Biol. Reprod. **63:** 82–88.
24. JOHNSTON, H., P.J. BAKER, M. ABEL *et al.* 2004. Regulation of Sertoli cell number and activity by follicle-stimulating hormone (FSH) and androgen during postnatal development in the mouse. Endocrinology **145:** 318–329.
25. TAN, K.A., K. DE GENDT, N. ATANASSOVA *et al.* 2005. The role of androgens in Sertoli cell proliferation and functional maturation: studies in mice with total or Sertoli cell-selective ablation of the androgen receptor. Endocrinology **146:** 2674–2683.
26. ATANASSOVA, N.N., M. WALKER, C. MCKINNELL *et al.* 2005. Evidence that androgens and oestrogens, as well as follicle-stimulating hormone, can alter Sertoli cell number in the neonatal rat. J. Endocrinol. **184:** 107–17.
27. JANNINI, E.A., S. ULISSE & M. D'ARMIENTO. 1995. Thyroid hormone and male gonadal function. Endocrine Rev. **16:** 443–459.
28. MARSHALL, G.R., F. JOCKENHOVEL, D. LUDECKE *et al.* 1986. Maintenance of complete but quantitatively reduced spermatogenesis in hypophysectomized monkeys by testosterone alone, Acta Endocrinol. (Copenh.) **113:** 424–431.
29. WEINBAUER, G.F., H.M. BEHRE, U. FINGSCHEIDT *et al.* 1991. Human follicle-stimulating hormone exerts a stimulatory effect on spermatogenesis, testicular size, and serum inhibin levels in the gonadotropin-releasing hormone antagonist-treated non-human primate (*Macaca fascicularis*). Endocrinology **129:** 1831–1839.
30. ZHENGWEI, Y., N,G. WREFORD, S. SCHLATT *et al.* 1998. Acute and specific impairment of spermatogonial development by GnRH antagonist-induced gonadotrophin withdrawal in the adult macaque (*Macaca fascicularis*). J. Reprod. Fertil. **112:** 139–147.
31. O'DONNELL, L., A. NARULA, G. BALOURDOS *et al.* 2001. Impairment of spermatogonial development and spermiation after testosterone induced gonadotropin suppression in adult monkeys (*Macaca fascicularis*). J. Clin. Endocrinol. Metab. **86:** 1814–1822.
32. KLUIN, PH.M., M.F. KRAMER & D.G. DE ROOIJ. 1983. Testicular development in *Macaca irus* after birth. Int. J. Androl. **6:** 25–43.
33. DYM, M. & L. FENG. 2005. The role of Sertoli cell in spermatogonial stem cell fate. *In* Sertoli Cell Biology. M.K. Skinner & M.D. Griswold, Eds.: 303–315. Elsevier Academic Press. San Diego.
34. PLANT, T.M., V.L. GAY, G.R. MARSHALL *et al.* 1989. Puberty in monkeys is triggered by chemical stimulation of the hypothalamus. Proc. Natl. Acad. Sci. USA **86:** 2506–2510.
35. HONARAMOOZ, A., M.W. LI, M.C. PENEDO *et al.* 2004. Accelerated maturation of primate testis by xenografting into mice. Biol. Reprod. **70:** 1500–1503.
36. MCKINNELL, C., P.T. SAUNDERS, H.M. FRASER *et al.* 2001. Comparison of androgen receptor and oestrogen receptor beta immunoexpression in the testes of the common marmoset (*Callithrix jacchus*) from birth to adulthood: low androgen receptor immunoexpression in Sertoli cells during neonatal increase in testosterone concentrations. Reproduction **122:** 419–429.
37. MAJUMDAR, S.S., K. SARDA, S.A. OÑATE *et al.* 2004. Limited Sertoli cell expression of androgen and FSH receptors in the rhesus monkey during infancy may underlie the failure of spermatogonial stem cells to differentiate in the presence of an adult-like gonadotropin milieu at this stage of development. Presented at the 86th Annual Meeting of The Endocrine Society, New Orleans, LA, June 16-19, 2004 (No. P3-318).
38. SAEZ, J.M. 1994. Leydig cells: endocrine, paracrine, and autocrine regulation. Endocrine Rev. **15:** 574–626.

39. PRINCE, F.P. 2001. The triphasic nature of Leydig cell development in humans, and comments on nomenclature. J. Endocrinol. **168:** 213–216.
40. FOUQUET, J.P, N. MEUSY-DESSOLLE & D.C. DANG. 1984. Relationship between Leydig cell morphometry and plasma testosterone during postnatal development of the monkey, *Macaca fascicularis*. Reprod. Nutrition & Dev. **24:** 281–296.
41. BERCU, B.B., B.C. LEE, J.L. PINEDA *et al.* 1983. Male sexual development in the monkey. 1. Cross sectional analysis of pulsatile hypothalamic-pituitary testicular function. J. Clin. Endocrinol. Metab. **56:**1214–1226.
42. RAMASWAMY, S., C.R. POHL, G.R. MARSHALL *et al.* 2004. A switch from continuous to episodic testicular testosterone release in response to pulsatile LH stimulation in juvenile rhesus monkeys (*Macaca mulatta*). J. Endocrinol. **183:** 61–68.
43. RAMASWAMY, S. 2005. Pubertal augmentation in juvenile rhesus monkey testosterone production induced by invariant gonadotropin stimulation is inhibited by estrogen. J. Clin. Endocrinol. Metab. **90:** 5866–5875.
44. DE ROUX, N., E. GENIN, J-C. CAREL *et al.* 2003. Hypogonadotropic hypogonadism due to loss of function of the KiSS1-derived peptide receptor GPR54. Proc. Natl. Acad. Sci. USA **100:** 10972–10976.
45. SEMINARA, S.B., S. MESSAGER, E.E. CHATZIDAKI *et al.* 2003. The GPR54 gene as a regulator of puberty. New Engl. J. Med. **349:** 1614–1627.
46. SHAHAB, M., C. MASTRONARDI, S.B. SEMINARA *et al.* 2005. Increased hypothalamic GPR54 signaling: a potential mechanism for initiation of puberty in primates. Proc. Natl. Acad. Sci. USA **102:** 2129–2134.
47. MURPHY, K.G. 2005. Kisspeptins: Regulators of metastasis and the hypothalamic-pituitary-gonadal axis. J. Neuroendocrinol. **17:** 519–525.
48. IRWIG, M.S., G.S. FRALEY & J.T. SMITH *et al.* 2004. Kisspeptin activation of gonadotropin releasing hormone neurons and regulation of KiSS-1 mRNA in the male rat. Neuroendocrinol. **80:** 264–272.
49. SHIBATA, M., R.B. GIBBS, M. SHAHAB *et al.* 2005. GnRH neurons in the peripubertal male rhesus monkey (*Macaca mulatta*) express GPR54: implication for the control of primate puberty. Presented at the 87th Annual Meeting of The Endocrine Society, San Diego, CA., June 4-7, 2005. (No. P1-98)

The mRNA Encoding τCstF-64 Is Expressed Ubiquitously in Mouse Tissues

ZANE HUBER, ROBERTO R. MONAREZ, BRINDA DASS, AND CLINTON C. MACDONALD

Department of Cell Biology & Biochemistry, Texas Tech University Health Sciences Center, Lubbock, Texas 79430, USA

ABSTRACT: Polyadenylation is a process of endonucleolytic cleavage of the mRNA, followed by addition of up to 250 adenosine residues to the 3′ end of the mRNA. Polyadenylation is essential for eukaryotic mRNA expression, and CstF-64 is a subunit of the CstF polyadenylation factor that is required for accurate polyadenylation. We discovered that there are two forms of the CstF-64 protein in mammalian male germ cells, one of which (CstF-64) is expressed in all tissues, the other of which (τCstF-64) is expressed only in male germ cells and in brain (albeit at significantly lower levels in the brain). Therefore, we were surprised to find that, using reverse transcription-PCR, cDNA cloning, and RNA blot analyses, τCstF-64 mRNA was expressed at higher levels in brain than in testis. Also, τCstF-64 mRNA was expressed at lower but detectable levels in all tissues tested, including epididymis, heart, kidney, liver, lung, muscle, ovary, spleen, thymus, and uterus. These results suggest the hypothesis that τCstF-64 mRNA is regulated at the translational or post-translational level.

KEYWORDS: polyadenylation; spermatogenesis; male germ cells; brain; mRNA processing; CstF

INTRODUCTION

Eukaryotic mRNA transcription is a complex stepwise process that involves transcriptional initiation, elongation, and termination.[1,2] Coordinately with transcription, the nascent mRNA is capped at its 5′ end, introns are removed by splicing, and its 3′ end is cleaved and polyadenylation occurs.[3] Polyadenylation is the process of endonucleolytic cleavage of the mRNA at a specific site followed by addition of up to 250 adenosine residues to the newly formed 3′ end. Accurate and efficient polyadenylation is essential for gene expression and is an important controller of translation and mRNA stability. The polyadenylation mechanism involves four steps: recognition of the correct site of cleavage/polyadenylation, endonucleolytic cleavage of the pre-mRNA at that site, addition of poly(A) to the newly formed 3′ end, and degradation of the remaining downstream portion of the pre-mRNA transcript.

ABBREVIATIONS: CstF: cleavage stimulation factor; CPSF: cleavage and polyadenylation specificity factor; UTR: untranslated region; SSC: standard saline citrate; qRT-PCR: quantitative reverse transcription-mediated polymerase chain reaction.

Address for correspondence: Clinton C. MacDonald, Department of Cell Biology & Biochemistry, Texas Tech University Health Sciences Center, 3601 4th Street, Lubbock, TX 79430. Voice: 806-743-2703; fax: 806-743-2990.
clint.macdonald@ttuhsc.edu

**Ann. N.Y. Acad. Sci. 1061: 163–172 (2005). © 2005 New York Academy of Sciences.
doi: 10.1196/annals.1336.017**

Several factors are required for polyadenylation. Two of these, the cleavage and polyadenylation specificity factor (CPSF), with four polypeptide subunits, and the cleavage stimulation factor (CstF), with three subunits, are important for the recognition and cleavage steps.[4–6] Over many years, CstF has been implicated in regulation of pre-mRNA polyadenylation in several systems. In particular, RNA binding of the 64,000 M_r subunit of CstF, CstF-64 (gene name, *Cstf2*) has been shown to regulate[7] or to correlate with[8–20] polyadenylation efficiency of specific substrates. Other studies showed that CstF-64 binds to a U- or GU-rich sequence downstream of most eukaryotic polyadenylation signals.[9,10,21,22] Furthermore, CstF-64 was shown to be essential for vertebrate[23] and invertebrate[24] viability. Together, these data suggest that CstF-64 is a key regulatory component of the polyadenylation machinery in most, if not all, eukaryotic cells.

Male germ cells express a second form of CstF-64 called τCstF-64. The τCstF-64 protein is expressed predominantly in meiotic and postmeiotic male germ cells[25] and to a smaller degree in brain, spleen, and thymus;[25,26] in rats, τCstF-64 is also expressed in the liver.[26] While CstF-64 is X-linked, τCstF-64 is encoded by a different gene, *Cstf2t*, which is on chromosome 19 in mice[27] and chromosome 10 in humans.[28] We have hypothesized that τCstF-64 is expressed in male germ cells, because *Cstf2* is on the X chromosome and is therefore inactivated during male meiosis.[29,30] This leads to the necessity of an auxiliary copy of *Cstf2* that can be expressed during meiosis, in a manner similar to that described for phosphoglycerokinase (*Pgk1*, *Pgk2*), pyruvate dehydrogenase E1a subunit (*Pdha1*, *Pdha2*), centrin (*Cetn2, Cetn1*), heterogeneous ribonucleoprotein G (HNRNPG, HNRNPG-T), Utp14a and Utp14b (*Utp14a, Utp14b*), and others.[30,31] This places τCstF-64 in a class of retroposons that are involved in male germ cell–specific development and X chromosome inactivation. Although the τCstF-64 protein is expressed at high levels in testis and at lower levels in brain, spleen, and thymus, we were surprised to find that expression of the τCstF-64 mRNA was more widespread. Reverse transcription-PCR (RT-PCR)–based cloning experiments revealed that the τCstF-64 mRNA could be cloned from a wide variety of tissues (including whole blood, liver, and muscle) from both males and females from a variety of eutherians (B. Dass, unpublished) (TABLES 1 and 2). Therefore, using RNA blotting and quantitative RT-PCR (qRT-PCR), we examined τCstF-64 mRNA levels in several tissues from mice and rats. Much to our surprise, significant levels of τCstF-64 mRNA were detected in every tissue examined, although τCstF-64 protein has only been reported in a few tissues. Therefore, we propose that the majority of the regulation of τCstF-64 expression is translational or post-translational. If translational, there must exist translational control elements in the τCstF-64 mRNA, possibly in its 3′-untranslated region.

MATERIALS AND METHODS

Animals Used in This Study

In all studies, animal protocols were approved by the Institutional Animal Care and Use Committee and National Institutes of Health guidelines for animal care were followed. Adult CD-1 mice were anesthetized with CO_2 and killed by cervical dislocation before tissue collection.

Antibodies and Protein Immunoblot Assays

Testicular tubules were prepared by decapsulating freshly harvested mouse testes in phosphate-buffered saline and washing to remove interstitial cells. Nuclei were prepared by centrifugation over a sucrose pad, boiled and sonicated in loading buffer, separated by 10% sodium dodecyl sulfate (SDS)-polyacrylamide gel electrophoresis, and transferred to polyvinylidene difluoride membranes for immunoblotting.[25] Membranes were stained with Ponceau S and incubated 1 to 16 hours in block solution (Tris-buffered saline with 10% nonfat dry milk). Monoclonal antibodies 3A7 and 6A9 were prepared against the human CstF-64 but specifically recognize the somatic CstF-64 and τCstF-64 protein isoforms, respectively, in mice.[25–27] Antibodies were used at a dilution of 1:50. Membranes were incubated with primary antibody in block followed by incubation with horseradish peroxidase-coupled sheep anti-mouse immunoglobulin G (1:10,000). After washing in Tris-buffered saline, immunoreactive bands were visualized by chemiluminescence using the Pierce SuperSignal™ kit (Pierce Biotechnology, Inc., Rockford, IL).

RNA Blot ("Northern") Analysis

Tissues were collected from adult male CD-1 mice (Charles River Laboratories, Wilmington, MA), frozen in liquid nitrogen, and stored at –80°C. Alternatively, tissues were collected into RNAlater® (Ambion, Inc., Austin, TX) and stored at 4°C. Total RNA was prepared from frozen tissues using TRIzol (Invitrogen Corporation, Carlsbad, CA) according to the manufacturer's directions. RNA integrity was estimated by examination of mobilities of 18S and 28S rRNA after ethidium bromide staining of formaldehyde gels.[32]

RNA blots were prepared as described previously,[33] except that transfer to nylon membranes was by capillary blotting overnight in 10 × SSC (standard saline citrate). For probe, DNA was isolated from a gel-purified 850-bp EcoRI fragment representing the 5′ end of the mouse τCstF-64 cDNA.[27] [^{32}P]dCTP-labeled probes were prepared by random priming using the Prime-It® II kit (Stratagene, La Jolla, CA). Prehybridization and hybridization were at 55°C in hybridization solution (5 × SSC, 50% formamide, 12.5 mM sodium phosphate, pH 6.6, 0.1% SDS, 0.2 mg/ml sonicated salmon sperm DNA, 0.4 mg/ml yeast tRNA). After 16 to 24 hours of hybridization, filters were washed four times at room temperature in 1 × SSC, 0.1% SDS and two to three times in 0.1 × SSC, 0.1% SDS at 65°C, followed by autoradiography with an intensifying screen at –80°C.

RT-PCR and qRT-PCR

Total RNA was isolated from mouse tissues using the TRIzol reagent, and cDNA was made using SuperScript™ II reverse transcriptase according to the manufacturer's recommendations (Invitrogen Corporation, Carlsbad, CA). Conditions for qRT-PCR (or "real-time" qRT-PCR) using the LightCycler® II apparatus (Roche Diagnostics Corporation, Indianapolis, IN) were optimized empirically for each primer set, and the number of cycles was confirmed to be within the linear range.[34] Using qRT-PCR, we amplified the mRNAs for *Cstf2t* and the ribosomal protein S16 as a control. The primers and PCR conditions were as follows: Cstf2t forward primer, 5′-GCAAAGCCAGGTAACTCCACAA-3′, and reverse primer, 5′-CCTGCATCTTC-

CTCACCTTAC-3′; expected product, 446 bp; $MgCl_2$ buffer, 4 mM; Roche Light Cycler settings of denaturing at 94°C for 30 seconds, annealing at 59°C for 0 seconds, and elongation at 72°C for 20 seconds; all slopes at 20. Also used was the following: S16 forward primer, 5′-CGCTGCAGTCCGTGCAGGTCTT-3′, and reverse primer, 5′-TCCAAACTTTTTGGATTCGCAGCG-3′; expected product, 383 bp; $MgCl_2$ buffer, 3 mM; Roche Light Cycler settings of denaturing at 94°C for 0 seconds, annealing at 60°C for 5 seconds, and elongation at 72°C for 20 seconds; all slopes at 20. All experiments were performed at least three times using RNA from different animals. The data in FIGURE 3 are normalized from averages of the raw gene-specific data divided by the averages of S16 data for each tissue.

RESULTS

τCstF-64 Protein Is Expressed in Many Mouse Tissues

The CstF-64 polyadenylation protein was expressed in all tissues examined.[25,26] Previously, we described a variant form of the CstF-64 polyadenylation protein that was expressed only in mouse testis and brain[25] and, oddly, in rat liver.[26] We named this protein τCstF-64.[27] FIGURE 1 shows the tissue distribution of CstF-64 and τC stF-64 in mouse tissues. The CstF-64 isoform was expressed in HeLa cells (panel A, lane 1) and in all mouse tissues examined (lanes 2–8). The τCstF-64 isoform was expressed at low levels in brain (panel B, lane 2) and at higher levels in testis (lane 7). τCstF-64 was undetectable in any other mouse tissues in this experiment (lanes 3–6

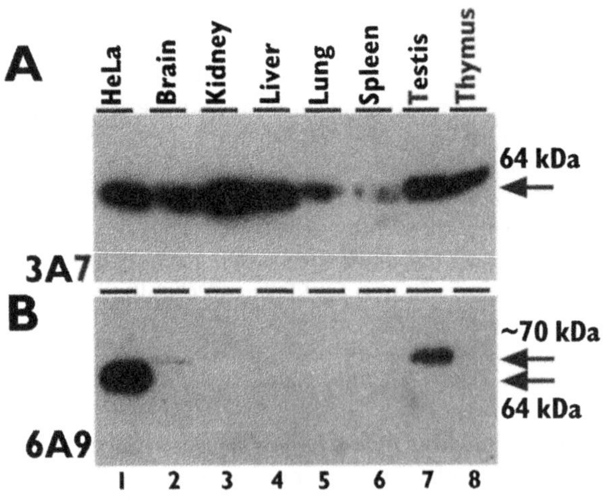

FIGURE 1. Protein immunoblot detection of two forms of CstF-64 in mouse tissues. **(A and B)** HeLa nuclear extract (*lane* 1), nuclei from indicated mouse tissues (lanes 2–6 and 8), or whole testicular tubules (*lane* 7) were examined by protein immunoblotting. Immunoblots were incubated with either the 3A7 (**A**) or 6A9 (**B**) monoclonal antibodies (the 3A7 monoclonal antibody detects the somatic CstF-64, while the 6A9 antibody detects τCstF-64[25]). *Arrows* indicate approximate sizes of the CstF-64 (64 kDa) and τCstF-64 (~70 kDa) proteins. (From Wallace *et al.*[25] Reproduced by permission.)

TABLE 2. Species in which τCstF-64 mRNA has been found in nontestis RNA

Species	Gender	Tissue(s)
Eptesicus fuscus	male	brain
Macaca mulatta	female	whole blood
Spermophilus tridecemlineatus	female	brain, liver

NOTE: Using degenerate oligonucleotide primers that distinguish CstF-64 from τCstF-64 sequences (B. Dass, unpublished), cDNAs representing τCstF-64 were cloned from RNA from the indicated tissues from each of the indicated species.

and 8). This suggests that the CstF-64 polyadenylation protein functions in all tissues, but that the τCstF-64 protein only functions appreciably in testis and in brain.

τCstF-64 mRNA Is Expressed in Many Mouse Tissues

In attempting to clone the cDNA encoding τCstF-64 from several different animal species (B. Dass, unpublished), we noticed that the mRNA for τCstF-64 was present in several tissues beyond testis and brain and in animals of both sexes (TABLES 1 and 2). TABLE 1 lists the species in which the τCstF-64 cDNA was cloned from RNA from testis or germ cell tissues, consistent with the high levels of expression of the τCstF-64 protein in testis. However, as shown in TABLE 2, the τCstF-64 cDNA was also found in nontesticular tissues and from animals of both genders where the τCstF-64 protein was not expressed. This suggested that the τCstF-64 mRNA might be expressed in a greater range of tissues than was the protein. On the basis of these and other similar results, we decided to examine the distribution of the τCstF-64 mRNA in multiple mouse tissues

Whole-cell RNA from mouse tissues was prepared, separated by denaturing 1.2% agarose gel electrophoresis, blotted to nylon membranes, and hybridized with probes representing the N-terminal RNA-binding domain and adjacent regions of τCstF-64 (FIG. 2A). For comparison of loading, the separated RNA was stained with ethidium bromide and photographed (FIG. 2B). In every tissue examined in both mice (FIG. 2A) and rats (not shown), some τCstF-64-hybridizing RNA (about 4.5 kb; arrow) was detected (FIG. 2A, lanes 1–8). Testis RNA (where τCstF-64 protein expression was reported to be greatest[25,26]) showed the greatest intensity of signal, although some of the variation in signal may have been due to uneven probe hybridization (e.g., lane 7). However, RNA from other tissues such as brain (where τCstF-64 is also expressed, although at levels lower than in testis), thymus, heart, spleen, and kidney (where τCstF-64 protein expression was low or unmeasurable[26]) also showed visible intensity of τCstF-64-hybridizing signal. In contrast, an earlier report showed that CstF-64 mRNA was at exceedingly low levels in all tissues except testis.[33]

τCstF-64 mRNA Levels Are Highest in Brain and Testis But Are Detectable in All Tissues Examined

Because we could detect τCstF-64 mRNA in several mouse tissues by RNA blotting, including several that have not been shown to express τCstF-64 protein, we decided to examine τCstF-64 mRNA levels more accurately using qRT-PCR.[34] RNAs

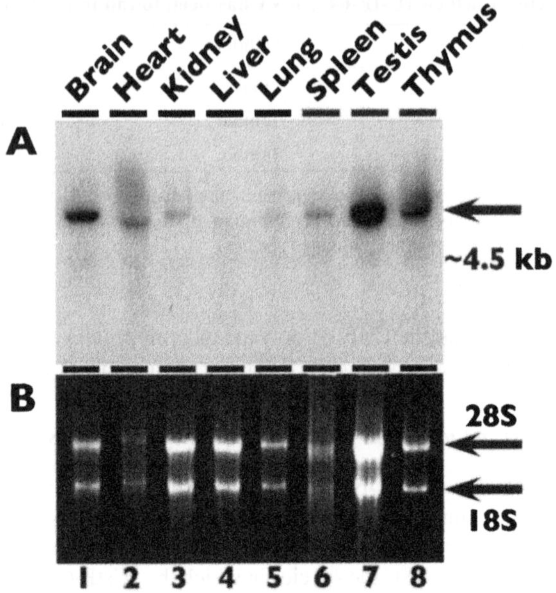

FIGURE 2. RNA blot analysis of τCstF-64 mRNA expression in mouse tissues. Approximately 10 mg of RNA from each of the indicated mouse tissues was separated by 1.2% agarose formaldehyde gel electrophoresis, transferred to a nylon membrane, and hybridized to a τCstF-64-specific cDNA probe (**A**). The ethidium bromide-stained agarose gel for comparison of loading. *Lanes* 1–8 include RNA from each of the tissues, as indicated. *Arrows* indicate the mobilities of τCstF-64 mRNA (**A**) or 28S or 18S rRNAs (**B**).

from the indicated tissues were isolated from adult male CD-1 mice, except for ovary and uterus, which were from adult CD-1 female mice. For comparison, we also included RNA from testes of mice in which the entire coding region of the τCstF-64 gene, *Cstf2t*, had been deleted by targeted disruption (B. Dass *et al.*, unpublished). Note that *Cst2t* mRNA is undetectable in testes of knockout mice (FIG. 3).

The τCstF-64 mRNA amount is greater in brain RNA (FIG. 3, 0.00325 copies/copy of S16 mRNA) than in testis (0.00236 copies/copy of S16). This does not reflect protein expression, since τCstF-64 expression is much greater in testis than in brain.[25,26] Next in rank is heart, in which τCstF-64 mRNA is approximately 26% of that in brain (0.00085 copies/copy of S16). Other amounts vary from 0.0005 (epididymis) to 0.0001 (lung) copies/copy of S16. Again, note that these amounts are not reflective of protein expression, since we have reported low but detectable τCstF-64 protein in spleen and thymus,[26] but not in the other tissues assayed here.

DISCUSSION

It is well-established that the τCstF-64 protein is expressed in a highly restricted set of tissues, primarily testis and brain.[25,26] However, we show here that the mRNA

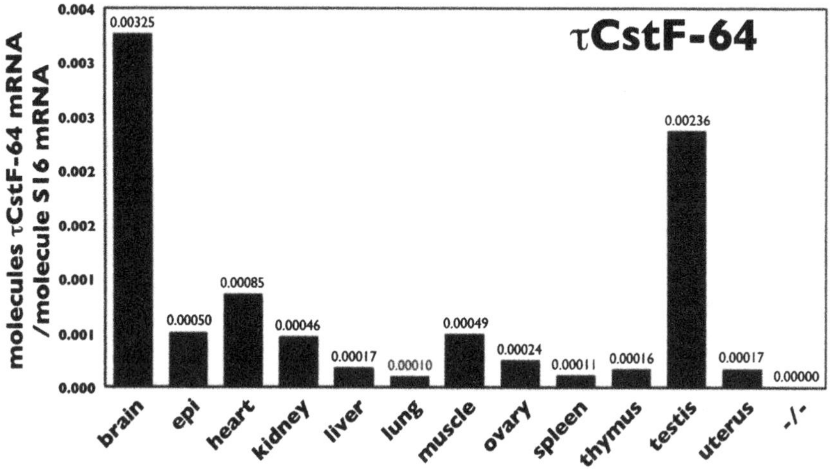

FIGURE 3. qRT-PCR analysis of τCstF-64 mRNA expression in mouse tissues. qRT-PCR was performed using RNA from CD-1 mice for each of the tissues indicated (epi, epididymis). The symbol –/– indicates qRT-PCR from testis RNA of a mouse homozygous for a targeted gene disruption of τCstF-64 (*Cstf2t*$^{-/-}$) (B. Dass, unpublished data). Values are expressed as the number of molecules of τCstF-64 mRNA divided by the number of molecules of ribosomal protein S16 mRNA determined for each tissue.

encoding τCstF-64 is not so highly restricted. We found the highest levels of τCstF-64 mRNA in brain and testis (in which the τCstF-64 protein is found). However, we also found moderate levels of τCstF-64 mRNA in some tissues (heart, skeletal muscle, kidney, and epididymis) and lower levels in several other tissues (liver, lung, ovary, spleen, thymus, and uterus). None of these tissues has measurable levels of τCstF-64 protein in mice, though liver, spleen, and thymus express τCstF-64 protein in rats. Because the protein expression levels do not correlate with the mRNA levels, we propose that expression of the τCstF-64 protein is controlled by a translational element, the mechanism for which we have not discovered.

Messenger RNA processing is an essential set of steps in eukaryotic gene expression. In addition to being required for assembly of exons within the mature mRNA, the machineries for capping, splicing, and polyadenylation are all required for efficient transcription.[1,2,35] Alternative mRNA processing is also essential to ensure the diversity of mRNA processing required for eukaryotic development and physiology. Two tissues with the highest degree of alternative mRNA processing are brain and testis.[36–39] While it is obvious there is a need for diversity of gene expression in the brain,[40] it is less obvious why such diversity is required for testicular germ cell function.[37,38] It has been speculated that diversity of gene expression in male germ cells is to provide functions of genetic diversity through translational control mechanisms,[41] but there are few data to support these and similar theories. We speculate, however, that alternative members of core proteins involved in gene expression, such as τCstF-64, are key contributors to genetic diversity in these two tissues.

Why, then, is τCstF-64 mRNA expressed in such a wide range of tissues if the protein is not? Why is τCstF-64 mRNA higher in brain than testis, though the protein expression is weighed heavily in the opposite direction? We are not certain of the answers to these questions, and we may not be able to answer them until the mechanism of translational or post-translational control is uncovered. One possibility is that the promiscuous expression of τCstF-64 mRNA is a consequence of its history as an expressed retroposon.[28] It has been speculated that retroposon expression in the testis is increased because of general mechanisms that increase transcription from heterochromatic DNA.[31,42] However, this would suggest that many retroposed genes would behave similarly, which has not been reported. A second possibility is that τCstF-64 mRNA is promiscuously expressed because it is located inside an intron of another gene, *Prkg1*, a type I cGMP-dependent protein kinase.[43] Because *Prkg1* has a wide range of expression, *Cstf2t* is likely in a euchromatic region of the chromosome and may be expressed widely. There are two arguments that temper this possibility, however: first, another retroposed gene with germ cell expression, *Utp14b*, is within the intron of a gene, *Facl3*, but does not show a similar wide range of expression.[44] Second, targeted disruption of *Cstf2t* by replacement with a gene encoding resistance to G418 does not alter *Prkg1* expression, suggesting that *Cstf2t* and *Prkg1* are transcriptionally independent (B. Dass, unpublished data). Furthermore, none of these speculations addresses the likely translational or post-translational mechanism that must repress τCstF-64 expression in all but a few tissues. It seems unlikely we will be able to answer these questions satisfactorily until we have investigated more closely the mechanism of τCstF-64's transcriptional and translational control.

ACKNOWLEDGMENTS

We would like to acknowledge the members of the MacDonald laboratory for their support and insightful discussions. This work was supported by Grant R01 HD37109-04 from the National Institutes of Health. Roberto Monarez was supported by a minority student supplement to R01 HD37109-04.

REFERENCES

1. Bentley, D. 1999. Coupling RNA polymerase II transcription with pre-mRNA processing. Curr. Opin. Cell Biol. **11:** 347–351.
2. Minvielle-Sebastia, L. & W. Keller. 1999. mRNA polyadenylation and its coupling to other RNA processing reactions and to transcription. Curr. Opin. Cell Biol. **11:** 352–357.
3. Zhao, J., L. Hyman & C. Moore. 1999. Formation of mRNA 3′ ends in eukaryotes: mechanism, regulation, and interrelationships with other steps in mRNA synthesis. Microbiol. Mol. Biol. Rev. **63:** 405–445.
4. Takagaki, Y., J.L. Manley, C.C. MacDonald *et al.* 1990. A multisubunit factor CstF is required for polyadenylation of mammalian pre-mRNAs. Genes Dev. **4:** 2112–2120.
5. Bienroth, S., E. Wahle, C. Suter-Crazzolara & W. Keller. 1991. Purification of the cleavage and polyadenylation factors involved in the 3′-processing of messenger RNA precursors. J. Biol. Chem. **266:** 19768–19776.
6. Colgan, D.F. & J.L. Manley. 1997. Mechanism and regulation of mRNA polyadenylation. Genes Dev. **11:** 2755–2766.
7. Takagaki, Y., R.L. Seipelt, M.L. Peterson & J.L. Manley. 1996. The polyadenylation factor CstF-64 regulates alternative processing of IgM heavy chain pre-mRNA during B cell differentiation. Cell **87:** 941–952.

8. WILUSZ, J. & T. SHENK. 1988. A 64 kDa nuclear protein binds to RNA segments that include the AAUAAA polyadenylation motif. Cell **52:** 221–228.
9. MANN, K.P., E.A. WEISS & J.R. NEVINS. 1993. Alternative poly(A) site utilization during adenovirus infection coincides with a decrease in the activity of a poly(A) site processing factor. Mol. Cell. Biol. **13:** 2411–2419.
10. MACDONALD, C.C., J. WILUSZ & T. SHENK. 1994. The 64-kilodalton subunit of the CstF polyadenylation factor binds to pre-mRNAs downstream of the cleavage site and influences cleavage site location. Mol. Cell. Biol. **14:** 6647–6654.
11. BAGGA, P.S., L.P. FORD, F. CHEN & J. WILUSZ. 1995. The G-rich auxiliary downstream element has distinct sequence and position requirements and mediates efficient 3′ end pre-mRNA processing through a trans-acting factor. Nucleic Acids Res. **23:** 1625–1631.
12. EDWALDS-GILBERT, G. & C. MILCAREK. 1995. Regulation of poly(A) site use during mouse B-cell development involves a change in the binding of a general polyadenylation factor in a B-cell stage-specific manner. Mol. Cell. Biol. **15:** 6420–6429.
13. MCGREGOR, F., A. PHELAN, J. DUNLOP & J. B. CLEMENTS. 1996. Regulation of herpes simplex virus poly(A) site usage and the action of immediate-early protein IE63 in the early-late switch. J. Virol. **70:** 1931–1940.
14. BAGGA, P.S., G.K. ARHIN & J. WILUSZ. 1998. DSEF-1 is a member of the hnRNP H family of RNA-binding proteins and stimulates pre-mRNA cleavage and polyadenylation in vitro. Nucleic Acids Res. **26:** 5343–5350.
15. MARTINCIC, K., R. CAMPBELL, G. EDWALDS-GILBERT, *et al.* 1998. Increase in the 64-kDa subunit of the polyadenylation cleavage stimulatory factor during the G0 to S phase transition. Proc. Natl. Acad. Sci. USA **95:** 11095–11100.
16. MOREIRA, A., Y. TAKAGAKI, S. BRACKENRIDGE, *et al.* 1998. The upstream sequence element of the C2 complement poly(A) signal activates mRNA 3′ end formation by two distinct mechanisms. Genes Dev. **12:** 2522–2534.
17. EVANS, D., I. PEREZ, M. MACMORRIS, *et al.* 2001. A complex containing CstF-64 and the SL2 snRNP connects mRNA 3′ end formation and trans-splicing in *C. elegans* operons. Genes Dev. **15:** 2562–2571.
18. VERALDI, K.L., G.K. ARHIN, K. MARTINCIC, *et al.* 2001. hnRNP F influences binding of a 64-kilodalton subunit of cleavage stimulation factor to mRNA precursors in mouse B cells. Mol. Cell. Biol. **21:** 1228–1238.
19. CHENNATHUKUZHI, V.M., S. LEFRANCOIS, C. R. MORALES *et al.* 2001. Elevated levels of the polyadenylation factor CstF 64 enhance formation of the 1kB testis brain RNA-binding protein (TB-RBP) mRNA in male germ cells. Mol. Reprod. Dev. **58:** 460–469.
20. YAO, Y., L. SONG, Y. KATZ & G. GALILI. 2002. Cloning and characterization of *Arabidopsis* homologues of the animal CstF complex that regulates 3′ mRNA cleavage and polyadenylation. J. Exp. Botany **53:** 2277–2278.
21. PÉREZ CAÑADILLAS, J.M. & G. VARANI. 2003. Recognition of GU-rich polyadenylation regulatory elements by human CstF-64 protein. EMBO J. **22:** 2821–2830.
22. DEKA, P., P.K. RAJAN, J.M. PÉREZ-CAÑADILLAS & G. VARANI. 2005. Protein and RNA dynamics play key roles in determining the specific recognition of GU-rich polyadenylation regulatory elements by human CstF-64 protein. J. Mol. Biol. **347:** 719–733.
23. TAKAGAKI, Y. & J.L. MANLEY. 1998. Levels of polyadenylation factor CstF-64 control IgM heavy chain mRNA accumulation and other events associated with B cell differentiation. Mol. Cell **2:** 761–771.
24. MACMORRIS, M., C. BROCKER & T. BLUMENTHAL. 2003. UAP56 levels affect viability and mRNA export in *Caenorhabditis elegans*. RNA **9:** 847–857.
25. WALLACE, A.M., B. DASS, S. E. RAVNIK *et al.* 1999. Two distinct forms of the 64,000 M_r protein of the cleavage stimulation factor are expressed in mouse male germ cells. Proc. Natl. Acad. Sci. USA **96:** 6763–6768.
26. WALLACE, A.M., T. DENISON, E.N. ATTAYA & C.C. MACDONALD. 2004. Developmental differences in expression of two forms of the CstF-64 polyadenylation protein in rat and mouse. Biol. Reprod. **70:** 1080–1087.
27. DASS, B., K.W. MCMAHON, N.A. JENKINS, *et al.* 2001. The gene for a variant form of the polyadenylation protein CstF-64 is on chromosome 19 and is expressed in pachytene spermatocytes in mice. J. Biol. Chem. **276:** 8044–8050.

28. DASS, B., L. MCDANIEL, R.A. SCHULTZ, *et al.* 2002. The gene CSTF2T encoding the human variant CstF-64 polyadenylation protein τCstF-64 is intronless and may be associated with male sterility. Genomics **80:** 509–514.
29. HANDEL, M.A., P.A. HUNT, M.C. KOT, *et al.* 1991. Role of sex chromosomes in the control of male germ-cell differentiation. Ann. N.Y. Acad. Sci. **637:** 64–73.
30. HANDEL, M.A. 2004. The XY body: a specialized meiotic chromatin domain. Exp. Cell Res. **296:** 57–63.
31. WANG, P.J., J.R. MCCARREY, F. YANG & D.C. PAGE. 2001. An abundance of X-linked genes expressed in spermatogonia. Nat. Genet. **27:** 422–426.
32. MACDONALD, C.C. & D.L. WILLIAMS. 1993. RNase H oligodeoxyribonucleotide-directed mRNA purification (ROMP) of apoII mRNA. Nucleic Acids Res. **21:** 765–766.
33. DASS, B., E.N. ATTAYA, A.M. WALLACE & C.C. MACDONALD. 2001. Overexpression of the CstF-64 and CPSF-160 polyadenylation protein messenger RNAs in mouse male germ cells. Biol. Reprod. **64:** 1722–1729.
34. LIVAK, K.J., S.J. FLOOD, J. MARMARO, *et al.* 1995. Oligonucleotides with fluorescent dyes at opposite ends provide a quenched probe system useful for detecting PCR product and nucleic acid hybridization. PCR Methods Applic. **4:** 357–362.
35. KORNBLIHTT, A.R., M. DE LA MATA, J.P. FEDEDA, *et al.* 2004. Multiple links between transcription and splicing. RNA **10:** 1489–1498.
36. EDWALDS-GILBERT, G., K.L. VERALDI & C. MILCAREK. 1997. Alternative poly(A) site selection in complex transcription units: means to an end? Nucleic Acids Res. **25:** 2547–2561.
37. WALKER, W.H., F.J. DELFINO & J.F. HABENER. 1999. RNA processing and the control of spermatogenesis. Front. Horm. Res. **25:** 34–58.
38. VENABLES, J.P. 2002. Alternative splicing in the testes. Curr. Opin. Genet. Dev. **12:** 615–619.
39. YEO, G., D. HOLSTE, G. KREIMAN & C.B. BURGE. 2004. Variation in alternative splicing across human tissues. Genome Biol. **5:** R74.
40. GRAVELEY, B.R., A. KAUR, D. GUNNING *et al.* 2004. The organization and evolution of the dipteran and hymenopteran Down syndrome cell adhesion molecule (Dscam) genes. RNA **10:** 1499–1506.
41. KLEENE, K.C. 2001. A possible meiotic function of the peculiar patterns of gene expression in mammalian spermatogenic cells. Mech. Dev. **106:** 3–23.
42. JONES, M.H., Y. ZHANG, K.N. Tirosvoutis *et al.* 1997. Chromosomal assignment of 311 sequences transcribed in human adult testis. Genomics **40:** 155–167.
43. ORSTAVIK, S., V. NATARAJAN, K. TASKEN, *et al.* 1997. Characterization of the human gene encoding the type I alpha and type I beta cGMP-dependent protein kinase (PRKG1). Genomics **42:** 311–318.
44. ROHOZINSKI, J. & C.E. BISHOP. 2004. The mouse juvenile spermatogonial depletion (jsd) phenotype is due to a mutation in the X-derived retrogene, mUtp14b. Proc. Natl. Acad. Sci. USA **101:** 11695–11700.

Drivers of Germ Cell Maturation

K.L. LOVELAND,[a,c] C. HOGARTH,[a,c] S. MENDIS,[a] A. EFTHYMIADIS,[b,c] J. LY,[b,c] C. ITMAN,[a,c] S. MEACHEM,[d] C.W. BROWN,[e] AND D.A. JANS[b,c]

[a]*Monash Institute of Medical Research, Monash University, Melbourne, Australia*

[b]*Department of Biochemistry and Molecular Biology, Monash University, Melbourne, Australia*

[c]*Australian Research Council Centre of Excellence in Biotechnology and Development, Australia*

[d]*Prince Henry's Institute of Medical Research, Melbourne Australia*

[e]*Department of Molecular and Human Genetics and Department of Pediatrics, Baylor College of Medicine, Houston, Texas 77030, USA*

ABSTRACT: Spermatogenesis requires progression of germ line stem cells through a precisely ordered differentiation pathway to form spermatozoa. Diverse and dynamic signals from the transforming growth factor-β (TGF-β) superfamily influence many stages of germ cell development. For example, interactions between several TGF-β superfamily ligands (bone morphogenetic proteins, activin, and glial-derived neurotrophic growth factor [GDNF]) appear to govern the onset of spermatogenesis, and we are exploring how germ cells interpret these competing signals. We examined the *in vivo* impact of activin on testis development using two mouse models, the *inhba*$^{-/-}$ mouse (which lacks the gene encoding the activin A subunit and dies at birth) and BK mice, with *inhbb* (encoding the activin βB subunit) replacing *inhba* (which survive to adulthood and show delayed fertility onset in males). Distinct effects on Sertoli cell and germ cell populations during fetal and early postnatal development were measured. We recognize that specific proteins, including downstream targets of TGF-β signals, such as Smads, must move into the nucleus to implement the gene transcription changes required for development. We hypothesized that changes at the level of cellular nuclear transport machinery may be required to mediate this. Examination of proteins involved in classical nuclear import, the importins, revealed that each importin has a developmentally regulated expression pattern in male germ cells. Because each importin binds a selected range of cargo proteins and mediates their nucleocytoplasmic passage, our findings suggest that each importin ferries cargo required for discrete stages of spermatogenesis.

KEYWORDS: activin; importins; spermatogenesis; nuclear transport; Sertoli cells; Smads; stereology; nuclear import; TGF-β superfamily

Address for correspondence: Kate Lakoski Loveland, Monash Institute of Medical Research, 27-31 Wright Street, Clayton, Victoria 3168 Australia. Voice: 61-3-9594-7418; fax: 61-3-9594-7111.

Kate.loveland@med.monash.edu.au

Ann. N.Y. Acad. Sci. 1061: 173–182 (2005). © 2005 New York Academy of Sciences. doi: 10.1196/annals.1336.018

INTRODUCTION

Spermatogenesis involves a precisely regulated sequence of differentiation steps, with adult fertility achieved through the integration of a succession of hormonal and local factors driving both somatic and germ cell maturation, with accompanying changes in gene expression characteristic of each developmental stage.[1,2] Since this paradigm of coordinated development is not unique to the mammalian testis, it is of value to explore the impact of key regulatory molecules and processes identified in other areas of biology on testis development and sperm production. We have been investigating two distinct regulators of cell behavior: the regulated expression and actions of transforming growth factor β (TGF-β) superfamily members and the family of nuclear transport importin proteins. Each of these appears to undergo regulation in expression that coincides with discrete stages of male germ cell maturation, and we have used this information to guide our efforts to uncover their functional significance.

THE TGF-β SUPERFAMILY

The TGF-β superfamily of ligands, encompassing a large set of proteins with diverse functions, has recently been shown to exert control over mammalian male germ cell differentiation at multiple steps of differentiation.[3,4] These ligands are characterized by a conserved set of seven cysteines, and they are classically understood to form homo- and heterodimers, each with a distinct capacity to provoke a biological outcome. Because most family members have both separate and shared components, signaling outputs from these ligands are likely to be strongly influenced by competing signals from other family members (FIG. 1). TGF-β superfamily ligands typically signal upon binding to heterotetrameric transmembrane receptor proteins with intracellular serine/threonine kinase activities; the bone morphogenetic proteins (BMPs) and activins share receptor subunits, while the TGF-βs employ a distinct set of receptor proteins. A large number of extracellular and transmembrane inhibitors govern ligand binding, and many of these inhibitors modulate the activity of more than one family member. For example, the soluble protein, follistatin, can reduce both activin and BMP actions, and it has differential effects on members within each of these subfamilies.[5,6] The membrane-bound inhibitor, Bambi (for BMP and *a*ctivin *m*embrane-*b*ound *i*nhibitor), demonstrated to inhibit signaling through TGF-β, BMP, and activin, has been described as a promiscuous inhibitor of this superfamily.[7]

In response to ligand binding, signal transduction to the nucleus is typically mediated through phosphorylation of receptor-activated Smad proteins (R-Smads), with activin and TGF-β both capable of activating Smad2 or 3 and BMPs activating Smad1, 5, or 8.[8] Each activated R-Smad must bind the common Smad4 to gain nuclear access, and this complex binds to gene promoter elements with specificity governed by the existing complement of modifying proteins. The inhibitory Smads, Smad6 and 7, can prevent signal transduction by these ligands through more than one mechanism. Hence, the allotment of receptor subunits, inhibitory molecules, Smads, and transcriptional regulators will collectively determine the response of a particular cell to the diverse signals that may be simultaneously imparted by TGF-β

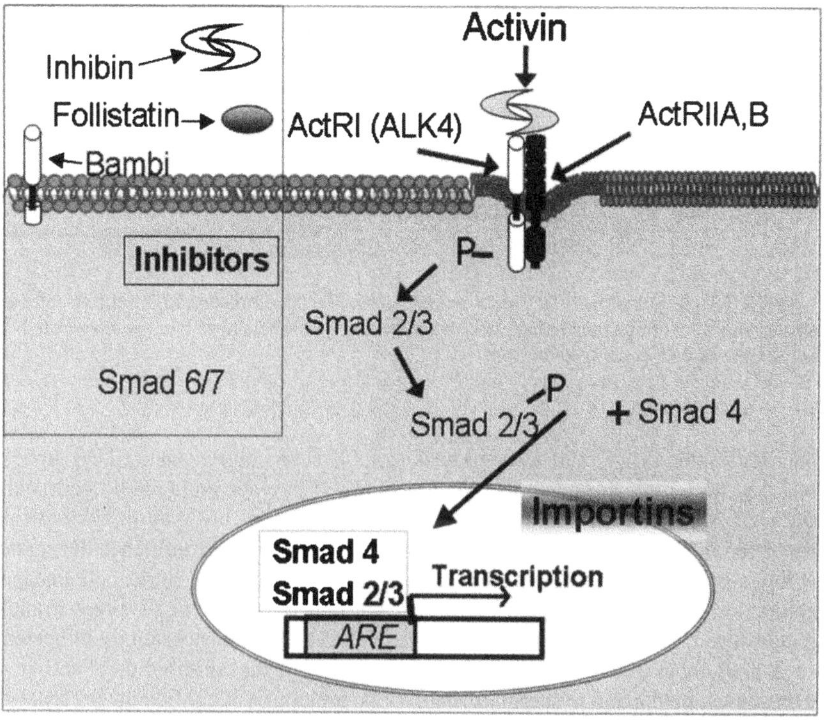

FIGURE 1. Summary of classical signaling pathway for TGF-β superfamily ligands, shown here for activin binding to its type II and type I serine/threonine kinase receptors. Extracellular binding of dimeric ligand leads to phosphorylation of either Smad2 and/or -3, enabling it to interact with Smad4. This complex then enters the nucleus via importin-mediated transport to activate transcription of specific downstream genes. Examples of various activin signaling inhibitors are depicted in the left-hand panel as soluble extracellular (follistatin and inhibin), membrane-bound (Bambi), and intracellular activin inhibitors (Smad6 and 7). See text for references.

superfamily members. Smad-independent signaling has also been documented,[9] providing additional avenues for complex interactions between TGF-β family ligand signals and a host of other signaling pathways.

The pervasive influence of this superfamily ranges from stem cells and embryonic development to governance of wound healing, inflammation, immune responses, and cancer progression.[10] As a result, they have attracted intense interest from many disciplines from which informative *in vivo* and *in vitro* models have emerged.

THE TGF-β SUPERFAMILY AND TESTIS DEVELOPMENT: ROLES FOR ACTIVIN

The development of transgenic and knockout mouse models has decisively demonstrated the requirement of TGF-β superfamily signaling for (1) allocation of cells to the germ line (*BMP4*, *BMP8b*, *Smad1*, and *Smad5* knockout mice), (2) Müllerian

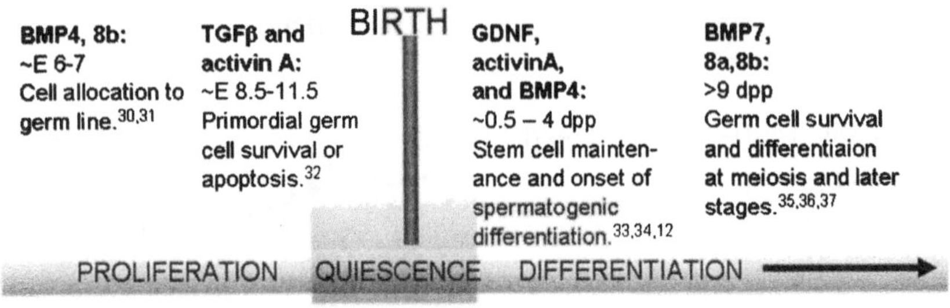

FIGURE 2. Timeline of key steps in male germ line development, highlighting selected points where functions for distinct TGF-β superfamily members have been demonstrated. E, embryonic day; dpp, day postpartum.

duct regression in the male (*Amhr* knockout), (3) stem cell renewal (*GDNF* knockouts and transgenic animals), and (4) maturation of meiotic and haploid germ cells (*ACVR2, BMP8b,* and *BMP7* knockouts) (reviewed in Ref. 11). Additional information from *in vitro* studies has revealed several potential sites of influence during development of the male gonad and gamete. Both somatic and germ cell lineages appear to be directly affected.[3,4] FIGURE 2 presents a timeline of key steps in male gonad development, with some points of influence of TGF-β superfamily indicated.

A main focus of our research has been to investigate the contribution of activin to testis development and to spermatogenesis. We previously identified the presence of the activin βA subunit protein in the quiescent germ cells that are present in the rat testis at birth.[12] On the basis of the timing of activin βA subunit mRNA synthesis (encoded by the *inhba* gene), we understand that this complement of germ cell activin is synthesized in the fetal testis and is stored until the onset of spermatogenesis. Thus, within 3 days after birth in the rat, as these cells re-enter the cell cycle and migrate to the basement membrane of the seminiferous cord, activin protein disappears from the cytoplasm of these cells. As they transform into spermatogonia, they begin to synthesize two antagonists of activin, follistatin and Bambi.[12,13] Our functional analysis with cultured rat testis fragments indicates that exogenous follistatin and follicle-stimulating hormone promote cellular changes that mimic the onset of spermatogenesis, and this is blocked by addition of activin.[12] We interpreted these data to indicate that activin participates in governing the growth arrest of germ cells that are formed in the fetal testis.

This hypothesis was tested by examining testes from mice which lack the gene encoding the activin βA subunit ($inhba^{-/-}$).[14] Our comparison of wild-type versus $inhba^{-/-}$ mice showed there was no significant difference between these genotypes in terms of body weight or testis weight (TABLE 1). However, the application of stereological analysis to precisely enumerate Sertoli cell and germ cell numbers in these two groups revealed that, while the absence of activin βA did not affect Sertoli cell numbers, the number of germ cells present was approximately doubled. These findings indicate that activin may act to limit germ cell proliferation in the fetal testis.

As indicated in FIGURE 2, additional roles for activin in governing postnatal testis development have been demonstrated through *in vivo* and *in vitro* studies. Both germ

TABLE 1. Phenotype of *inhibin* βA$^{-/-}$ mice at day of birth

	Wild type		Inhibin βA$^{-/-}$	
	Mean	SEM	Mean	SEM
Pooled testis weight (mg)	6.8	0.429	8.917	1.939
Body weight (g)	1.412	0.05	1.362	0.051
Total no. of Sertoli cells/testis (× 10 ^{6}cells)	1.438	0.263	1.395	0.247
Total no. of germ cells/testis (× 10^{6} cells)	0.156	0.026	0.405	0.066

NOTE: Data are presented from n = 6 per group. Total germ cell numbers are significantly different between groups (P <0.01; unpaired t test).

cells and Sertoli cells have been shown to alter their behavior in response to culture periods with this factor. Our recent survey of activin βA and βB and inhibin α subunit mRNA and proteins throughout rat postnatal testis development illustrated their changing levels,[15] with a general picture emerging to show that these are present at relatively high levels in the juvenile animal compared to the adult. Of particular interest to us and others is the demonstrated cooperative impact of activin on the stimulation of rat postnatal Sertoli cell proliferation during a discrete period immediately before they undergo terminal differentiation and exit the cell cycle.[16–18] These data suggest that local production of activin, which is relatively high in this phase of testis development, may influence the final Sertoli cell numbers that exist in the adult. This is understood to be a key determinant of male fertility, as each adult Sertoli cell supports maturation of a limited number of germ cells at a time,[19] and factors reducing Sertoli cell proliferation prior to terminal differentiation would be expected to compromise fertility.

To examine the *in vivo* role of activin in the postnatal testis, we have turned to analysis of the BK mouse strain. This strain was created by replacement of the mature domain of the *Inhba* gene with that of the closely related *Inhbb* gene to produce the *Inhba*$^{\mathrm{tm1Zuk}}$ mouse, also termed BK/BK. These mice survive to adulthood and show delayed fertility onset in males. This animal produces the activin βB subunit at sites and levels as measured for the activin βA subunit in wild-type animals, and the insertion rescued the phenotype of neonatal lethality seen in the *inhba*$^{-/-}$ mice.[20] In the BK/BK animals, the replacement of activin βA with the less bioactive activin βB protein led to a significant reduction in both body weight and testis weight at days 7 and 14 post partum, and our preliminary data indicate that Sertoli and germ cell numbers are both decreased. This finding indicates that the impact of activin A on postnatal testis development is distinct from that in the fetal testis, in which Sertoli cell numbers do not appear to be affected. Reflecting the common understanding that reduction in the number of terminally differentiated Sertoli cells will lead to survival of a reduced number of germ cells,[21] we predict that the germ cell numbers will be reduced in the BK/BK mice in proportion to the number of Sertoli cells present.

In addition to this documentation of the functional consequences of regulated expression of activin and its inhibitors, we and others have recently observed that the key intracellular signaling components for TGF-β superfamily signaling, Smads, are also developmentally regulated in the mouse testis.[22] Our *in situ* hybridization anal-

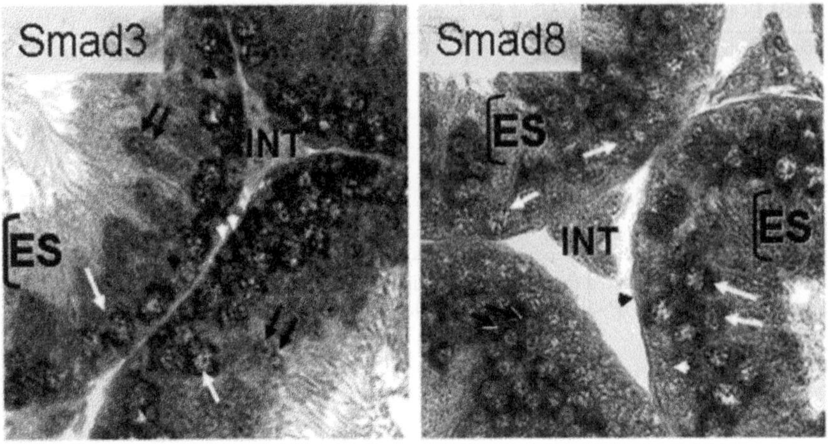

FIGURE 3. Detection of *Smad3* and *Smad8* mRNAs in adult mouse testis by *in situ* hybridization with a digoxigenin-labeled cRNA reveals that each Smad is produced in germ cells at different stages of maturation. The *Smad3* mRNA signal is most intense in spermatogonia and spermatocytes, weak in round spermatids, and weak to not detectable in elongated spermatids or somatic cells. The *Smad8* mRNA signal is most intense in spermatocytes and round spermatids and weak or not detected in other cell types. *Black arrowhead*, Sertoli cell nucleus; *white arrowhead*, spermatogonium; *white arrows*, spermatocyte; *black arrows*, round spermatid; *ES with bracket*, elongated spermatids; *INT*, interstitium.

ysis of adult mouse testis illustrates that, while *Smad3* mRNA is most readily observed in spermatogonia and spermatocytes of the adult testis, the signal for *Smad8* mRNA is most intense in spermatocytes and round spermatids (FIG. 3).

Collectively, these data demonstrate that differential expression of genes encoding TGF-β superfamily ligands, inhibitors, and transcription factors are regulated throughout testis development. This phenomenon is a paradigm of developmental biology in many systems, and we have extended our investigations into the mechanisms that govern testis development and spermatogenesis by examining how these changes in gene expression may be effected. To do this, we turned to examine how access of transcription factors and other nuclear proteins might be influencing developmental switches, such as those illustrated by male gamete maturation.

REGULATED NUCLEAR TRANSPORT IN DEVELOPMENT

Movement of proteins between the cell cytoplasm and nucleus is determined by their capacity to pass through the nuclear pore complexes, the portal of the nuclear envelope which is composed of over 40 distinct nucleoporin proteins.[23] Active transport is required for movement of proteins of 50 kDa or greater, and this typically requires their interaction with specific components of the cellular nuclear transport machinery. Members of the importin family of transport proteins mediate nucleocytoplasmic transport through mechanisms common in a wide diversity of species, from *Caenorhabditis elegans* and *Drosophila* through to mouse and human.[24] Multiple α forms and β forms exist, each with a distinct set of functional properties, including different binding affinities for different cargo proteins (reviewed in Refs. 25 and 26). In classical nuclear import, the importin α protein recognizes a specific nuclear localization signal on a cargo protein, while a heterodimer of importin α/β1 is required for nuclear pore translocation. In addition, nuclear transport may be mediated by an importin α molecule alone (distinct from importin β1). In both types of transport, it is a sequence of transient interactions between the importin α protein and various nucleoporins which mediates translocation. The complex is dissociated inside the nucleus when GTP-bound Ran binds to importin β. We have recently reviewed the many points of testis development that are governed by regulated nuclear transport.[27]

We hypothesized that the readily apparent developmental switches of spermatogenesis are mediated through the regulated expression of nuclear transport factors acting in concert with key nuclear proteins which also undergo regulated expression. While many examples of regulated transcription factor/nuclear proteins are known,[1]

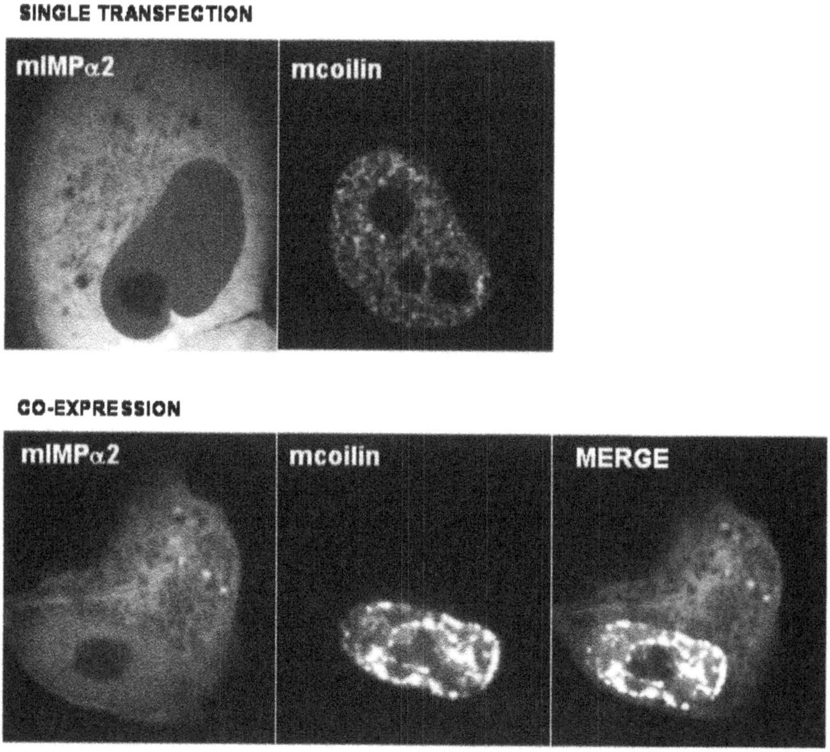

FIGURE 4. Intracellular colocalization of coilin and importin α2 illustrated in a heterologous cell type. Vero monkey kidney cells were transfected with constructs encoding mouse coilin and importin α2, each with a different-colored fluorescent tag. Coilin, which mainly localizes to the nucleoplasm (*top*), is observed in nuclear speckles (coiled bodies) and within the nucleoli (*bottom*) when coexpressed with importin α2, indicating the two proteins interact in this model system.

there is a paucity of information available about developmentally regulated expression of specific nuclear transport machinery. To redress this, we performed a comprehensive survey of the five known murine importin α genes[27] and two importin β genes,[38] determining their expression pattern in the juvenile and adult testis using a combination of *in situ* hybridization, Northern blotting, and reverse transcription-PCR. This survey demonstrated that each importin examined has a different pattern of expression, thereby supporting our hypothesis that nuclear transport of specific cargoes, mediated through regulated expression of nuclear transport proteins, is a feature of discrete stages of spermatogenesis.

We have extended these findings by performing yeast two-hybrid analysis to identify binding partners for specific importin proteins in the adult mouse testis. To date our screen with full-length importin α2 protein has detected at least eight different candidate cargo proteins, each of which has been validated in yeast. For one of these, coilin, a component of coiled/Cajal bodies,[28] the interaction has been further validated in a mammalian cell line (FIG. 4). Coiled bodies are small, round nuclear inclusions that can associate with histone genes. Since coilin can bind to U7 small nuclear ribonucleoprotein, it may facilitate histone pre-mRNA processing by recruiting RNA processing factors to the coiled bodies.[29] Using the yeast two-hybrid approach, we expect to identify interactions between proteins that make up the nuclear transport proteins and their cargo, which are of functional importance during the highly regulated switches that characterize spermatogenesis.

CONCLUSION

Investigations of activin biology have revealed its capacity to influence both fetal and neonatal rodent testis development. Our ongoing analyses of *in vivo* models indicate that local activin production is precisely regulated and differentially affects germ cell and somatic cell maturation. We have previously demonstrated the upregulation of activin inhibitors specifically at the onset of spermatogenesis. However, our emerging knowledge of the complexity of TGF-β superfamily signaling interactions leads us to predict that a complex local network of factors governs this process. In exploring the potential role of regulated nuclear import in driving spermatogenesis, we have illustrated shifting patterns of the importin α gene expression that accompanies male germ cell development. This information can be used to identify proteins that may participate in the key developmental switches that characterize the precisely ordered processes required for production of fertile spermatozoa.

ACKNOWLEDGEMENTS

We would like to thank Professor David de Kretser for his many contributions to the study of testicular biology that have inspired our investigations of activin biology. Funding for this work has been provided in Australia by the NHMRC as fellowships to K.L.L. (143792) and D.A.J. (334013), program funding to K.L.L. (143710); and program funding to S.M. (241000); the Australian Research Council (K.L.L., C.H., C.I., A.H., J.L., and D.A.J.), and Monash University Postgraduate Scholarships (C.H. and C.I.). Funding to C.W.B. was provided by HD01156 and HD27823, the Robert Wood Johnson Foundation, and a Burroughs Wellcome Fund Career Award in the Biomedical Sciences.

REFERENCES

1. EDDY, E.M. 2002. Male germ cell gene expression. Rec. Prog. Horm. Res. **57:** 103–128.
2. DE KRETSER, D.M., J.J. BUZZARD, Y. OKUMA, *et al.* 2004. The role of activin, follistatin and inhibin in testicular physiology. Mol. Cell. Endocrinol. **225:** 57–64.
3. LOVELAND, K.L. & G.R. HIME. 2005. TGFβ superfamily members in spermatogenesis: setting the stage for fertility in mouse and *Drosophila.* Cell Tissue Res. **322:** 141–146.
4. LOVELAND, K.L. & D.M. ROBERTSON. 2005. The TGFβ superfamily in Sertoli cell biology. *In* Sertoli Cell Biology. M. Griswold & M. Skinner, Eds. :227–247. Elsevier Science. New York.
5. GLISTER, C., C.F. KEMP & P.G. KNIGHT. 2004. Bone morphogenetic protein (BMP) ligands and receptors in bovine ovarian follicle cells: actions of BMP-4, -6 and -7 on granulosa cells and differential modulation of Smad-1 phosphorylation by follistatin. Reproduction **127:** 239–254.
6. SCHNEYER, A., A. SCHOEN, A. QUIGG & Y. SIDIS. 2003. Differential binding and neutralization of activins A and B by follistatin like-3 (FSTL-3/FSRP/FLRG). Endocrinology **144:** 1671–1674.
7. ONICHTCHOUK, D., Y.G. CHEN, R. DOSCH, *et al.* 1999. Silencing of TGF-β signalling by the pseudoreceptor BAMBI. Nature **401:** 480–485.
8. MIYAZAWA, K., M. SHINOZAKI, T. HARA, *et al.* 2002. Two major Smad pathways in TGF-β superfamily signaling. Genes Cells **7:** 1191–1204.
9. DERYNCK, R. & Y.E. ZHANG. 2003. Smad-dependent and Smad-independent pathways in TGF-β family signaling. Nature **425:** 577–583.
10. CHANG, H., C.W. BROWN & M.M. MATZUK. 2002. Genetic analysis of the mammalian transforming growth factor-β superfamily. Endocr. Rev. **23:** 787–823.
11. JORGEZ, C., Y.-C. LIN & M.M. MATZUK. 2003. Genetic manipulations to study reproduction. Mol. Cell Endocrinol. **234:** 127–135.
12. MEEHAN, T., S. SCHLATT, M.K. O'BRYAN, *et al.* 2000. Activin A and follistatin in the developing rat testis: localisation and regulation of Sertoli and germ cell development. Dev. Biol. **220:** 225–237.
13. LOVELAND, K.L., M. BAKKER, T. MEEHAN, *et al.* 2003. Expression of *Bambi* in juvenile and adult rat tissues is widespread and regulated in male germ cells. Endocrinology **144:** 4180–4186.
14. MATZUK, M.M, T.R. KUMAR, A. VASSALLI, *et al.* 1995. Functional analysis of activins during mammalian development. Nature **374:** 354–356.
15. BUZZARD J.J., K.L. LOVELAND, M.K. O'BRYAN *et al.* 2004. Changes in circulating and testicular levels of inhibin A and B and activin A during post-natal development in the rat. Endocrinology **145:** 3532–3541.
16. BOITANI, C., M. STEFANINI, A. FRAGALE & A.R. MORENA. 1995. Activin stimulates Sertoli cell proliferation in a defined period of rat testis development. Endocrinology **136:** 5438–5444.
17. FRAGALE, A., R. PUGLISI, A.R. MORENA, *et al.* 2001. Age-dependent activin receptor expression pinpoints activin A as a physiological regulator of rat Sertoli cell proliferation. Mol. Hum. Reprod. **7:** 1107–1114.
18. BUZZARD, J.J., P.G. FARNWORTH, D.M. DE KRETSER, *et al.* 2003. Proliferative phase Sertoli cells display a developmentally regulated response to activin in vitro. Endocrinology **144:** 474–483.
19. ORTH, J.M., G.L. GUNSALUS & A.A. LAMPERTI. 1988. Evidence from Sertoli cell-depleted rats indicates that spermatid number in adults depends on numbers of Sertoli cells produced during perinatal development. Endocrinology **122:** 787–794.

20. BROWN, C.W., D.E. HOUSTON-HAWKINS, T.K. WOODRUFF & M.M. MATZUK. 2000. Insertion of *Inhbb* into the *Inhba* locus rescues the *Inhba-null* phenotype and reveals new activin functions. Nat. Genet. **25:** 453–455.
21. PRINT, C. & K.L. LOVELAND. 2000. Germ cell suicide: recent insights into the regulation of spermatogenesis by apoptosis. Bioessays **22:** 423–430.
22. XU, J., A.R. BEYER, W.H. WALKER & E.A. MCGEE. 2003. Developmental and stage-specific expression of Smad2 and Smad3 in rat testis. J. Androl. **24:** 192–200.
23. JANS, D.A., C.Y. XIAO & M.H. LAM. 2000. Nuclear targeting signal recognition: a key control point in nuclear transport? BioEssays **22:** 532–544.
24. HOGARTH, C., C. ITMAN, D.A. JANS & K.L. LOVELAND. 2005. Regulated nucleocytoplasmic transport in spermatogenesis: a driver of cellular differentiation? BioEssays **27:** 1011–1025.
25. GOLDFARB, D.S., A.H. CORBETT, D.A. MASON, *et al.* 2004. Importin alpha: a multipurpose nuclear-transport receptor. Trends Cell Biol. **14:** 505–514.
26. HAREL, A. & D.J. FORBES. 2004. Importin beta: conducting a much larger cellular symphony. Mol. Cell **16:** 319–330.
27. HOGARTH, C., S. CALANNI, D.A. JANS & K.L. LOVELAND. 2005. Importin α mRNAs have distinct expression profiles during spermatogenesis. Dev. Dyn. **235:** 253–262.
28. ANDRADE, L.E.C., E.K.L. CHAN, I. RASKA, *et al.* 1991. Human autoantibody to a novel protein of the nuclear coiled body: immunological characterization and cDNA cloning of p80-coilin. J. Exp. Med. **173:** 1407–1419.
29. BELLINI, M. & J.G. GALL. 1998. Coilin can form a complex with the U7 small nuclear ribonucleoprotein. Mol. Biol. Cell **9:** 2987–3001.
30. LAWSON, K.A., N.R. DUNN, B.A. ROELEN, *et al.* 1999. *Bmp4* is required for the generation of primordial germ cells in the mouse embryo. Genes Dev. **13:** 424–436.
31. YING, Y., X.M. LIU, A. MARBLE, *et al.* 2000. Requirement of Bmp8b for the generation of primordial germ cells in the mouse. Mol. Endocrinol. **14:** 1053–1063.
32. RICHARDS, A.J., G.C. ENDERS & J.L. RESNICK. 1999. Activin and TGFβ limit murine primordial germ cell proliferation. Dev. Biol. **207:** 470–475.
33. PELLEGRINI, M., P. GRIMALDI, P. ROSSI, *et al.* 2003. Developmental expression of BMP4/ALK3/SMAD5 signaling pathway in the mouse testis: a potential role of BMP4 in spermatogonia differentiation. J. Cell Sci. **116:** 3363–3372.
34. NAGANO, M., B.Y. RYU, C.J. BRINSTER, *et al.* 2003. Maintenance of mouse male germ line stem cells in vitro. Biol. Reprod. **68:** 2207–2214.
35. ZHAO, G.Q., K. DENG, P.A. LABOSKY, *et al.* 1996. The gene encoding bone morphogenetic protein 8B is required for the initiation and maintenance of spermatogenesis in the mouse. Genes Dev. **10:** 1657–1669.
36. ZHAO, G.Q., L. LIAW & B.L.M. HOGAN. 1998. Bone morphogenetic protein 8A plays a role in the maintenance of spermatogenesis and the integrity of the epididymis. Development **125:** 1103–1112.
37. ZHAO, G.Q., Y.X. CHEN, X.M. LIU, *et al.* 2001. Mutation in *Bmp7* exacerbates the phenotype of *Bmp8a* mutants in spermatogenesis and epididymis. Dev. Biol. **240:** 212–222.
38. LOVELAND, K.L., C. HOGARTH, A. SZCZEPNY, *et al.* 2005. Developmental regulation of importins β1 and β3 in rodent testis. Biol. Reprod. (Eub ahead of print.)

Role of Nuclear Receptors in *INSL3* Gene Transcription in Leydig Cells

JACQUES J. TREMBLAY[a,b] AND NICHOLAS M. ROBERT[a]

[a]*Ontogeny-Reproduction Research Unit, CHUL Research Centre, Ste-Foy, Québec, Canada*

[b]*Centre for Research in Biology of Reproduction, Department of Obstetrics and Gynecology, Faculty of Medicine, Université Laval, Ste-Foy, Québec, Canada*

ABSTRACT: Insulin-like 3 (INSL3) is a hormone produced by testicular Leydig cells throughout life. During embryonic life it regulates an essential step of testicular descent, whereas in adults it acts as a male germ cell survival factor. Despite the importance of INSL3 for male sex differentiation and function, very little is known regarding the molecular mechanisms that regulate its expression. So far, the nuclear receptor SF-1 is the only transcription factor known to regulate the mouse Insl3 promoter in Leydig cells. In order to further our understanding of the transcriptional regulation of *INSL3* expression, we have isolated the human INSL3 promoter and tested the effects of the nuclear receptors SF-1, LRH-1, and Nur77 on its activity in Leydig cells. In transfections assays, all three nuclear receptors activated the human INSL3 promoter but especially Nur77, which acted through a novel regulatory element. Thus, the human INSL3 promoter constitutes a novel target for the orphan nuclear receptor Nur77.

KEYWORDS: male sex differentiation; testis descent; germ cell survival; NGFI-B; NR4A1; NBRE; relaxin-like factor

INTRODUCTION

Insulin-like 3 (INSL3), also known as relaxin-like factor, belongs to the insulin-IGF-relaxin family of growth factors and hormones.[1,2] In males, it is specifically produced by testicular Leydig cells from the onset of gonadal development throughout adulthood, whereas in females it is only secreted by luteal cells of the adult ovary.[3,4]

INSL3 plays important roles in male and female reproductive function. In males, INSL3 is involved in two very different, but critical, processes. During fetal life, it regulates an essential step of testicular descent as revealed by gene inactivation experiments in the mouse: Insl3$^{-/-}$ null mice had bilateral cryptorchid testes located high in the abdominal cavity close to the kidneys.[5,6] In adult life, INSL3 was recent-

Address for correspondence: Dr. Jacques J. Tremblay, Ontogeny-Reproduction, Room T1-49, CHUL Research Centre, 2705 Laurier Blvd., Ste-Foy, Québec, Canada G1V 4G2. Voice: 418-656-4141 ext. 46254; fax: 418-654-2765.
Jacques-J.Tremblay@crchul.ulaval.ca

Ann. N.Y. Acad. Sci. 1061: 183–189 (2005). © 2005 New York Academy of Sciences.
doi: 10.1196/annals.1336.020

ly shown to act as a survival factor for testicular germ cells.[7] In adult females, INSL3 is involved in oocyte maturation.[7]

Despite its crucial role in mammalian reproductive function, very little is known about the mechanisms that regulate *INSL3* expression. To date, the nuclear receptor SF-1 is the only transcription factor that has been reported to regulate the mouse Insl3 promoter. Three binding sites for SF-1 were identified in the proximal region of mouse and rat Insl3 promoters.[8,9] The involvement of SF-1 in mouse Insl3 expression was also assessed *in vivo*. In Leydig cell-specific Sf1 null mice, Jeyasuria *et al.* showed that testicular descent was affected, but it was not completely abrogated like in Insl3$^{-/-}$ mice.[5,6,10] The testes partially descended to the level of the bladder in Leydig cell-specific Sf1 null mice, indicating that these animals retain the ability to produce Insl3, albeit at a lower level.[10] Therefore, other transcription factors are involved in the initiation and maintenance of INSL3 expression.

Nur77 (NGFI-B, TR3, NR4A1) is the prototypic member of the NR4A family of orphan nuclear receptors, which also includes Nurr1 (NR4A2) and Nor1 (NR4A3). Transcription factors belonging to the NR4A family are known to be immediate-early response factors in several tissues, including steroidogenic Leydig cells.[11,12] In these cells, Nur77 is expressed at low levels, and its expression is rapidly and robustly induced in response to LH and cAMP analogs.[11,12] In the past few years, members of the NR4A family of nuclear receptors, particularly Nur77 and Nurr1, have received increased attention as novel regulators of basal and hormone-induced gene transcription in testicular Leydig cells. So far, Nur77 has been shown to activate the promoters of several genes involved in testosterone biosynthesis in Leydig cells. These include the human HSD3B2, rat Cyp17, mouse HSD3B1, and mouse StAR promoters.[11,13,14]

Here we report the results of a study on the role of three nuclear receptors, SF-1, LRH-1, and Nur77, in human INSL3 (hINSL3) promoter activity in Leydig cells. Although the hINSL3 promoter was activated by all three nuclear receptors, Nur77 was a stronger activator than SF-1 and LRH-1.

RESULTS AND DISCUSSION

Activation of hINSL3 Promoter Activity by Nuclear Receptors

Our lab is interested in the role played by the orphan nuclear receptor Nur77 (NGFI-B/NR4A1) in Leydig cell gene expression. Nur77 generally binds DNA as a monomer to an element called the Nur77-binding response element (NBRE) (AAAAGGTCA) which is highly related to the SF-1 binding site.[15] Since the rodent Insl3 promoter contains three SF-1 binding sites, we hypothesized that Nur77 could also regulate this promoter. As a first step to test this hypothesis, we have isolated a 1.1-kb promoter fragment of the *hINSL3* gene. Through sequence analysis, we found that the hINSL3 promoter contains only one (at bp −56) of the three SF-1 binding sites found in the rodent INSL3 promoter. Interestingly, we have identified a novel consensus sequence for the binding of Nur77 at position bp −100 that is conserved across species. To test the possibility that Nur77 could be involved in hINSL3 transcription, we performed transient transfections in MA-10 Leydig cells, a homologous cell line known to endogenously express INSL3.[3] As shown in FIGURE 1, we

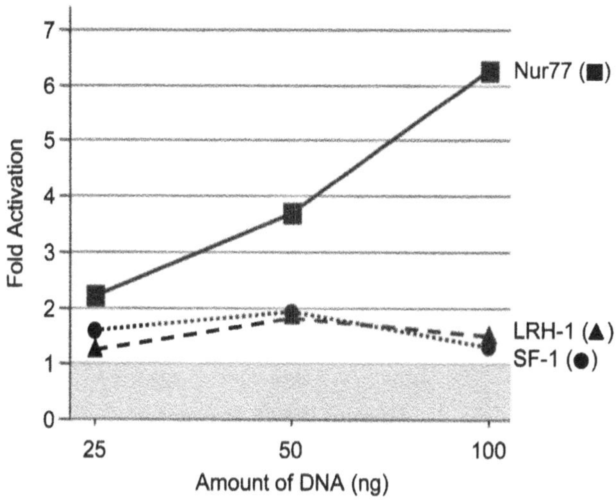

FIGURE 1. Activation of the hINSL3 promoter by nuclear receptors. MA-10 mouse Leydig cells were cotransfected with a bp −1137 to bp +11 hINSL3 promoter construct and either an empty expression vector or increasing doses (25, 50, and 100 ng) of expression vectors for Nur77 (*squares*), SF-1 (*circles*), and LRH-1 (*triangles*), as indicated. The *grey zone* corresponds to the activity of the hINSL3 reporter in the presence of the empty expression vector (control), which was arbitrarily set at 1. Results are shown as the fold activation over control.

found that Nur77 could activate the hINSL3 promoter fragment in a dose-dependent manner (up to 6-fold), whereas SF-1 and LRH-1, another nuclear receptor closely related to SF-1, were poor activators (1.8-fold). These results suggest a novel role for Nur77 in hINSL3 transcription. This is also supported by the fact that Nur77 is coexpressed with INSL3 within the mammalian testis and ovary as well as in Leydig cell lines.[11,12,16,17]

Nur77 Specifically Acts through a Novel NBRE

Sequence analysis of the hINSL3 promoter revealed the presence of a putative NBRE at bp −100. To test whether Nur77-dependent activation of the hINSL3 promoter was mediated through this NBRE, we generated a series of 5′-progressive deletions of the hINSL3 promoter. They were then tested for Nur77-responsiveness in MA-10 cells (FIG. 2). A deletion construct to bp −322 that retained the putative NBRE at bp −100 and the SF-1 element at bp −56 was still activated by Nur77. Deletion to bp −93, which removed the NBRE while retaining the SF-1 element (bp −56), however, completely abrogated transactivation of the hINSL3 promoter by Nur77. Thus, these results indicate that the Nur77-dependent activation of the hINSL3 promoter requires the region from bp −322 to −93, which contains the putative NBRE. The SF-1 element at bp −56 is not required for hINSL3 promoter activation by Nur77. The importance of this novel NBRE in Nur77-mediated hINSL3 promoter activation was confirmed by site-directed mutagenesis in the context of the bp −1137 hINSL3 reporter; mutation of the bp −100 NBRE (AAAAGGTCA →

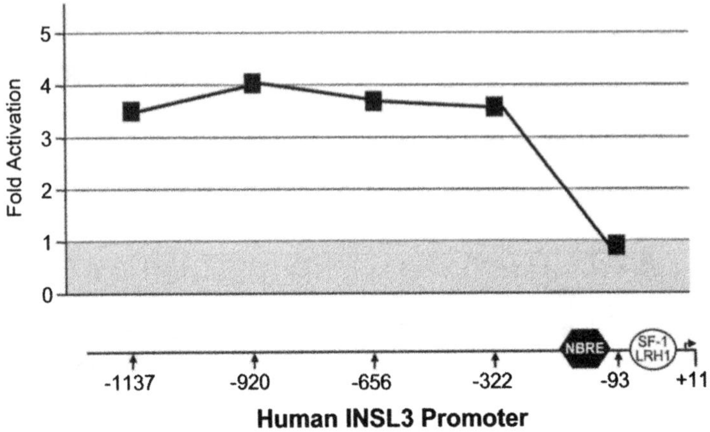

FIGURE 2. Mapping of the Nur77-responsive element in the hINSL3 promoter. MA-10 Leydig cells were cotransfected with various 5′ deletion constructs of the hINSL3 promoter (the 5′-end point of each construct is indicated under the graph) with either 100 ng of an empty expression (*grey zone*) or an expression vector for Nur77 (*square*). The positions of the two nuclear receptor binding sites are indicated (*octagon* for Nur77 and *circle* for SF-1). Results are shown as the fold activation over control.

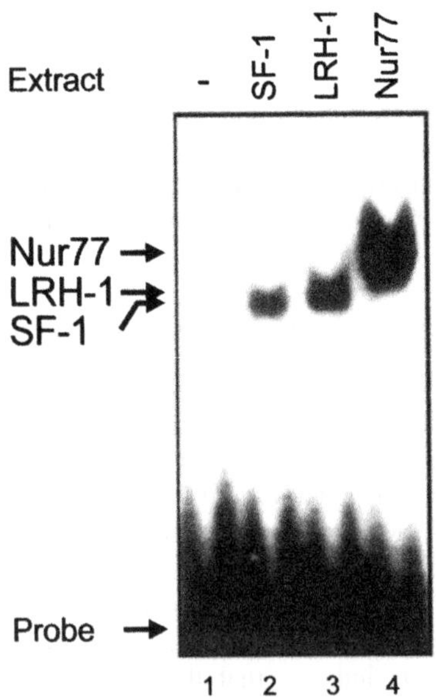

FIGURE 3. The novel NBRE element is a Nur77-specific binding site. EMSA was used to assess the binding of *in vitro*-produced Nur77 to a double-stranded ^{32}P-labeled oligonucleotide corresponding to the NBRE element at bp −100 of the hINSL3 promoter. Binding of *in vitro*-synthesized SF-1 and LRH-1 was also assessed.

AAAATTTCA) completely abolished hINSL3 promoter activation by Nur77 in MA-10 cells (data not shown).

The DNA binding of Nur77 to the novel NBRE element (bp −100) was next assessed by electromobility shift assay (EMSA). Consistent with the specific requirement for the NBRE element in Nur77-mediated activation of the hINSL3 promoter (FIG. 2), *in vitro*-translated Nur77 bound with high affinity to the NBRE element (FIG. 3, lane 4). On the other hand, SF-1 and LRH-1 could only bind with low affinity to this element (FIG. 3, lanes 2 and 3). This is consistent with the fact that although Nur77 and SF-1/LRH-1 bind as monomers to similar DNA elements, they have different nucleotide requirements in the 5′ extension of the binding site.[15] For instance, Nur77 prefers AAAAGGTCA, whereas SF-1/LRH-1 binds with high affinity to TCAAGGTCA.[15] Thus, these results indicate that the novel NBRE at bp −100 of the hINSL3 promoter is preferentially bound by Nur77.

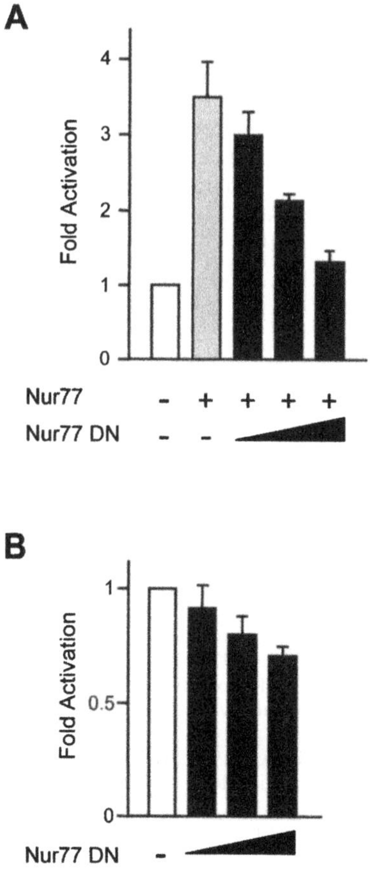

FIGURE 4. A Nur77 dominant negative competitor represses hINSL3 promoter activity in Leydig cells. **(A)** A truncated Nur77 protein acts as a dominant negative competitor. MA-10 Leydig cells were cotransfected with a bp −1137 to bp +11 hINSL3 promoter construct and either an empty expression vector (Ctl; *open bar*) or an expression vector (100 ng) for Nur77 in the absence (*grey bar*) or the presence of increasing doses of an expression vector encoding a truncated Nur77 protein (20, 50, and 100 ng; *solid triangles*). Results are shown as the fold activation over control. **(B)** MA-10 Leydig cells were cotransfected with the bp −1137 to bp +11 hINSL3 promoter construct and either an empty expression vector (Ctl; *open bar*) or increasing doses of an expression vector encoding the Nur77 dominant negative protein (20, 50, and 100 ng; *solid triangles*). Results are shown as relative activity (± standard error of the mean); the activity of the hINSL3 reporter cotransfected with the control (empty) vector was arbitrarily set to 1.

Modulation of Endogenous Nur77 Activity in Leydig Cells Affects hINSL3 Promoter Activity

To assess further the role of Nur77 in hINSL3 expression, we have generated a truncated Nur77 protein (amino acids 252–601) that lacks the N-terminal activation domain. Since the truncated Nur77 protein retains its zinc finger region, it could still bind DNA. However, since the truncated Nur77 lacked a transactivation domain, it could no longer activate the hINSL3 promoter. The Nur77 truncated protein antagonized the transactivation mediated by the wild-type Nur77 protein (FIG. 4A). Thus, the truncated Nur77 protein behaves as a dominant negative competitor. When overexpressed in MA-10 Leydig cells, the Nur77 dominant negative protein decreased, in a dose-dependent manner, the activity of the wild-type bp −1137 hINSL3 promoter, indicating that endogenous Nur77 is required for full promoter activity in these cells (FIG. 4B).

CONCLUSION

In the present study, we have identified the promoter of the human *INSL3* gene, encoding a key hormone essential for male and female reproductive function, as a novel target for Nur77. Nur77-mediated activation of the hINSL3 promoter is entirely dependent on a novel NBRE located at bp −100 and does not require the SF-1 element at bp −56. The Nur77 orphan nuclear receptor has received increasing consideration in recent years as an important regulator of steroidogenic enzyme-encoding genes (Cyp17, HSD3B2, and aromatase) in Leydig cells.[11,14,18] Thus, our data add to the list of Nur77 targets in Leydig cells a hormone-encoding gene (INSL3) and provide further evidence for a role for Nur77 in mammalian reproductive function.

ACKNOWLEDGMENTS

J.J.T. is a New Investigator of the Canadian Institutes of Health Research (CIHR). This work was supported by CIHR grant number MOP-62752 to J.J.T.

REFERENCES

1. ADHAM, I.M., E. BURKHARDT, M. BENAHMED & W. ENGEL. 1993. Cloning of a cDNA for a novel insulin-like peptide of the testicular Leydig cells. J. Biol. Chem. **268:** 26668–26672.
2. PUSCH, W., M. BALVERS & R. IVELL. 1996. Molecular cloning and expression of the relaxin-like factor from the mouse testis. Endocrinology **137:** 3009–3013.
3. BALVERS, M., A.N. SPIESS, R. DOMAGALSKI *et al.* 1998. Relaxin-like factor expression as a marker of differentiation in the mouse testis and ovary. Endocrinology **139:** 2960–2970.
4. ZIMMERMANN, S., P. SCHOTTLER, W. ENGEL & I.M. ADHAM. 1997. Mouse Leydig insulin-like (Ley I-L) gene: structure and expression during testis and ovary development. Mol. Reprod. Dev. **47:** 30–38.
5. NEF, S. & L.F. PARADA. 1999. Cryptorchidism in mice mutant for Insl3. Nat. Genet. **22:** 295–299.

6. ZIMMERMANN, S., G. STEDING, J.M. EMMEN *et al.* 1999. Targeted disruption of the Insl3 gene causes bilateral cryptorchidism. Mol. Endocrinol. **13:** 681–691.
7. KAWAMURA, K., J. KUMAGAI, S. SUDA *et al.* 2004. Paracrine regulation of mammalian oocyte maturation and male germ cell survival. Proc. Natl. Acad. Sci. USA **101:** 7323–7328.
8. KOSKIMIES, P., J. LEVALLET, P. SIPILA *et al.* 2002. Murine relaxin-like factor promoter: functional characterization and regulation by transcription factors steroidogenic factor 1 and DAX-1. Endocrinology **143:** 909–919.
9. ZIMMERMANN, S., A. SCHWARZLER, S. BUTH *et al.* 1998. Transcription of the Leydig insulin-like gene is mediated by steroidogenic factor-1. Mol. Endocrinol. **12:** 706–713.
10. JEYASURIA, P., Y. IKEDA, S.P. JAMIN *et al.* 2004. Cell-specific knockout of steroidogenic factor 1 reveals its essential roles in gonadal function. Mol. Endocrinol. **18:** 1610–1619.
11. MARTIN, L.J. & J.J. TREMBLAY. 2005. The human 3β-hydroxysteroid dehydrogenase/Δ^5-Δ^4 isomerase type 2 promoter is a novel target for the immediate early orphan nuclear receptor Nur77 in steroidogenic cells. Endocrinology **146:** 861–869.
12. SONG, K.H., J.I. PARK, M.O. LEE *et al.* 2001. LH induces orphan nuclear receptor Nur77 gene expression in testicular Leydig cells. Endocrinology **142:** 5116–5123.
13. HONG, C.Y., J.H. PARK, R.S. AHN *et al.* 2004. Molecular mechanism of suppression of testicular steroidogenesis by proinflammatory cytokine tumor necrosis factor alpha. Mol. Cell. Biol. **24:** 2593–2604.
14. ZHANG, P. & S.H. MELLON. 1997. Multiple orphan nuclear receptors converge to regulate rat P450c17 gene transcription: novel mechanisms for orphan nuclear receptor action. Mol. Endocrinol. **11:** 891–904.
15. WILSON, T.E., T.J. FAHRNER & J. MILBRANDT. 1993. The orphan receptors NGFI-B and steroidogenic factor 1 establish monomer binding as a third paradigm of nuclear receptor-DNA interaction. Mol. Cell. Biol. **13:** 5794–5804.
16. GE, R.S., Q. DONG, C.M. SOTTAS *et al.* 2005. Gene expression in rat Leydig cells during development from the progenitor to adult stage: a cluster analysis. Biol. Reprod. **72:** 1405–1415.
17. PARK, J.I., H.J. PARK, H. CHOI *et al.* 2001. Gonadotropin regulation of NGFI-B messenger ribonucleic acid expression during ovarian follicle development in the rat. Endocrinology **142:** 3051–3059.
18. WU, Y., S. GHOSH, Y. NISHI *et al.* 2005. The orphan nuclear receptors NURR1 and NGFI-B modulate aromatase gene expression in ovarian granulosa cells: a possible mechanism for repression of aromatase expression upon luteinizing hormone surge. Endocrinology **146:** 237–246.

Analysis of the Human Sperm Proteome

DANIEL S. JOHNSTON,[a] JOE WOOTERS,[b] GREGORY S. KOPF,[a]
YONGCHANG QIU,[b] AND KENNETH P. ROBERTS[c]

[a]*Contraception, Women's Health & Musculoskeletal Biology, Woman's Health Research Institute, Wyeth Research, Collegeville, Pennsylvania, USA*

[b]*Biological Technologies, Molecular Profiling and Biomarker Discovery, Wyeth Research, Cambridge, Massachusetts, USA*

[c]*Department of Urologic Surgery, University of Minnesota, Minneapolis, Minnesota, USA*

ABSTRACT: As part of our effort to identify putative protein targets for the development of male contraceptives, we performed an in-depth proteomic analysis of human sperm by liquid chromatography and tandem mass spectrometry. Motile sperm were collected from a single fertile individual and fractionated into detergent-soluble and detergent-insoluble fractions. Sodium dodecyl sulfate–polyacrylamide gel electrophoresis separation of these fractions, followed by manual cutting of the gel, yielded 35 gel sections for each fraction to include proteins across the full range of electrophoretic mobility. Proteomic analysis of these gel sections identified more than 1,760 proteins with high confidence, with 1,350 proteins identified in the soluble fraction, 719 identified in the insoluble fraction, and 309 identified in both fractions. This characterization of the human sperm proteome provides a high-resolution, physiologically relevant index of the proteins that comprise human sperm.

KEYWORDS: sperm; proteome; nonhormonal contraception; male gamete

INTRODUCTION

While numerous forms of contraception are available worldwide, it is clear that there is no single method that fulfills the requirements, preferences, and intentions of all individuals during their reproductive lifetime.[1] The most effective contraceptive option for men, vasectomy, is more accurately described as sterilization, with an approximately 50% chance of reversibility by vasovasostomy.[2,3] The remaining options for male-initiated methods of contraception are estimated to account for 37% of all contraceptive use,[4] even though choices are limited to withdrawal and condoms. A clear gap exists in the market for safe, effective, convenient, and reversible male-based contraception, a fact that has attracted the attention of the pharmaceutical industry. Hormone-based approaches to male contraception have been investigated for the past 30 years[5,6] and have recently gained significant attention

Address for correspondence: Dr. Daniel S. Johnston, Wyeth Research, 500 Arcola Rd., N3169, Collegeville, PA 19426, USA. Voice: 484-865-2050; fax: 484-865-9367.
johnstd2@wyeth.com

**Ann. N.Y. Acad. Sci. 1061: 190–202 (2005). © 2005 New York Academy of Sciences.
doi: 10.1196/annals.1336.021**

from both the media and the scientific community.[7–10] However, the potential for untoward effects associated with hormone-based therapeutics, including acne, weight gain, decrease in testicular volume, decrease in high-density lipoprotein cholesterol,[11] oiliness of skin, and increased hematocrit and hemoglobin concentrations,[12] has driven research towards hormone-independent approaches to male contraception (reviewed in Refs. 13 to 15). This innovative strategy aims to target specific gene products that are required for the production or function of fertile sperm without affecting hormonal balance or somatic processes.

One of the most obvious tissues under investigation for the identification of male-based contraceptive target proteins is the testis. The objective of a contraceptive that acts in the testis would be to disrupt the coordinated process of spermatogenesis to the point where the testis either produces fertilization-incompetent spermatozoa or does not produce a sufficient number of spermatozoa for male fertility. Gene products necessary for meiosis[16–19] or male germ cell-specific events, including formation of the acrosome, histone–protamine exchange, assembly of the flagellum, and membrane organization,[14,20,21] have been shown to be necessary for quantitatively or qualitatively normal spermatogenesis. The recent transcriptional profiling of distinct populations of mouse testicular cells has revealed testis-specific gene products for which expression is limited to the meiotic and postmeiotic phases of spermato-genesis[22] and will facilitate identification of additional testicular targets. Contraceptive agents that could target these male germ cell–specific events may have the lowest chance of disrupting somatic cell processes, with the benefit of eliminating side effects.

A second approach to male-based contraception is the disruption of epididymal function by interfering with post-testicular sperm maturation (reviewed in Refs. 23 to 25). During their transit through the epididymis, sperm acquire forward motility and the ability to undergo capacitation, both of which are required for normal fertility.[26,27] Theoretically, contraceptive agents could act directly on the epididymis to prevent normal sperm transport through the tubule or perturb the luminal environment and prevent sperm maturation.[24,28] As in the testis, transcriptional profiling studies of the mouse epididymis[29] have begun the process of cataloging the gene products that are involved in producing mature sperm (expression data are available at <http://mrg.genetics.washington.edu>). While such studies provide insight into the gene products that may be present in the reproductive organs during spermatogenesis and epididymal maturation, they are unable to verify which of these persist in the ejaculated sperm.

The commonality of both testis- and epididymis-based contraceptive approaches is the eventual disruption of sperm function, suggesting that a viable approach to developing a male-based, nonhormonal contraceptive is to target proteins that are present in the ejaculated sperm cell. Sperm represent a unique, terminally differentiated single-cell population and are both transcriptionally and translationally inactive. Sperm cannot respond to their environment by *de novo* gene expression or protein synthesis but rather are restricted to using the protein repertoire present at the time of ejaculation to carry out all of the processes required for fertilization (capacitation, hyperactivated motility, and acrosomal exocytosis, and interactions with the uterus, oviduct, zona pellucida, or egg plasma membrane).[30–44] Because of the transcriptional/translational inactivity of sperm, a cataloging and subsequent bioinformatic analysis of the human sperm proteome represents an opportunity to obtain a global understanding of sperm function.

METHODS

Preparation of Human Sperm Proteins

A semen sample was produced by masturbation from a normal fertile individual with normal semen parameters and collected into a sterile sample container. Fertility was established by paternity within the past 12 months. The semen sample was incubated at 37°C until liquefaction was complete (~15 minutes). Sperm number was determined by hemacytometer. One milliliter of liquefied semen (~300 million sperm) was overlaid onto Isolate® (Irvine Scientific, Santa Ana, CA) and centrifuged at 300 × *g* for 20 minutes to select for viable, motile sperm. The pellet was washed by resuspending in BWW medium (Irvine Scientific) and pelleted at 1,000 × *g* for 5 minutes. The sperm were resuspended in 1 ml of 40 mM OBG (1-*O*-*n*-octyl-beta-D-glucopyranoside) and extracted for 1 hour with gentle rocking. The OBG-soluble fraction (membrane and cytosolic proteins) was separated by centrifugation from the insoluble components at 12,000 × *g* for 10 minutes. The pellet was washed twice with 1 mM Tris (pH 7.2) and resuspended in sodium dodecyl sulfate (SDS) sample buffer. The OBG-soluble supernatant was buffer-exchanged to 1 mM Tris, pH 7.2, and equilibrated in 1× SDS sample buffer.

One-Dimensional SDS-Polyacrylamide Gel Electrophoresis (1D SDS-PAGE)

Samples were run on 8- to 18% Laemmli gels and stained with the SilverQuest kit (Invitrogen, Carlsbad, CA). Each lane was cut into 35 bands based on the individual size of the major protein bands to preserve the original resolution of the gel separation. Proteins in these gel bands were then digested with trypsin using a robot (DigestPro; Abimed, Analysen-Technik GmbH, Langenfeld, Germany).

Nano-LC-MS/MS

The digest solutions were analyzed on a PicoFrit column (New Objective, Woburn, MA) packed with Jupiter C_{18} (5 μm; 75 μm × 12 cm), coupled to a linear ion trap mass spectrometer (LTQ; ThermoFinnigan, San Jose, CA). Peptides were eluted with a 70-minute convex gradient using 0.4% acetic acid–0.005% heptafluorobutyric acid as a modifier. The fragment ion spectra (tandem mass spectrometry [MS/MS] data) were acquired in an MS data-dependent mode in which each MS acquisition was followed by four MS/MS acquisitions of the first four most intense peptide ions in the prior MS spectrum. The dynamic exclusion and isotope exclusion functions in the acquisition software provided by the manufacturer were also employed to increase the number of peptide ions that were analyzed. Typically, more than 10,000 fragmentation ion spectra were collected from each liquid chromatography-MS/MS (LC-MS/MS) analysis. A total of 570,000 MS/MS spectra were collected from the entire analysis.

Protein Identification

Fragmentation spectra were matched against the human, mouse, and rat subsets of the nonredundant NCBI protein database (NCBInr) using the Sequest software package (ThermoQuest). Peptide sequences were loaded into an Oracle database and compiled into protein consensus groups using in-house software called Dynammid. A proprietary scoring scheme was used to obtain a confidence value for each protein. The confidence value embodies the X_{corr}, rank, and number of unique sequences as returned by the Sequest search engine and was computed according to the following equation:

$$PIDscore = \sum_{n=1}^{k} \left(\frac{X_{corr}}{Rank}\right) d ,$$

where *PIDscore* is the Protein Identification score, *k* is the number of peptides satisfying the X_{corr}, rank, and DeltaCn selected by the user and *d* is the number of distinct peptide sequences represented in a given protein. For this computation, only peptides exceeding an X_{corr} of 2.2 and proteins with two or more distinct peptide sequences were considered. The X_{corr} parameter has been analyzed statistically,[45] and the analysis indicates that an X_{corr} of 2.2 represents a 99% confidence level for 1+ and 2+ peptides and exceeds 96% for 3+ peptides. Multiple distinct hits to a protein are expected to contribute additional confidence in a multiplicative fashion, as is reflected in the PID score equation above. Generally, proteins with scores of >100 would be represented by four to six unique sequences and considered very high confidence hits and would be used without further validation. A total of 1,760 sperm proteins were detected with this level of confidence.

Functional Characterization

Annotation of protein cellular localization and biological function were performed using the Gene Ontology tool at the NIH DAVID website <http://david.niaid.nih.gov/david/version2/index.htm>.

RESULTS AND DISCUSSION

Cataloging and Functional Annotation of the Human Sperm Proteome

Many of the proteins present on ejaculated sperm are synthesized in the accessory sex glands, acquired long after the morphological differentiation of the spermatozoon. The methodology used in the current study aims to identify proteins that persist on sperm regardless of their originating organ and selects for viable, motile sperm washed of excess seminal plasma. To provide a degree of subcellular context to the peptides that would be identified, proteins soluble in OBG were extracted from the sperm, yielding two fractions (detergent soluble and insoluble) for separation by SDS-PAGE and LC-MS/MS as described in the METHODS section above.

The detergent-soluble and -insoluble fractions yielded 1,350 and 719 proteins with high confidence, respectively (FIG. 1). Three hundred and nine identified proteins were represented in both fractions (15%), yielding a total of 1,760 distinct proteins. All identified proteins were subjected to gene ontology analysis to categorize the list of proteins based on biological function. As expected, the detergent-soluble fraction is rich in cytoplasmic proteins, signaling molecules, and membrane receptors (FIG. 2A), while the detergent-insoluble fraction is enriched for cytostructural

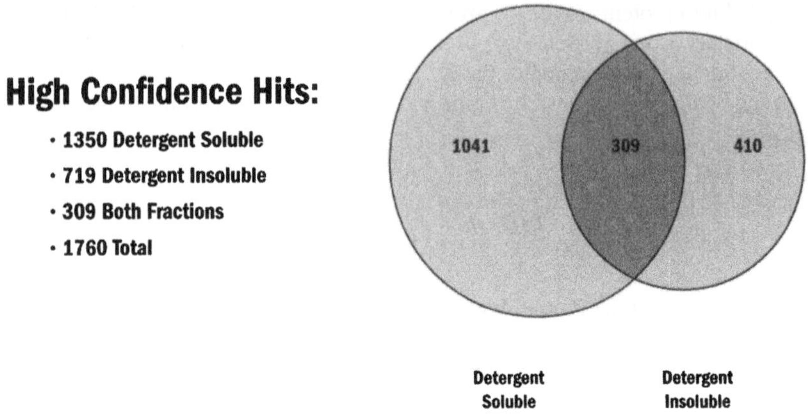

FIGURE 1. Ven diagram showing the number of proteins identified uniquely in the detergent-soluble or detergent-insoluble fractions or in both fractions.

and nuclear chromatin binding proteins (FIG. 2B). The proteome is also highly populated by proteins involved in protein folding and degradation. Heat-shock proteins 70 and 90 are both abundant, and more than 120 identified proteins contain protease domains.

In both fractions, the greatest number of proteins were characterized as either novel proteins or proteins for which no gene ontology is available (FIG. 2A and B). Overall, novel and uncharacterized proteins are more abundant in the detergent-soluble fraction, with many being described only as "putative" or "predicted" translation products. Manual annotation of the proteins for which no gene ontology is available revealed an unusually large number of proteins related to venoms (wasp and snake-like), developmental proteins, heavy metal binding proteins, and archaebacteria-like proteins. This cataloging and functional annotation of the human sperm proteome is the first step in understanding the machinery that sperm possess to facilitate movement through the female reproductive tract and complete fertilization. For the purposes of developing nonhormonal male contraceptives, these findings could provide a number of putative drug targets when their function is elucidated.

Evaluation of Tissue-Specific Expression

Our primary reason for investigating the human sperm proteome was to identify previously uncharacterized proteins that were exclusively expressed in the male reproductive tract. These proteins are of highest interest as targets for male contraceptives. To this end, proteins identified from the proteomic analysis were screened electronically for tissue specificity and selectivity by comparing them to the Wyeth Research databases of U133 (human), MOE430 (mouse), and RAE230 (rat) Affymetrix Gene Chip profiling data from hundreds of intact or microdissected tissues as well as various cell lines. Using these data, we have assembled a list of the proteins whose expression is restricted to the male reproductive tract and whose function may therefore be sperm-specific. One example is zona pellucida binding protein 2

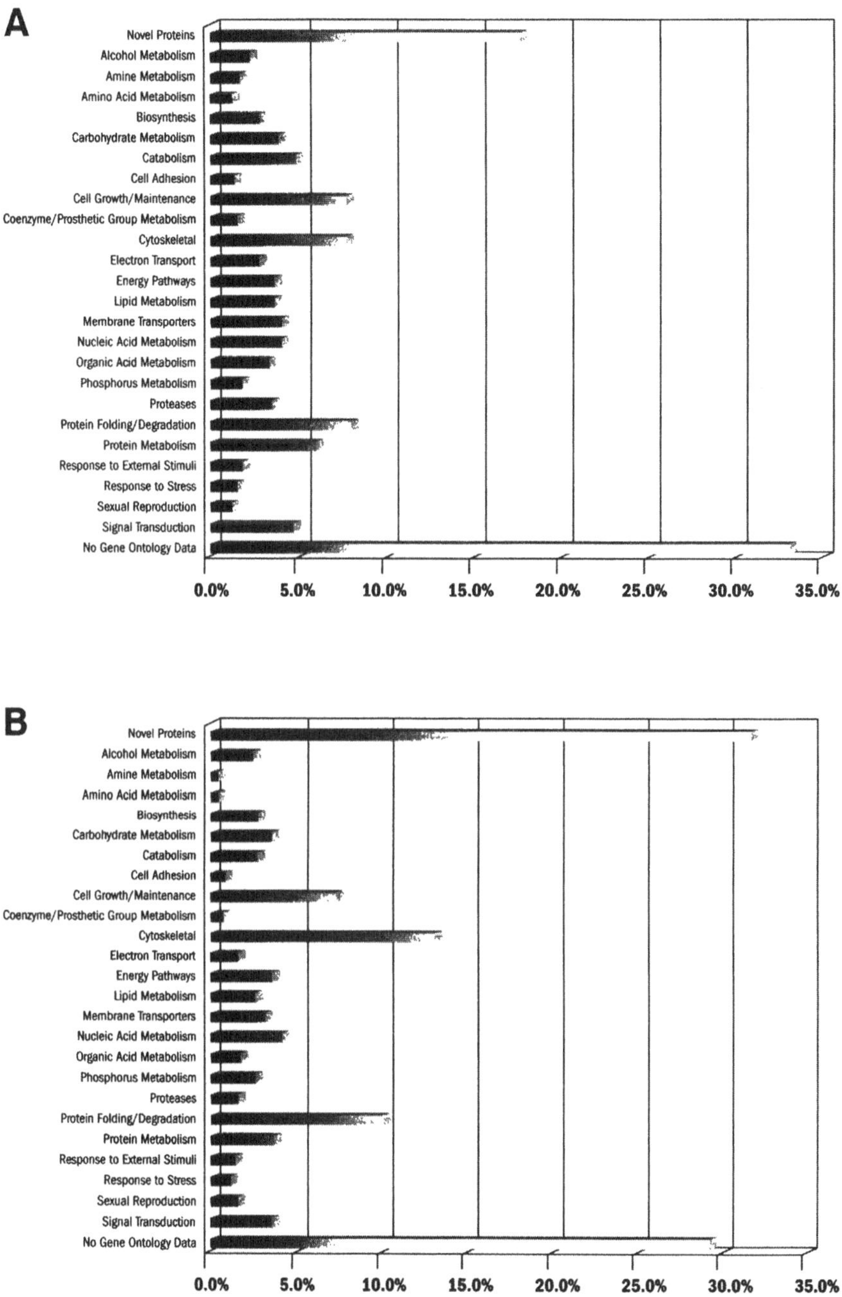

FIGURE 2. Functional composition of the proteins identified in the analysis. **(A)** Functional composition of proteins identified in the soluble fraction. **(B)** Functional composition of compounds identified in the insoluble fraction.

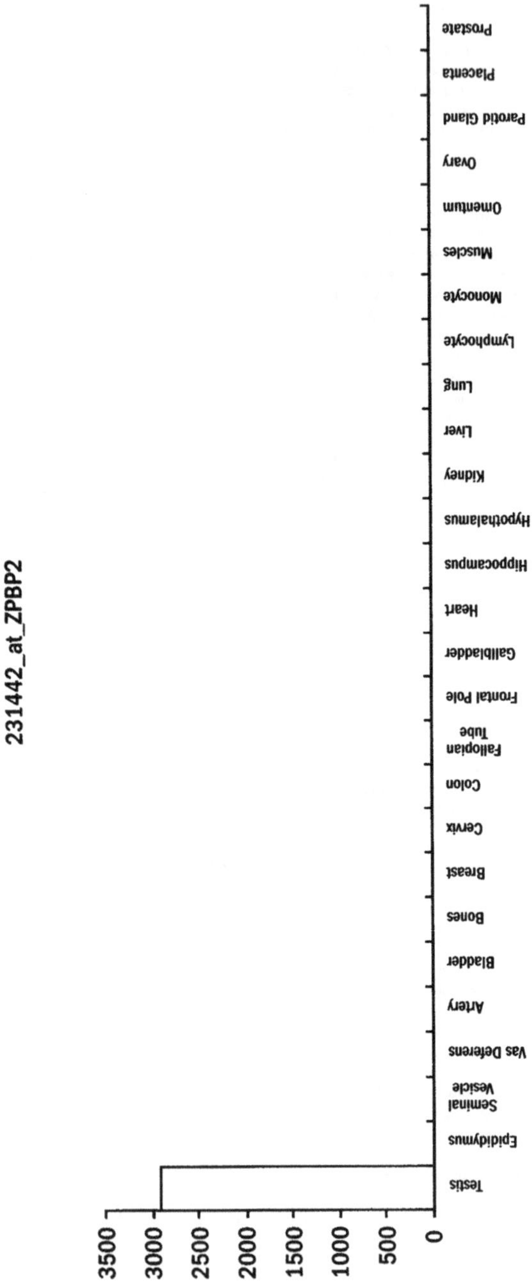

FIGURE 3. Tissue expression profile of zona pellucida binding protein 2. Affymetrix identifier shown in center.

(FIG. 3), for which expression was observed only in the testis. The tissue distributions of these putative targets will be further analyzed using more sensitive methods to authenticate the expression patterns seen by electronic Northern analyses.

Detailed Examination of Individual Proteins Identified in the Analysis

The MS/MS output not only allows for the identification of a protein as present in either the OBG-soluble or OBG-insoluble fraction of the sperm, but it may also be queried to provide detailed information for each protein. An example of this analysis is demonstrated in FIGURE 4 using the data generated for the epididymal protein CRISP-1, starting with the numbers of total and unique tryptic peptides identified in both the soluble and insoluble fractions (FIG. 4A). In the case of CRISP-1, the combined OBG-soluble and -insoluble fractions yielded 12 distinct spectra a total of 36 times. The number of unique spectra for any protein partially depends on the number of lysine and arginine residues that are available for digestion by trypsin. Post-translational modifications, whether biologically relevant (phosphorylation, glycosylation) or experimentally induced (oxidation of methionine, linkage of acrylamide to cysteine), may also influence the number of unique spectra obtained for a particular protein. Each of the peptides identified for a protein may be displayed within the context of the known amino acid sequence. For CRISP-1, identified peptides represent the majority of the protein, including the amino and carboxy termini (FIG. 4B, in grey). However, distinct coverage gaps exist in two regions of the protein. The first, between amino acids 76 and 147, is not detected because there are no lysine or

A. Number of Peptides Identified

	Sol	Insol	Total
# of Peptides	15	21	36
Unique Peptides	12	[illegible]	12

B.

MLSMKKKSARDQFNKLV
TDLPNVQEEIVNIHNALR
RRVVPPASNMLKMSWS
EEAAQNARIFSKYCDMT
ESNPLERRLPNTFCGEN
MHMTSYPVSWSSVIGV
WYSESTSFKHGEWTTTD
DDITTDHYTQIVWATSYLI
GCAIASCRQQGSPRYLY
VCHYCHEGNDPETKNEP
YKTGVPCEACPSNCEDK
LCTNPCIYYDEYFDCDIQ
VHYLGCNHSTTILFCKAT
CLCDTEIK

C.

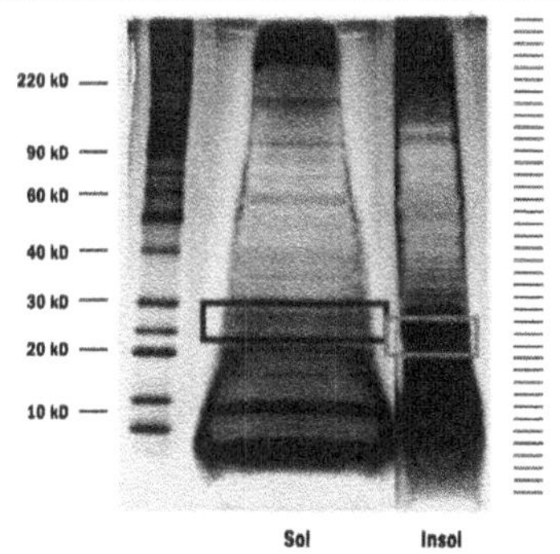

FIGURE 4. Proteomic information acquired for CRISP-1. (**A**) The number of total and unique peptides identified in the soluble and insoluble fractions. (**B**) Mapping of the identified peptides to the amino acid sequence of human CRISP-1. (**C**) Electrophoretic mobility of the CRISP-1 proteins in the detergent-soluble and detergent-insoluble fractions.

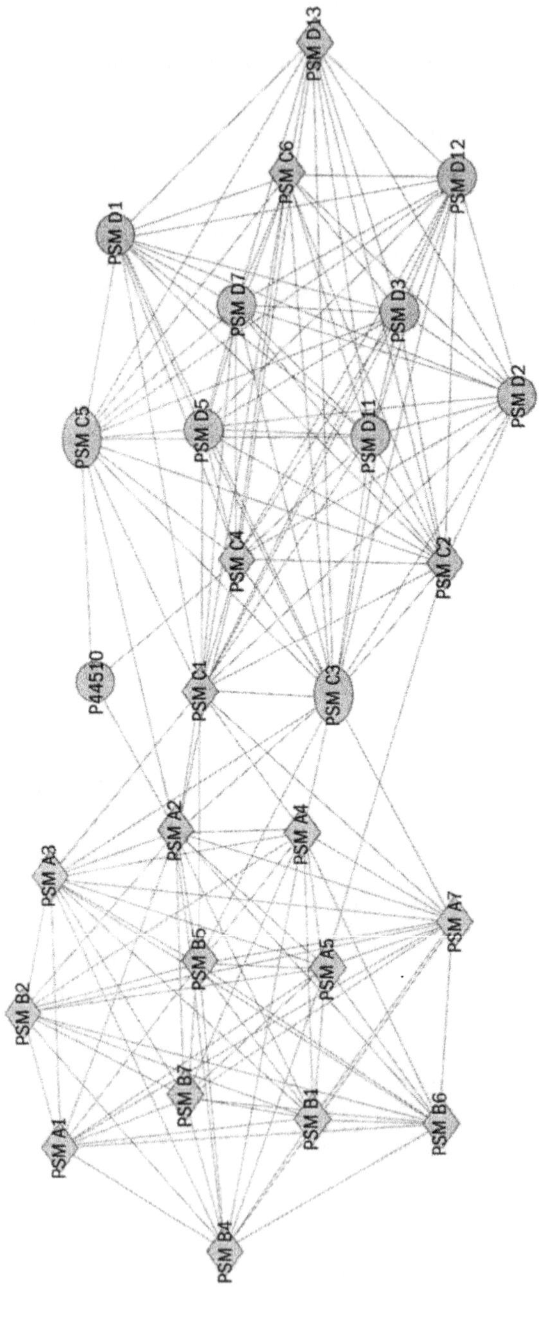

FIGURE 5. Visualization of each of the proteins that comprise a 26S proteosome. *Shaded circles* around each gene identifier indicate that that protein was identified in the analysis.

arginine residues and the resulting peptide is too large to be detected by mass spectrometry analysis. The second region, between residues 170 and 218, has several putative glycosylation sites, and we predict that the lysine and arginine residues in this region were not cleaved because of the presence of glycosylation in these regions.

In addition to identifying which peptides are detected for a given protein, the data may be used to correlate the peptide with the gel band from which it was cut to allow for an estimation of the SDS-PAGE mobility of the whole protein. For CRISP-1 (FIG. 4C), one of the proteins present in both detergent-soluble and detergent-insoluble fractions, the peptides in the detergent-soluble fraction originated from between 28 and 32 kDa. Interestingly, CRISP-1 peptides in the OBG-insoluble fragment came from the region of the gel with an apparent mass of between 26 and 30 kDa, but the cause or consequence of this subtle difference remains unknown. When applicable, this analysis may be used to identify proteins that have dramatically unexpected mobility patterns by SDS-PAGE and to support hypotheses regarding proteolysis or post-translational modifications, such as glycosylation or transglutamination. When used in combination for detailed analysis of individual proteins, recognition of unique spectra, peptide coverage mapping, and correlation of peptides with gel mobility provide basic, yet valuable, information to aid in the generation and testing of hypotheses regarding protein function.

Use of Proteomic Data To Identify Protein Networks

One of the goals at the outset of this analysis was to be able to utilize the proteomic data to perform pathway analyses). The set of 1,760 proteins identified from human sperm can be uploaded into either public databases, such as the Database for Annotation, Visualization and Integrated Discovery (DAVID; <www.http://apps1.niaid.nih.gov/david/>, or proprietary databases, such as that of Ingenuity Pathway Systems. These programs analyze the list of identified proteins and identify biological pathways and protein assemblies that are highly represented and presumed to be active. For example, when the sperm proteome data are uploaded, all of the proteins of the glycolytic pathway that directly participate in the conversion of glucose to pyruvate are present. This analysis also identifies each of the 27 proteins that comprise a functional 26S proteosome (FIG. 5). The proteosome inhibitors MG-132 and lactacystin inhibit *in vitro* fertilization in the boar by causing the sperm to fail to penetrate the zona pellucida.[46] The current study provides the first definitive evidence that all 27 proteins constituting the 26S proteosome are present in sperm.

Analyses such as these allow the identification of highly represented protein networks that could facilitate the selection of targets for contraception development. It is inherent in our strategy to identify contraceptive targets of highly populated pathways and protein assemblies that are both active and important in overall sperm production, maturation, and function and, therefore, potential targets for pharmaceutical perturbation or modulation.

CONCLUSIONS

This characterization of the human sperm proteome provides a high-resolution, physiologically relevant index of 1,760 of the approximately 2,300 proteins that we predict to be present in human sperm. The data have a wide range of uses, including (1) identifying the majority of the protein complement that sperm possess to carry out fertilization, (2) providing detailed quantitative data on the amino acid composition and mobility of both OBG-soluble and OBG-insoluble proteins, (3) evaluating active biological pathways in sperm, and (4) assisting in the identification of novel targets for the development of nonhormonal male contraceptives. These data will eventually be made available to the research community via a public database and will be a powerful resource for future studies relating to the function of male gametes.

REFERENCES

1. HARRISON, P.F. & A. ROSENFIELD. 1996. Contraceptive Research and Development. National Academy Press. Washington, D.C.
2. BANERJEE, A.K. & A. SIMPSON. 1992. Reversing vasectomy. BMJ **304:** 1130.
3. KAY, D.J., V. CLIFTON, J.S. TAYLOR & B. BOETTCHER. 1993. Anti-sperm antibodies and semen profiles in re-anastomosed men. Reprod. Fertil. Dev. **5:** 135–139.
4. DARROCH, J.E. 2000. The pill at 40—a new look at a familiar method: the pill and men's involvement in contraception. Fam. Plann. Perspect. **32:** 90–91.
5. EWING, L.L. & B. ROBAIRE. 1978. Endogenous antispermatogenic agents: prospects for male contraception. Annu. Rev. Pharmacol. Toxicol. **18:** 167–187.
6. EWING, L.L., R.C. COCHRAN, R.J. ADAMS, *et al.* 1983. Testis function in rhesus monkeys treated with a contraceptive steroid formulation. Contraception **27:** 347–362.
7. ANAWALT, B.D., J.K. ARMORY, K.L. HERBST, *et al.* 2005. Intramuscular testosterone enanthate plus very low dosage oral levonorgestrel suppresses spermatogenesis without causing weight gain in normal young men: a randomized clinical trial. J. Androl. **26:** 405–413.
8. AMORY, J.K. 2004. Testosterone/progestin regimens: a realistic option for male contraception? Curr. Opin. Invest. Drugs **5:** 1025–1030.
9. MATTHIESSON, K.L., J.K. ARMORY, R. BERGER, *et al.* 2005. Novel male hormonal contraceptive combinations: the hormonal and spermatogenic effects of testosterone and levonorgestrel combined with a 5α-reductase inhibitor or gonadotropin-releasing hormone antagonist. J. Clin. Endocrinol. Metab. **90:** 91–97.
10. GUI, Y.L., C.H. HE, J.K. ARMORY, *et al.* 2004. Male hormonal contraception: suppression of spermatogenesis by injectable testosterone undecanoate alone or with levonorgestrel implants in Chinese men. J. Androl. **25:** 720–727.
11. NIESCHLAG, E. 2000. Male contribution to contraception. *In* Andrology: Male Reproductive Health and Dysfunction. E. Nieschlag & H.M. Behre, Eds.: 399–418. Springer. New York.
12. WU, F.C. 1996. Male contraception. Baillières Clin. Obstet. Gynaecol. **10:** 1–23.
13. DEPAOLO, L.V., B.T. HINTON & R.E. BRAUN. 2000. Male contraception: views to the 21st century, Bethesda, MD, USA, 9–10 September 1999. Trends Endocrinol. Metab. **11:** 66–69.
14. COOKE, H.J. & P.T. SAUNDERS. 2002. Mouse models of male infertility. Nat. Rev. Genet. **3:** 790–801.
15. LYTTLE, C.R. & G.S. KOPF. 2003. Status and future direction of male contraceptive development. Curr. Opin. Pharmacol. **3:** 667–671.
16. KNEITZ, B., P.E. COHEN, E. ARDIEVICH, *et al.* 2000. MutS homolog 4 localization to meiotic chromosomes is required for chromosome pairing during meiosis in male and female mice. Genes Dev. **14:** 1085–1097.
17. EDELMANN, W., P.E. COHEN, B. KNEITZ, *et al.* 1999. Mammalian MutS homologue 5 is required for chromosome pairing in meiosis. Nat. Genet. **21:** 123–127.
18. LIU, D., M.M. MATZUK, W.K. SUNG, *et al.* 1998. Cyclin A1 is required for meiosis in the male mouse. Nat. Genet. **20:** 377–380.

19. Yoshida, K., G. Kondoh, Y. Matsuda, *et al.* 1998. The mouse RecA-like gene Dmc1 is required for homologous chromosome synapsis during meiosis. Mol. Cell **1:** 707–718.
20. Martianov, I., G.M. Fimia, A. Dierich, *et al.* 2001. Late arrest of spermiogenesis and germ cell apoptosis in mice lacking the TBP-like TLF/TRF2 gene. Mol. Cell **7:** 509–515.
21. Roest, H.P., J. van Klaveren, J. de Wit, *et al.* 1996. Inactivation of the HR6B ubiquitin-conjugating DNA repair enzyme in mice causes male sterility associated with chromatin modification. Cell **86:** 799–810.
22. Shima, J.E., D.J. McLean, J.R. McCarrey & M.D. Griswold. 2004. The murine testicular transcriptome: characterizing gene expression in the testis during the progression of spermatogenesis. Biol. Reprod. **71:** 319–330.
23. Cooper, T.G. & C.H. Yeung. 1999. Recent biochemical approaches to post-testicular, epididymal contraception. Hum. Reprod. Update **5:** 141–152.
24. Cooper, T.G. 2002. The epididymis as a target for male contraception. *In* The Epididymis: from Molecules to Clinical Practice. B. Robaire & B.T. Hinton, Eds.: 483–502. Kluwer Academic/Plenum Publishers. New York.
25. Johnston, D.S., Y. Bai & G.S. Kopf. 2002. Utilization of human genome databases in strategies for contraceptive development. *In* Epididymis III: The Third International Conference on the Epididymis. T.T. Turner & B.T. Hinton, Eds. :256–270. The Van Doren Company. Charlottesville.
26. Robaire, B. & B.T. Hinton. 2002. The Epidiydmis: From Molecules to Clinical Practice. Kluwer Academic/Plenum Publishers. New York.
27. Kopf, G.S., P.E. Visconti & H. Galentino-Homer. 1999. Capacitation of the mammalian spermatozoon. Adv. Dev. Biochem. **5:** 83–107.
28. Breton, S., P.J. Smith, B. Liu & D. Brown. 1996. Acidification of the male reproductive tract by a proton pumping (H^+)-ATPase. Nat. Med. **2:** 470–472.
29. Johnston, D.S., S.A. Jelinsky, H.J. Bang, *et al.* 2005. The mouse epididymal transcriptome: transcriptional profiling of segmental gene expression in the epididymis. Biol. Reprod. **73:** 404–413.
30. Parmentier, M., F. Libert & S. Schurmans. 1992. Expression of members of the putative olfactory receptor gene family in mammalian germ cells. Nature **355:** 453–455.
31. Vanderhaeghen, P., S. Schurmans, G. Vassart & M. Parmentier. 1993. Olfactory receptors are displayed on dog mature sperm cells. J. Cell Biol. **123:** 1441–1452.
32. Vanderhaeghen, P., S. Schurmans, G. Vassart & M. Parmentier. 1997. Specific repertoire of olfactory receptor genes in the male germ cells of several mammalian species. Genomics **39:** 239–246.
33. Spehr, M., G. Gisselmann, A. Poplawski, *et al.* 2003. Identification of a testicular odorant receptor mediating human sperm chemotaxis. Science **299:** 2054–2058.
34. Bahat, A., I. Tur-Kaspa, A. Gakamsky, *et al.* 2003. Thermotaxis of mammalian sperm cells: a potential navigation mechanism in the female genital tract. Nat. Med. **9:** 149–150.
35. Eisenbach, M. 1999. Sperm chemotaxis. Rev. Reprod. **4:** 56–66.
36. Evans, J.P. & H.M. Florman. 2002. The state of the union: the cell biology of fertilization. Nat. Cell Biol. **4**(Suppl.)**:** s57–s63.
37. Primakoff, P. & D.G. Myles. 2002. Penetration, adhesion, and fusion in mammalian sperm-egg interaction. Science **296:** 2183–2185.
38. Inoue, N., M. Ikawa, A. Isotani & M. Okabe. 2005. The immunoglobulin superfamily protein Izumo is required for sperm to fuse with eggs. Nature **434:** 234–238.
39. Buck, J., M.L. Sinclair, L. Schapal, *et al.* 1999. Cytosolic adenylyl cyclase defines a unique signaling molecule in mammals. Proc. Natl. Acad. Sci. USA **96:** 79–84.
40. Zippin, J.H., L.R. Levin & J. Buck. 2001. CO_2/HCO_3^--responsive soluble adenylyl cyclase as a putative metabolic sensor. Trends Endocrinol. Metab. **12:** 366–370.
41. Ren, D., B. Navarro, G. Perez, *et al.* 2001. A sperm ion channel required for sperm motility and male fertility. Nature 413: 603–609.
42. Quill, T.A., D. Ren, D.E. Clapham & D.L. Garbers. 2001. A voltage-gated ion channel expressed specifically in spermatozoa. Proc. Natl. Acad. Sci. USA **98:** 12527–12531.

43. FUKAMI, K., K. NAKAO, T. INOUE, *et al.* 2001. Requirement of phospholipase Cδ4 for the zona pellucida-induced acrosome reaction. Science **292:** 920–923.
44. NOLAN, M.A., D.F. BABCOCK, G. WENNEMUTH, *et al.* 2004. Sperm-specific protein kinase A catalytic subunit Cα2 orchestrates cAMP signaling for male fertility. Proc. Natl. Acad. Sci. USA **101:** 13483–13488.
45. QIAN, W.J., T. LIU, M.E. MONROE, *et al.* 2005. Probability-based evaluation of peptide and protein identifications from tandem mass spectrometry and SEQUEST analysis: the human proteome. J. Proteome Res. **4:** 53–62.
46. SUTOVSKY, P., G. MANANDHAR, T.C. MCCAULEY, *et al.* 2004. Proteasomal interference prevents zona pellucida penetration and fertilization in mammals. Biol. Reprod. **71:** 1625–1637.

Male Gamete Contributions to the Embryo

RYUZO YANAGIMACHI

The Institute for Biogenesis Research, University of Hawaii Medical School, Honolulu, Hawaii 96822, USA

ABSTRACT: During normal fertilization, plasma membranes of a spermatozoon and an oocyte mingle to form a mosaic plasma membrane of a zygote. This may contribute to the polyspermy block of the zygote. Sperm tail components (mitochondria, axonema, and accessory fibers) that enter the oocyte are "digested" without playing major roles in embryo development. The proximal centrosome adjacent to the sperm nucleus may become the center of the sperm aster that brings the male and female pronuclei to the center of the zygote, but it may not be essential for embryonic development per se. Whether sperm RNAs contribute to embryonic development is the subject of controversy. The nucleus is the most important sperm component that enters the oocyte. It is known that 7–15% of the spermatozoa of fertile men are chromosomally abnormal. The proportion of the spermatozoa with subchromosomal abnormalities (including damage at the DNA level) is expected to be even higher. The majority of embryos and fetuses with genomic abnormalities are aborted before reaching term. Structurally abnormal spermatozoa are not necessarily genomically abnormal, even though the incidence of genomically abnormal spermatozoa is higher among structurally abnormal than normal spermatozoa. In mammals, certain genes (estimated to be about 100) in the spermatozoon and oocyte must be "imprinted" in a gender-specific manner to warrant normal embryonic development.

KEYWORDS: sperm; oocyte; chromosome; fertilization; genetic; epigenetic

INTRODUCTION

In some animal species reproduction is carried out exclusively by females. Parthenogenetic beetles and lizards are examples. A Hawaiian lizard, *Lepidodactylus lugubris*, reproduces parthenogenetically,[1] but nonfertile males appear once in a while,[2] suggesting that this species was originally bisexual. The bisexual mode of reproduction emerged during evolution and has been maintained in most animals, including mammals. The reason for this is still not fully understood, but this process may accelerate the rate of adaptation in evolution while avoiding irreversible accumulation of detrimental mutations in the face of a constantly changing, competitive environment. Although mammals can reproduce by females alone via artificial

Address for correspondence: Ryuzo Yanagimachi, Ph.D., Institute for Biogenesis Research, University of Hawaii Medical School, 1960 East-West Road, Honolulu, HI 96822. Voice: 808-956-8746; fax: 808-956-7316.
yana@hawaii.edu

Ann. N.Y. Acad. Sci. 1061: 203–207 (2005). © 2005 New York Academy of Sciences.
doi: 10.1196/annals.1336.022

parthenogenesis[3] or cloning,[4] all individuals thus produced have little or no genetic diversity. They may all perish when physical and biological environments change drastically or when serious infectious diseases spread. Bisexual reproduction is a masterpiece of Mother Nature that reduces the chance of the extinction of species.

FATE AND FUNCTIONS OF SPERM COMPONENTS ENTERING THE OOCYTE DURING FERTILIZATION

During the acrosome reaction, which occurs immediately before fertilization, the plasma membrane covering the anterior segment of the acrosome, together with the acrosomal contents, is shed from the main body of the fertilizing spermatozoon. The plasma membrane over the equatorial segment of the acrosome first fuses with the oocyte plasma membrane.[5,6] The remaining sperm plasma membrane in the head and tail then mingles with the oocyte plasma membrane to become a "mosaic" membrane covering the surface of the zygote. Intermingling of the sperm and oocyte plasma membranes may contribute to the polyspermy block of the zygote.[7,8] Claims that the point of sperm–oocyte fusion affects the first cleavage plane and that the first cleavage plane tends to be perpendicular to the embryo-aembryo axis of the blastocyst[9-11] (and consequently the dorso-ventral axis of the postimplantation embryo) have not been confirmed. Either the plasma membrane components or the intracellular components in the sperm head trigger activation of the oocyte.[12–14] The strongest candidate of the sperm-borne oocyte-activating factor is at present phospholipase C zeta.[15–17] In many species the proximal centrosome becomes the center of the sperm aster, which brings the male and female pronuclei to the center of the zygote,[18] but the sperm centrosome per se may not be essential for embryonic development.[19] Whether sperm RNAs contribute to embryonic development[20] is controversial. The intrinsic nuclear envelope of the spermatozoon disappears very quickly after sperm entry into the oocyte.[21] The nuclear envelopes of both male and female pronuclei are produced *de novo*.[21] Perinuclear materials and the inner acrosomal membrane in the sperm head as well as mitochondria, axonema, and accessory fibers in the sperm tail are "digested" by multivesicular bodies, lysosomes,[22] and ubiquitin-mediated proteosomes[23–25] without contributing to embryonic development.

The nuclei of fully mature mammalian spermatozoa are unusual in that they resist rather harsh chemical and physical stresses, perhaps because of cross-linking of nuclear protamines by disulfide bonds.[26] Snap-freezing of (mouse) spermatozoa without cryoprotection or freeze-drying “kills” the spermatozoa by disrupting sperm membranes and various organelles, but sperm nuclei remain capable of participating in normal embryonic development provided they are injected microsurgically into oocytes.[26–28] Thus, the life and death of spermatozoa is not synonymous with that of sperm nuclei. The spermatozoa with structurally abnormal heads and/or tails are commonly infertile, yet some, if not all, are able to participate in normal embryonic development when they are injected into oocytes.[29,30] Although the incidence of chromosomally (genomically) abnormal spermatozoa is significantly higher among structurally abnormal than normal spermatozoa,[31–33] it is still astonishing that many grossly abnormal spermatozoa are chromosomally (genomically) normal.

CONSEQUENCE OF FERTILIZATION BY SPERMATOZOA WITH GENETIC AND EPIGENETIC ERRORS

In humans about 20–37% of naturally ovulated oocytes of fertile women and about 7–15% of spermatozoa of fertile men have abnormal chromosomes; about 40% of zygotes after natural conceptions are estimated to be chromosomally abnormal.[33,34] The frequency of zygotes with chromosomal and/or subchromosomal abnormalities is expected to be even higher. It is amazing that the incidence of newborn babies with congenital abnormalities after natural conception (2.7%)[35] does not differ greatly from that after assisted reproduction, such as *in vitro* fertilization (IVF) and intracytoplasmic sperm injection (ICSI).[36,37] This must be due to a stringent maternal selection against abnormal embryos/fetuses during pregnancy. Reported higher incidences of congenital malformations after IVF and ICSI compared with natural conception are likely due to maternal characteristics (such as age) and the genetic backgrounds of couples and not to any aspects of the technical procedure itself.[37,38]

At least in mammals some genes (estimated to be about 100) in male and female gametes must be "imprinted" properly in a parent-specific manner for normal embryonic development.[39,40] Genomic imprinting (at least DNA methylation) is established during gametogenesis.[41,42] Errors in genomic imprinting, for example, due to defective imprinting centers, may result in abnormal development or even death of embryos and fetuses[43,44] or diseases of offspring after birth.[45] At present there are no noninvasive methods to detect genetic and epigenetic errors in maturing or mature gametes. Preimplantation diagnosis (isolation of one or a few blastomeres from a cleaving embryo and analysis of the genomic makeup) is the only method to accurately detect genetic and epigenetic errors in future offspring. A day may come when genetic and epigenetic errors in developing and mature gametes are corrected prior to fertilization.

ACKNOWLEDGMENTS

I thank Drs. Thomas Huang, Kazuto Morozumi, Steve Ward, and Stefan Moisyadi of the University of Hawaii and Dr. Keith Latham of Temple University for their invaluable help in the preparation of the manuscript. Studies by my research group were funded by the National Institute of Child Health and Human Development, the Katherine and Harold Castle Foundation, and the University of Hawaii RTRF Chancellor's Fund.

REFERENCES

1. Cuellar, O. & A.G. Kulge. 1972. Natural parthenogenesis in the gekkonid lizard *Lepiodactylus lugubris*. J. Genet. **61:** 14–26.
2. Brown, S.G. & S. Murphy-Walker. 1966. Behavioral interactions between rare male phenotype and female unisexual *Lepidodactylus lugubris*. Herpetol. J. **6:** 69–73.
3. Kono, T., Y. Obata, Q. Wu *et al.* 2004. Birth of parthenogenetic mice that can develop toadulthood. Nature **428:** 860–864.
4. Wilmut, I., N. Beaujean, P.A. de Sousa, *et al.* 2002. Somatic cell nuclear transfer. Nature **419:** 583–586.

5. MOORE, H.D.M. & J.M. BEDFORD. 1978. Ultrastructure of the equatorial segment of hamster spermatozoa during penetration of oocytes. J. Ultrastruct. Res. **62:** 110–117.
6. YANAGIMACHI, R. 1994. Mammalian fertilization. *In* The Physiology of Reproduction. E. Knobil & J.D. Neill, Eds.: 189–317. Raven Press. New York.
7. MALESZEWSKI, M., Y. KIMURA & R. YANAGIMACHI. 1996. Sperm membrane incorporation into oolemma contributes to the oolemma block to polyspermy: evidence based on intracytoplasmic injection experiments in the mouse. Mol. Reprod. Dev. **44:** 256–259.
8. PRZYNSKA, B., M. MALESZEWSKI & D. MALUCHNIK. 1996. Mouse oocytes penetrated by sperm at GV or GVBD stage lose the ability to fuse with spermatozoa. Zygote **4:** 123–128.
9. PITROWSKA, K. & M. ZERNICKA-GOETZ. 2001. Role for sperm in spatial patterning of the early mouse embryo. Nature **409:** 517–521.
10. PLUSA, B., K. PIOTROWSKA, AND M. ZEMICKA-GOETZ. 2002. Sperm entry position provides a surface marker for the first cleavage plane of mouse zygote. Genesis **32:** 193–198.
11. ZERNICKA-GOETZ, M. 2004. First cell fate decision and spatial patterning in the early mouse embryo. Semin. Cell. Dev. Biol. **15:** 563–572.
12. FOLTZ, K.R. 1995. Sperm-egg binding proteins. Int. Rev. Cytol. **163:** 249–303.
13. HOMA, S. & K. SWANN. 1994. A cytosolic sperm factor triggers calcium oscillations and membrane hyperpolarization in human oocytes. Hum. Reprod. **9:** 2356–2361.
14. KIMURA, Y., R. YANAGIMACHI, S. KURETAKE, *et al.* 1998. Analysis of mouse oocyte activation suggests the involvement of sperm perinuclear material. Biol. Reprod. **58:** 1407-1415.
15. SWANN, K., M.G. LARMAN, C.M. SAUNDERS & F.A. LAI. 2004. The cytosolic sperm factor that triggers Ca2+ oscillations and egg activation in mammals is a novel phospholipase C: PLC zeta. Reproduction **127:** 431–439.
16. KOUCHI, Z, K. FUKAMI, T. SHIKANO *et al.* 2004. Recombinant phospholipase C zeta has high Ca^{2+} sensitivity and induces Ca^{2+} oscillations in mouse eggs. J. Biol. Chem. **279:** 10408–10412.
17. KNOTT, J.G., M. KUROKAWA, R.A. FISSORE, *et al.* 2005. Transgenic RNA interference reveals role for mouse sperm phospholipase C zeta in triggering Ca^{2+} oscillations during fertilization. Biol. Reprod. **72:** 992–996.
18. SCHATTEN, G. 1994. The centrosome and its mode of inheritance: the reduction of the centrosome during gametogenesis and its restoration during fertilization. Dev. Biol. **165:** 299–235.
19. KIM, N.H, J.W. LEE, S.H. JUN, *et al.* 1998. Fertilization of porcine oocytes following intracytoplasmic spermatozoon or isolated sperm head injection. Mol. Reprod. Dev. **51:** 436–444.
20. OSTERMEIER, G.C., R.J. GOODRICH, J.S. MOLDENHAUER, *et al.* 2005. A suite of novel human spermatozoal RNAs. J. Androl. **26:** 70–74.
21. USUI, N., A. OGURA, Y. KIMURA & R. YANAGIMACHi. 1997. Sperm nuclear envelope: breakdown of intrinsic envelope and de novo formation in hamster eggs. Zygote **5:** 35–46.
22. Hirano, J. & Y. Hirao. 1988. Fate of sperm tail components after incorporation into the hamster egg. Gamete Res. **19:** 369–386.
23. SHITARA, H., H. KANEDA, A. SATO, *et al.* 2000. Selective continuous elimination of mitochondria injected into mouse egg from spermatids, but not from liver cells, occurs throughout embryogenesis. Genetics **156:** 1272–1284.
24. SUTOVSKY, P., G. MANANDHAR, A. WU, *et al.* 2000. Interaction of sperm perinuclear theca with oocytes: implications for oocyte activation, anti-polyspermy defense, and assisted reproduction. Microscop. Res. Tech. **61:** 362–378.
25. SUTOVSKY, P., K. VAN LEYEN, T. MCCAULEY, *et al.* 2000. Degradation of paternal mitochondria after fertilization: implications for heteroplasmic, assisted reproduction technologies and mtDNA inheritance. Reprod. Biomed. Online **8:** 24–33.
26. KANEKO, T, D.G. WHITTINGTON, J.W. OVERSTREET & R. YANAGIMACHI. 2003. Tolerance of the mouse sperm nuclei to freeze-drying depends on their disulfide status. Biol. Reprod. **69:** 1859–1862.

27. WAKAYAMA, T. & R. YANAGIMACHI. 1998. Development of normal mice from oocytes injected with freeze-dried spermatozoa. Nat. Biotechnol. **16:** 639–641.
28. WARD, M.A., T. KANEKO, H. KUSAKABE, *et al.* 2003. Long-term preservation of mouse spermatozoa after freeze-drying and freezing without cryoprotection. Biol. Reprod. **69:** 2100–2108.
29. BURRUEL, V.R., R. YANAGIMACHI & W.K. WHITTEN. 1996. Normal mice developed from oocytes injected with spermatozoa with grossly misshapen heads. Biol. Reprod. **55:** 709–714.
30. DE VOS, A., H. VAN DE VELDE, H. JORIS, *et al.* 2003. Influence of individual sperm morphology in fertilization, embryo morphology and pregnancy outcome of intracytoplasmic sperm injection. Fertil. Steril. **79:** 42–48.
31. LEE, J.D., Y. KANIGUCHI & R. YANAGIMACHI. 1996. Analysis of chromosome constitution of human spermatozoa with normal and aberrant head morphology after injection into mouse oocytes. Hum. Reprod. **11:** 1942–1946.
32. KISHIKAWA, H., H. TATENO & R. YANAGIMACHI. 1999. Chromosome analysis of mouse spermatozoa with normal and abnormal head morphology. Biol. Reprod. **61:** 809–812.
33. YANAGIMACHI, R. 2005. Intracytoplasmic injection of spermatozoa and spermatogenic cells: its biology and applications in human and animals. Reprod. Biomed. Online **10:** 247–286.
34. PLACHOT, M., A.M. JUNCA, J. MANDERBAUM, *et al.* 1998. Failure of embryonic development. Reprod. Nutr. Dev. **28:** 1781–1790.
35. ANTHONY, S., S.E. BUITENDIJK, C.A. DORREPAAL, *et al.* 2002. Congenital malformations in 4224 children conceived after IVF. Hum. Reprod. **17:** 2089–2095.
36. SCHLEGEL, P.N. 1999. Debate: is ICSI genetic time bomb? No: ICSI is safe and effective. J. Androl. **20:** 18–22.
37. LUDWIG, M. & A. KATRALINIC. 2002. Malformation rate in fetuses and children conceived after intracytoplasmic sperm injection (ICSI): results of a perspective cohort study. Reprod. Biomed. Online **5:** 171–178.
38. TOURNAYE, H. 2003. ICSI: a technique too far? Int. J. Androl. **26:** 63–69.
39. SURANI, M.A. 1998. Imprinting and the initiation of gene silencing in the germ cell line. Cell **93:** 309–312.
40. LATHAM, K.E. 1994. Epigenetic modification and reimprinting of the mammalian genome during development. Curr. Top. Dev. Biol. **43:** 1–49.
41. DAVIS, T.L., G.J. YANG, J.R. MCCARREY & M.S. BARTOLOMEI. 2000. The H19 methylation imprint is erased and reestablished differentially on the parental alleles during male germ cell development. Hum. Mol. Genet. **9:** 2885–2984.
42. LUCIFERO, D., M.R. MANN, M.S. BARTOLOMEI & J.M. TRASLER. 2004. Gene-specific timing and epigenetic memory in oocyte imprinting. Hum. Mol. Genet. **13:** 839–849.
43. BARTON, S.C., M.A. SURANI & M.L. NORRIS. 1984. Role of paternal and maternal genomes in mouse development. Nature **311:** 374–376.
44. MCGRATH, J. & D. SOLTER. 1984. Completion of mouse embryogenesis requires both the maternal and paternal genomes. Cell **37:** 179–183.
45. WALTER, J. & M. PAULSEN. 2003. Imprinting and disease. Semin. Cell. Dev. Biol. **14:** 101–110.

The Androgen Microenvironment of the Human Testis and Hormonal Control of Spermatogenesis

JONATHAN P. JAROW[a] AND BARRY R. ZIRKIN[b]

[a]*Department of Urology, Johns Hopkins University School of Medicine, Baltimore, Maryland 21287, USA*

[b]*Department of Biochemistry and Molecular Biology, Division of Reproductive Biology, Johns Hopkins University School of Public Health, Baltimore, Maryland 21205, USA*

ABSTRACT: **It is well established for both rat and man that the total testosterone concentration within the testis is far higher than that in serum. We know for the rat that intratesticular testosterone can be reduced by 50–60% without an adverse effect on spermatogenesis but that the required intratesticular testosterone concentration is still 10-fold greater than serum testosterone concentration. This kind of information, if available for the human, could prove invaluable for understanding and treating select men with infertility and in the development of male hormonal contraceptives. Unfortunately, we know little about the androgen content of intratesticular fluid within the human testis and nothing about the relationship between intratesticular androgens and human spermatogenesis. Using a newly developed minimally invasive technique for repetitive testicular sampling, our recent studies of the human have demonstrated that, as in the rat, there is a gradient between the concentration of testosterone in serum and within the testis; intratesticular testosterone levels were found to be 100-fold higher than serum testosterone levels in normal men. Using liquid chromatography tandem mass spectroscopy, we have shown that intratesticular 5α-dihydrotestosterone (DHT) levels are only 2% that of testosterone and, thus, despite greater affinity for the androgen receptor, intratesticular DHT is not significant in normal men. In order to assess how much of the testosterone within the human testis is bioactive, we adapted a highly sensitive recombinant protein mammalian cell-based bioassay to measure androgen bioactivity. The androgen bioactivity in the normal human testis is roughly two-thirds that of the total testosterone measurable by radioimmunoassay, despite the fact that the concentrations of the major androgen- binding proteins (sex hormone-binding globulin- and androgen-binding protein) are insufficient to account for this difference. This finding suggests that androgens may bind to other, as-yet-unknown molecules in the human testis. How, or if, this relates to spermatogenesis in the rat, or to man-to-man differences in the response to hormonal contraceptives, is not clear. We do not yet know how much testosterone is required within the human testis to either maintain or**

Address for correspondence: Jonathan P. Jarow, M.D., Department of Urology, Johns Hopkins University School of Medicine, 601 North Caroline Street, Baltimore, MD 21287. Voice: 410-955-3617; fax: 410-614-8487.
jjarow@jhmi.edu

**Ann. N.Y. Acad. Sci. 1061: 208–220 (2005). © 2005 New York Academy of Sciences.
doi: 10.1196/annals.1336.023**

restore quantitatively normal spermatogenesis because, as yet, experimental studies comparable to those performed in the rat have not been feasible for the human.

KEYWORDS: spermatogenesis; male infertility; testosterone

INTRODUCTION

Approximately 15% of couples have difficulty conceiving. A male factor is either the sole or contributory cause in more than 50% of these couples.[1] There are many known causes of male infertility, of which about 50% are potentially correctable with therapy. The most common among these is varicoceles.[2] The remaining known causes of male infertility, including genetic/chromosomal abnormalities, gonadotoxin exposure, testicular trauma, etc., are not amenable to therapy. For 25% of men evaluated, no identifiable cause of their abnormal semen analyses can be found, and these men currently are diagnosed with idiopathic male infertility. Men with idiopathic male factor infertility are likely to represent a heterogeneous population with a variety of unknown causes of infertility. Historically, such men have been treated with either empiric medical therapies in the hope of improving semen quality or assisted reproductive techniques (e.g., intrauterine insemination) to increase the chance of conception without improving semen quality.

With the introduction of *in vitro* fertilization (IVF) by intracytoplasmic sperm injection,[3] it is now possible for couples who previously were labeled "sterile" due to male factor infertility to have offspring with paternal genetic continuity. However, there are several disadvantages to this method of management. First, its cost is very high. Second, there is a very high incidence of multiple gestations (~40%) associated with IVF, which increases the cost (both monetary and human) of IVF over other methods of management. The costs are exacerbated because multiple gestations typically lead to low birth weight and an increased need for neonatal intensive care.[4] Third, the use of IVF for treatment of male factor infertility transfers all of the risks of intervention from the partners with an abnormality (men) to the healthy spouses (women). For example, ovarian hyperstimulation occurs in 5% of women undergoing IVF.[5] Fourth, standard medical practice argues for the desirability of establishing the etiology of a disorder before instituting therapy and for treating the underlying problem whenever possible.

SPERMATOGENESIS AND INTRATESTICULAR ANDROGENS

Spermatogenesis is a particularly complex, cyclic process that involves a number of different cell types interacting over time and space. In the rat, for example, any given cross section of the seminiferous tubule will contain one or two generations of spermatogonia, one or two generations of spermatocytes, and one or two generations of spermatids. The net result is continuous and asynchronous production of sperm. Regulation of spermatogenesis occurs via hormonal, paracrine, and autocrine signals. Given the complexity of signaling and of cell types and interactions, it is not surprising that our understanding of the mechanisms by which spermatogenesis is regulated is far from complete. Human spermatogenesis is equally complex, or perhaps even more so because of the non-linear arrangement of spermatogenesis in the seminiferous tubules. These, in combination with ethical concerns related to human experimentation, have made it very difficult to pursue studies of spermatogenesis regulation in man.

In all mammalian species studied to date, testosterone has been found to be the predominant intratesticular steroid. Turner and colleagues reported that, in the rat, the concentration of 5α-dihydrotestosterone (DHT) is only 5% that of testosterone.[6] The testosterone concentration within the rat testis, whether in the interstitial or seminiferous tubular compartment, has been shown to be far in excess (25- to 30-fold greater) of its concentration in blood serum. Local testosterone production by Leydig cells, in combination with a countercurrent mechanism in the pampiniform plexus, probably constitutes the primary mechanism for the concentration gradient between testicular and systemic testosterone concentrations under normal circumstances. Zirkin and associates have shown experimentally that the testosterone concentration gradient between serum and testis persists when testosterone is administered exogenously to animals under circumstances in which Leydig cell steroidogenesis is suppressed.[7] A possible explanation for this observation is that the testis may act as a reservoir for testosterone, perhaps due in part to the presence of androgen-binding proteins contained within.

In the adult rat, the normal total intratesticular testosterone concentration, 50 to 70 ng/ml,[6] is far higher than the concentration, 20 ng/ml, shown experimentally to be *required* for the quantitative maintenance of spermatogenesis[7] or for the quantitative restoration of spermatogenesis in animals rendered azoospermic by testosterone administration or by active immunization against GnRH.[8] The required concentration, 20 ng/ml, though less than 50% of the testosterone concentration that is normally present within the testis, nonetheless is 10-fold higher than the total testosterone concentration in normal rat blood serum (2 ng/ml). As yet, the reason intratesticular testosterone concentration must be so high to maintain or restore spermatogenesis is unknown. One possibility is that much of the total testosterone is bound and therefore unavailable to the Sertoli cell androgen receptors. Androgen-binding protein (ABP), a secretory product of Sertoli cells that itself is regulated by testosterone and follicle-stimulating hormone (FSH), is present in the mammalian testis.[9] The established role for ABP is as a binding/carrier protein for sex steroids, but it also has been suggested that ABP may regulate fertility and spermatogenesis. For example, in studies of transgenic mice that overexpress ABP,[10] spermatogenesis begins normally despite the excess ABP, but progressive structural and functional abnormalities occur in the testis in parallel with increasing amounts of ABP. In an in vitro study, Roberts and Zirkin found that ABP was capable of inhibiting the ability of testosterone to stimulate the transcription of a transfected reporter gene, presumably by preventing its access to the androgen receptor.[11] Excess ABP within the seminiferous tubules also may act by reducing the availability of testosterone as a substrate for 5α-reductase and aromatase, thus leading to reduced formation of dihydrotestosterone and estradiol, respectively, and, in some cases, infertility.[12] However, over expression of human sex hormone-binding globulin (SHBG) in the testes of transgenic mice was shown to have no effect on spermatogenesis.[13] As yet, therefore, the relationship of intratesticular binding molecules to the concentration of bioavailable androgens remains uncertain.

PERCUTANEOUS ASPIRATION OF THE HUMAN TESTIS

Earlier studies of the rat had shown that testosterone concentrations within the testicular interstitial and seminiferous tubular fluids are significantly higher than normal serum levels.[7] On the basis of these observations, we wished to conduct comparable studies in the human, that is, to establish the normal range of intratesticular testosterone in a group of fertile men. Until recently, it has not been feasible to relate intratesticular testosterone concentration to spermatogenesis in humans. A major problem with studying the hormonal regulation of human spermatogenesis has been the lack of noninvasive or minimally invasive methods suitable for obtaining intratesticular fluids. We recently reported the development of such a method, involving percutaneous aspiration of the testis.[14] Using this method we found that, as in the rat, the mean *total* testosterone concentration within the human testis, about 600 ng/ml, is well in excess of the testosterone concentration in serum (5 ng/ml). These results are consistent with previously conducted analyses of testosterone concentration in human testicular fluids obtained from men undergoing open surgery,[15–17] which have ranged from 350–750 ng/ml. Testosterone concentration also has been analyzed in testicular biopsies[18] with comparable results, that is, 600 ng/ml. Though open surgical aspiration and testis biopsy are reliable methods by which intratesticular fluids may be obtained, these approaches obviously represent invasive procedures that cannot be used for routine human studies, whereas percutaneous aspiration is by comparison a relatively non-invasive procedure that in

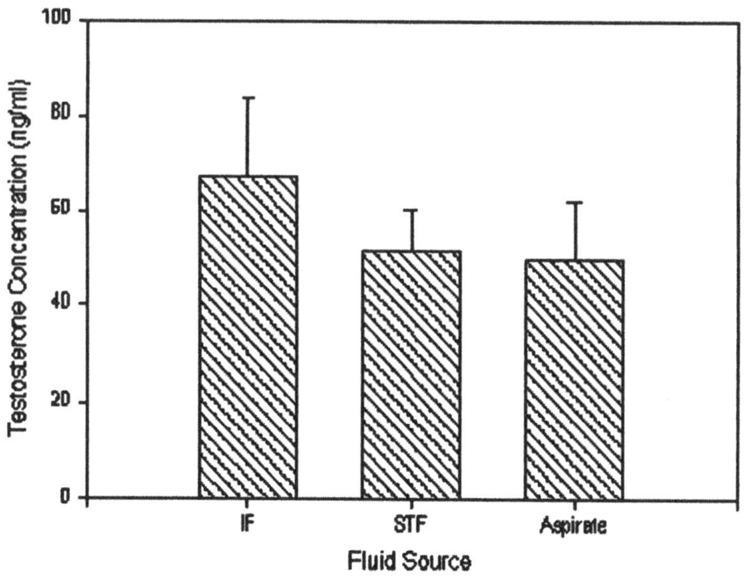

FIGURE 1. Intratesticular testosterone concentration in the rat.[14] The testosterone concentration of fluid obtained by percutaneous aspiration (*aspirate*) was compared to that of fluids from the interstitial compartment (*IF*) and seminiferous tubule compartment (*STF*) obtained by centrifugation methods. The testosterone concentration of the percutaneous aspirate most closely approximates that of the STF. *, $P < .5$ compared to STF.

additional studies we have shown can be performed on a repeated basis in the same individuals.

Preliminary studies in a rat model demonstrated that testicular aspiration could be safely performed. The sample obtained is comprised of both seminiferous tubular fluid and interstitial fluid, but the testosterone concentration most closely approximates that of the seminiferous tubular fluid (FIG. 1).[14] Subsequent studies in human volunteers undergoing vasectomy showed that testosterone concentration, measured by radioimmunoassay (RIA) after high-performance liquid chromatography (HPLC) separation, was the major androgen in the human testis, since dihydrotestosterone could not be detected in these small-volume samples. We found that percutaneous testicular aspiration, using a 19-gauge needle, could be used to aspirate testicular fluids safely and repeatedly.[14] Follow-up testicular ultrasounds did not reveal any deleterious effects to the aspirated testes. Testicular aspirations yielded 22 μl of fluid on average. The mean testosterone concentration in the aspirates, as measured by radioimmunoassay, was 609 ± 50 ng/ml, with testosterone clearly the dominant androgenic steroid; 5α-dihydrotestosterone (DHT) and 5α-androstane-3α, 17β-diol (3α-diol), separated by HPLC, were below the limits of detection by RIA. These results indicated that, as in the rat testis, testosterone is the major androgenic steroid in the human testis and that its concentration within the testis is considerably greater than that in human serum (about 5 ng/ml). The concentration of SHBG or ABP ranged from 3 to 21 nM, with a mean of 8.5 nM. Thus, the molar concentration of intratesticular testosterone, about 2,000 nM, greatly exceeded SHBG or ABP concentration, suggesting that much of the testosterone might be bioavailable. There was significant variation within and between individuals, comparable to that observed with serum testosterone concentrations (FIG. 2).

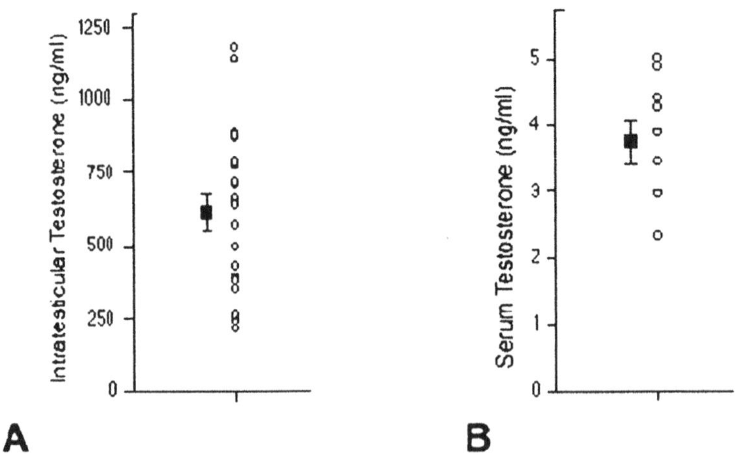

FIGURE 2. Intratesticular (**A**) and serum (**B**) testosterone concentrations in man. Testicular aspirates were obtained from 30 fertile volunteers, 9 of whom underwent bilateral testicular aspiration.[14] Blood serum was obtained from the latter 9 men at the time of testicular aspiration. Shown are individual measurements and the means ± the standard errors of the means.

TABLE 1. Testicular concentrations of T, DHT, 3α, and â-Diol and E2 of 10 normal male subjects measured by liquid chromatography tandem mass spectroscopy[24]

	Mean (± SE)	Range
Testosterone (ng/ml)	572 ±102	103–1,085
DHT (ng/ml)	13.4 ± 1.8	4.9–26
3α & â-diol (ng/ml)	12.3 ± 1.9	3.6–26
E2 (ng/ml)	15.7 ± 2.3	8–29

ANDROGEN ENVIRONMENT WITHIN THE HUMAN TESTIS

More-sensitive methods to detect sex steroids in these small-volume testis samples were needed, since on the basis of animal model studies we knew that other steroids were likely to be present. A rapid, sensitive, and specific method using liquid chromatography tandem mass spectrometry (LC/MS/MS) was developed for simultaneous determination of testosterone, DHT, estradiol (E2), and 3α-diol within human testicular fluid.[19] Using LC/MS/MS, we confirmed that the major androgenic steroid within the human testis indeed is testosterone, but that significant amounts of other steroids are also present, particularly DHT (TABLE 1).

After we observed that the intratesticular concentration of androgens was in great excess of the concentration of binding proteins for androgens, we next wished to assess the biological activity of the androgens within the testis. To do so, we used an established mammalian two-hybrid method to assay bioavailable androgen concentration in testicular fluids obtained by the percutaneous aspiration method.[20] The advantage of this assay is its exquisite sensitivity to active androgenic steroids. To perform the assay, the COS-7 cells were transfected with plasmids to express recombinant proteins of the N-terminus of the androgen receptor (AR) fused to the VP16 activation domain, the C-terminal ligand-binding domain of AR fused to the Gal4 DNA-binding domain, the AR coactivator ARIP3, and a firefly luciferase reporter under the regulation of five Gal4-binding sites (pG5-LUC). In brief, the specific binding of androgen alters the conformation of the AR ligand-binding domain fusion protein to promote its interaction with the AR N-terminal domain fusion protein. The binding of this activated complex of recombinant proteins, one of which contains the Gal4 DNA-binding domain, to one or more of the Gal4-binding sites of the luciferase reporter gene activates expression of the firefly luciferase enzyme. The sensitivity of the assay for androgen-mediated receptor transactivation is further enhanced by the coexpression of the androgen receptor coactivator, ARIP3, which interacts with and stabilizes the AR N- and C-terminal protein interactions, thus amplifying the expression of the firefly luciferase enzyme. A constitutively active *Renilla* luciferase plasmid is used to control for transfection efficiency. Firefly and *Renilla* luciferase enzyme activities were measured, and the results were quantified based on the ratio of firefly and *Renilla* luciferase activities for known and test samples. The values from samples containing known concentrations of added testosterone were used to plot a standard curve of relative luciferase activity to testosterone concentration. Utilizing the standard curve, the biologically active androgen in each test sample was determined based upon its relative luciferase activity. Bioactive androgen concentrations were quantified in serum (30 μl) and intratesticular fluid

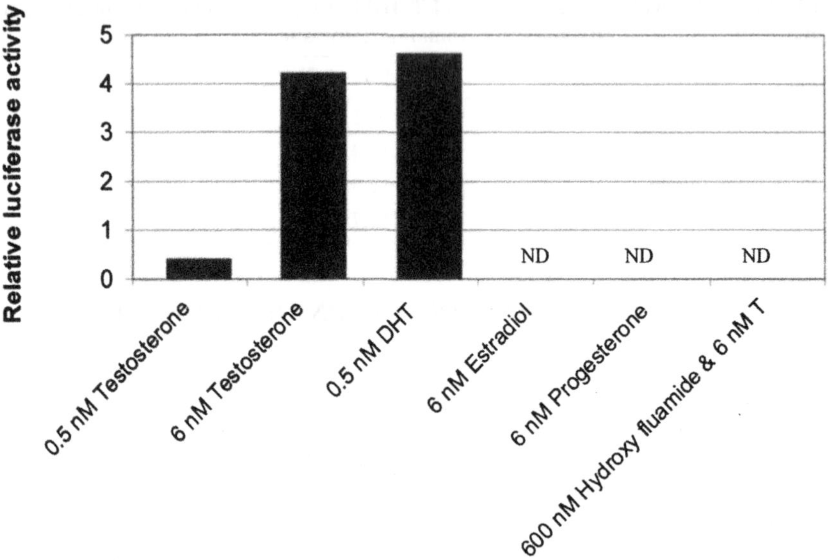

FIGURE 3. Specificity of the androgen bioactivity assay. Testosterone, estradiol, progesterone, or testosterone plus hydroxyflutamide were added to charcoal-stripped fetal calf serum.[22] N.D., not detectable.

(1 μl). The sensitivity of this assay is equivalent to 0.069 ± 0.03 nmol/liter of testosterone; inter- and intra-assay coefficients of variation are 23% and 6%, respectively. All samples were assayed simultaneously and run in quadruplicate.

The specificity of the assay was evaluated by measuring androgen bioactivity after adding various steroids to fetal calf serum stripped of endogenous steroids by absorption to dextran-coated charcoal. For this analysis, 0.5 nM testosterone, 6 nM testosterone, 0.5 nM dihydrotestosterone, 6 nM estradiol, 6 nM progesterone, and 6 nM testosterone plus the anti-androgen hydroxyflutamide (600 nM) were tested (FIG. 3). An approximate 10-fold increase in androgen bioactivity was seen with the addition of 6 nM testosterone, compared to 0.5 nM testosterone, and androgen bioactivity was completely suppressed by the presence of hydroxyflutamide. DHT had approximately 10-fold higher bioactivity than testosterone, as would be expected.[21] Neither estradiol nor progesterone demonstrated androgenic activity. In addition, testosterone-free human serum did not demonstrate detectable androgenic activity in this assay. In initial studies, the possibility that interfering substances might affect the results obtained with this assay was assessed by quantifying bioactive androgens in serial dilutions of human serum and human testicular fluid samples. This approach also tests whether dilution alters the free versus bound equilibrium. Assays were run in quadruplicate, with four different volumes tested. The results are shown in TABLE 2. Analysis of variance revealed no significant differences ($P > .05$) in androgen bioactivity with the different volumes of human serum or human testicular fluid.

In our initial studies using the androgen bioactivity assay, we studied the serum and intratesticular fluids of 22 fertile men.[22] The mean total (immunoreactive) serum

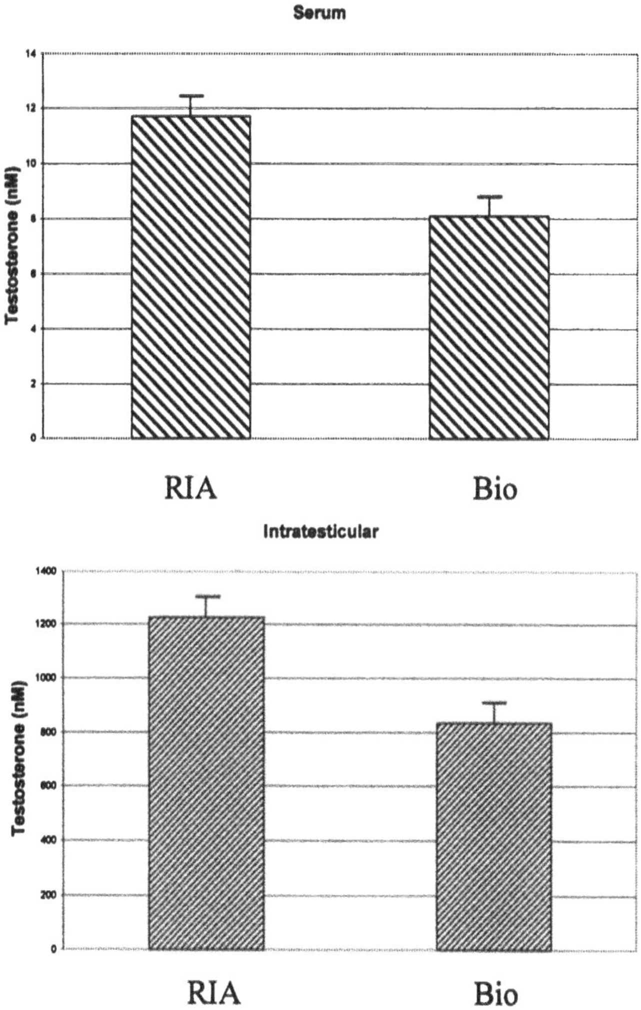

FIGURE 4. Comparison of the mean (± standard error of the mean) total concentration of testosterone, as measured by RIA, and androgen bioactivity (*Bio*) as measured by the bioactivity assay, in serum and intratesticular fluid of 22 fertile men.[22]

TABLE 2. Effect of sample volume on measurement of androgen bioactivity (mean ± SEM) in human serum and testicular fluid specimens[27]

Volume (μl)	Androgen bioactivity (nM)
Human serum	
10	9.26 ± 0.08
20	8.37 ± 0.83
30	7.46 ± 0.12
40	9.47 ± 0.53
Human testicular fluid	
1.0	1234 ± 19
1.5	1105 ± 48
2.0	856 ± 142
2.5	968 ± 161

testosterone concentration, measured by RIA, was 11.7 ± 0.7 nM, and the androgen bioactivity in the serum of these same subjects, measured by the androgen bioactivity assay, was 8.1 ± 0.5 nM. This represented a significant (P <.01), 30% decrease from the total concentration (FIG. 4). The total testosterone and bioactive androgen concentrations in intratesticular fluid were, in both cases, approximately 100-fold higher than the respective values in serum; the concentration of total (immunoreactive) testosterone was 1,226 ± 86 nM, and that of bioactive androgen was 835 ± 74 nM (FIG. 4). The 30% decrease in intratesticular bioactive androgen from immunoreactive intratesticular testosterone was significant (P <.01). The total testosterone and bioactive androgen concentrations in the serum of individual men were highly correlated (r = .624; P = .0019). In contrast, total testosterone and bioactive androgen concentrations in the intratesticular fluid did not correlate (r = .037; P = .87).

Thus, through the combination of RIA, liquid chromatography tandem mass spectrometry, and androgen bioactivity, we are now able to measure the sex steroid content and androgen bioactivity within the small-volume samples obtained by percutaneous aspiration of the testis in human subjects.

CLINICAL STUDIES

To understand the effect of administering hormones for contraception or infertility treatment, it is essential for far more to be understood about the relationship between androgen concentration within the human testis, whether total or bioavailable, and normal spermatogenesis. We have utilized the methods described above to assess the androgen microenvironment of the human testis in men undergoing hormonal contraception and hormonal therapy.[23,24] We studied a group of men receiving hormonal contraception utilizing weekly testosterone enanthate (TE) combined with oral daily levonorgestrel (LNG).[23] In this longitudinal study, baseline and follow-up (after 6 months of therapy) samples of serum and testicular fluid were obtained. The concentration of bioavailable androgen within the testis before treatment

TABLE 3. Serum hormones, intratesticular testosterone, and sperm count before and after 6 months of treatment with testosterone enanthate and levonorgestrel[28]

	Baseline	Month 6
Sperm count (million/ml)	65 ± 15	1.3 ± 1.3[a]
Serum T (nmol/L)	22.8 ± 2.0	28.7 ± 2.0
Intratesticular T (nmol/L)	822 ± 136	13 ± 4.5[a]
Intratesticular androgenic bioactivity (nmol/L)	583 ± 145	40 ± 22[a]
Serum LH (IU/L)	3.79 ± 0.80	0.08 ± 0.03[a]
Serum FSH (IU/L)	3.29 ± 0.67	0.10 ± 0.03

NOTE: Values are mean ± SEM.
[a]All but one individual (9 million/ml) was azoospermic.

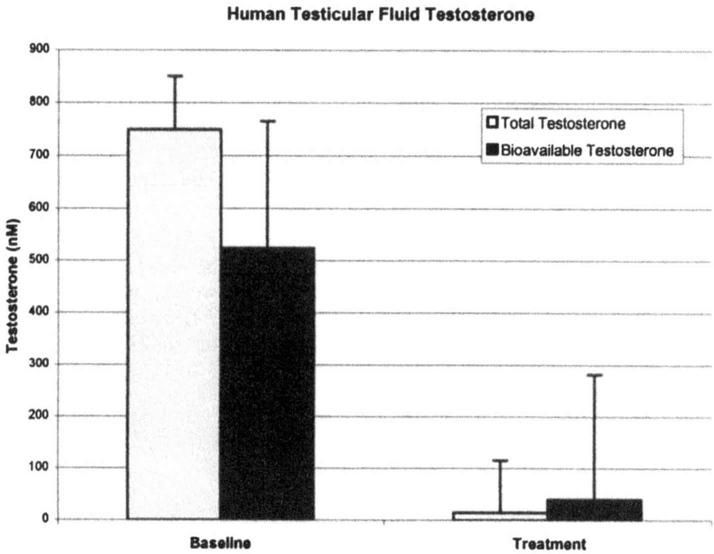

FIGURE 5. Comparison of total and bioavailable intratesticular testosterone (*HTF*) in normal men before and during hormonal contraceptive therapy (mean ± SD).[23]

was approximately 70% of the total concentration of testosterone (TABLE 3). Serum luteinizing hormone (LH) and FSH were markedly suppressed by the combination of testosterone and levonorgestrel, while serum testosterone concentration was maintained. Sperm counts declined, and all but one individual became azoospermic on this regimen. TE plus LNG dramatically suppressed intratesticular testosterone (98%) and intratesticular androgenic bioactivity (93%) to levels approximating that in serum (FIG. 5). Intratesticular testosterone levels comparable to normal serum testosterone concentrations were insufficient to support normal spermatogenesis. However, there was no correlation between either baseline or follow-up intratesticular

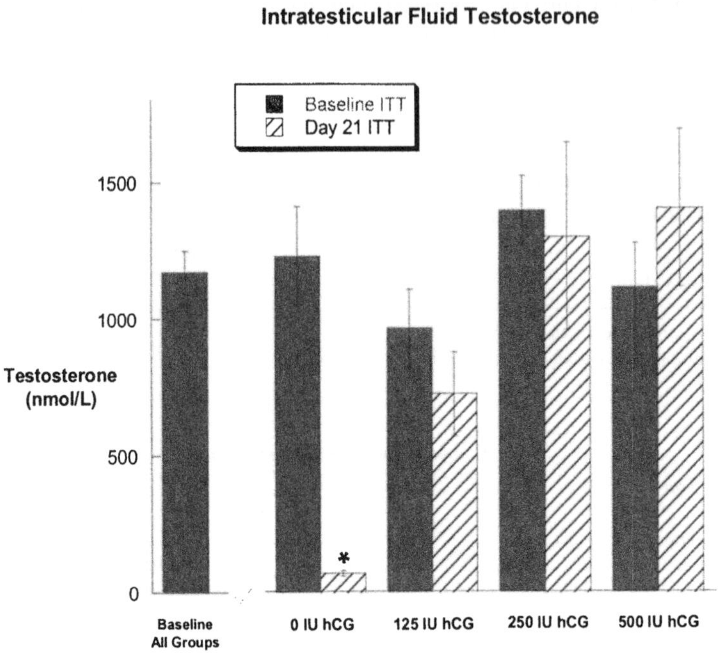

FIGURE 6. Intratesticular fluid testosterone (ITT) at baseline and after 3 weeks of therapy with testosterone enanthate, 200 mg intramuscular weekly, plus saline placebo or hCG (125, 250, or 500 IU) subcutaneously every other day.[25] *Gray bars* are mean baseline ITT levels. *Striped bars* are mean post-treatment ITT levels on day 21.

testosterone concentration and sperm count for the one patient who did not become azoospermic.

In order to assess the effects of hormonal therapy upon intratesticular testosterone, we measured the intratesticular testosterone concentration in a group of fertile volunteers receiving various dosages of human chorionic gonadotropin (hCG).[24] Endogenous testosterone production was suppressed, as demonstrated in our previous study,[25] by administration of weekly testosterone enanthate. We observed a direct relationship between HCG dose and intratesticular testosterone concentration (FIG. 6), with restoration of baseline levels with dosages as low as 125 units of hCG every other day. The dosages of hCG that effectively restored "normal" intratesticular testosterone concentrations in this study we re significantly lower (10%) than the dosages normally used in clinical practice.[25–28] Further study is needed to determine whether these dosages can be effectively used to treat male infertility and avoid some of the commonly observed adverse effects of supraphysiologic hCG doses, such as elevated estradiol levels and gynecomastia.[29]

SUMMARY

In summary, assessment of the testicular hormonal environment through percutaneous fluid aspiration has shown a similar testis-to-serum gradient of testosterone as noted in previous testicular biopsy studies in men and rats. Intratesticular testosterone is much higher than serum levels in normal young men. Intratesticular androgenic bioactivity is ~70% of intratesticular testosterone concentration as measured by RIA or mass spectroscopy. Using these methods, we have been able to document a profound reduction of both intratesticular testosterone concentration andandrogen bioactivity in volunteers receiving hormonal contraception by using a combination of testosterone enanthate and levonorgestrel. Suppression of intratesticular testosterone to levels comparable to those in serum was associated with near-complete suppression of spermatogenesis. Yet, not all men became azoospermic, suggesting that there is another mechanism for maintenance of low-level spermatogenesis in men resistant to standard hormonal contraceptive therapy. In addition, we have been able to directly monitor the testicular response to hormonal therapy, specifically hCG. We have found that very low doses of hCG maintain baseline levels of intratesticular testosterone concentration in men with gonadotropin withdrawal from exogenous testosterone administration. Lower doses of hCG may be as effective in treating male infertility due to hypogonadotropism as the higher doses used historically.

ACKNOWLEDGMENT

Dr. Jarow's and Dr. Zirkin's research is supported by National Institutes of Health Grant HD044258.

REFERENCES

1. Thonneau, P., S. Marchand, A. Tallec, *et al.* 1991. Incidence and main causes of infertility in a resident population (1,850,000) of three French regions (1988–1989). Hum. Reprod. **6:** 81–86.
2. Greenberg, S.H., L.I. Lipshultz & A.J. Wein. 1978. Experience with 425 subfertile male patients. J. Urol. **119:** 507–510.
3. Palermo, G.D., J. Cohen & Z. Rosenwaks. 1996. Intracytoplasmic sperm injection: a powerful tool to overcome fertilization failure. Fertil. Steril. **65:** 899–908.
4. Callahan, T.L., J.E. Hall, S.L. Ettner, *et al.* 1994. The economic impact of multiple-gestation pregnancies and the contribution of assisted-reproduction techniques to their incidence. N. Engl. J. Med. **331:** 244–249.
5. Bassil, S., P.A. Godin, S. Stallaert, *et al.* 1995. Ovarian hyperstimulation syndrome. A review. Assisted Reprod. Rev. **5:** 90–96.
6. Turner, T.T., C.E. Jones, S.S. Howards, *et al.* 1984. On the androgen microenvironment of maturing spermatozoa. Endocrinology **115:** 1925–1932.
7. Zirkin, B.R., R. Santulli, C.A. Awoniyi & L.L. Ewing. 1989. Maintenance of advanced spermatogenic cells in the adult rat testis: quantitative relationship to testosterone concentration within the testis. Endocrinology **124:** 3043–3049.
8. Awoniyi, C.A., R. Santulli, V. Chandrashekar, *et al.* 1989. Quantitative restoration of advanced spermatogenic cells in adult male rats made azoospermic by active immunization against luteinizing hormone or gonadotropin-releasing hormone. Endocrinology **125:** 1303–1309.
9. Joseph, D.R. 1994. Structure, function, and regulation of androgen-binding protein/sex hormone-binding globulin. Vitamins Hormones **49:** 197–280.

10. JOSEPH, D.R., D.A. O'BRIEN, P.M. SULLIVAN, *et al.* 1997. Overexpression of androgen-binding protein/sex hormone-binding globulin in male transgenic mice: tissue distribution and phenotypic disorders. Biol. Reprod. **56:** 21–32.
11. ROBERTS, K.P. & B.R. ZIRKIN. 1993. Androgen binding protein inhibition of androgen dependent transcription explains the high minimal testosterone concentration required to maintain spermatogenesis in the rat. Endocr. J. **1:** 41–47.
12. JEYARAJ, D.A., G. GROSSMAN, C. WEAVER & P. PETRUSZ. 2002. Dynamics of testicular germ cell proliferation in normal mice and transgenic mice overexpressing rat androgen-binding protein: a flow cytometric evaluation. Biol. Reprod. **66:** 877–885.
13. JANNE, M., H.K. DEOL, S.G. POWER, *et al.* 1998. Human sex hormone-binding globulin gene expression in transgenic mice. Mol. Endocrinol. **12:** 123–136.
14. JAROW, J.P., H. CHEN, T.W. ROSNER, *et al.* 2001. Assessment of the androgen environment within the human testis: minimally invasive method to obtain intratesticular fluid. J. Androl. **22:** 640–645.
15. ADAMOPOULOS, D., D.M. LAWRENCE, P. VASSILOPOULOS, *et al.* 1984. Hormone levels in the reproductive system of normospermic men and patients with oligospermia and varicocele. J. Clin. Endocrinol. Metab. **59:** 447–452.
16. SEALEY, J.E., M. GOLDSTEIN, T. PITARRESI, *et al.* 1988. Prorenin secretion from human testis: no evidence for secretion of active renin or angiotensinogen. J. Clin. Endocrinol.-Metab. **66:** 974–978.
17. SWERDLOFF, R.S. & P.C. WALSH. 1975. Pituitary and gonadal hormones in patients with varicocele. Fertil. Steril. **26:** 1006–1012.
18. MORSE, H.C., N. HORIKE, M.J. ROWLEY & C.G. HELLER. 1973. Testosterone concentrations in testes of normal men: effects of testosterone propionate administration. J. Clin. Endocrinol. Metab. **37:** 882–886.
19. ZHAO, M., S.D. BAKER, X. YAN, *et al.* 2004. Simultaneous determination of steroid composition of human testicular fluid using liquid chromatography tandem mass spectrometry. Steroids **69:** 721–726.
20. RAIVIO, T., J.J. PALVIMO, L. DUNKEL, *et al.* 2001. Novel assay for determination of androgen bioactivity in human serum. J. Clin. Endocrinol. Metab. **86:** 1539—1544.
21. RAIVIO, T., J.S. TAPANAINEN, P. KUNELIUS & O.A. JANNE. 2002. Serum androgen bioactivity during 5α-dihydrotestosterone treatment in elderly men. J. Androl. **23:** 919– 921.
22. JAROW, J.P., W.W. WRIGHT, T.R. BROWN, *et al.* 2005. Bioactivity of androgens within the testes and serum of normal men. J. Androl. **26:** 343—348.
23. COVIELLO, A.D., W.J. BREMNER, A.M. MATSUMOTO, *et al.* 2004. Intratesticular testosterone concentrations comparable with serum levels are not sufficient to maintain normal sperm production in men receiving a hormonal contraceptive regimen. J. Androl. **25:** 931–938.
24. COVIELLO, A.D., A.M. MATSUMOTO, W.J. BREMNER, *et al.* 2005. Low-dose human chorionic gonadotropin maintains intratesticular testosterone in normal men with testosterone-induced gonadotropin suppression. J. Clin. Endocrinol. Metab. **90:** 2595–2602.
25. LIU, P.Y., L. TURNER, D. RUSHFORD, *et al.* 1999. Efficacy and safety of recombinant human follicle stimulating hormone (Gonal-F) with urinary human chorionic gonadotrophin for induction of spermatogenesis and fertility in gonadotrophin-deficient men. Hum. Reprod. **14:** 1540–1545.
26. KLIESCH, S., H.M. BEHRE & E. NIESCHLAG. 1995. Recombinant human follicle-stimulating hormone and human chorionic gonadotropin for induction of spermatogenesis in a hypogonadotropic male. Fertil. Steril. **63:** 1326–1328.
27. LIU, L., S.M. BANKS, K.M. BARNES & R.J. SHERINS. 1988. Two-year comparison of testicular responses to pulsatile gonadotropin-releasing hormone and exogenous gonadotropins from the inception of therapy in men with isolated hypogonadotropic hypogonadism. J. Clin. Endocrinol. Metab. **67:** 1140–1145.
28. MATSUMOTO, A.M. & W.J. BREMNER. 1985. Stimulation of sperm production by human chorionic gonadotropin after prolonged gonadotropin suppression in normal men. J. Androl. **6:** 137–143.
29. ZITZMANN, M. & E. NIESCHLAG. 2000. Hormone substitution in male hypogonadism. Mol. Cell. Endocrinol. **161:** 73–88.

Sperm Chemotaxis

A Primer

E. HILDEBRAND AND U.B. KAUPP

Institut für Biologische Informationsverarbeitung, Forschungszentrum Jülich, 52425 Jülich, Germany

ABSTRACT: **Sperm become attracted by chemical substances that are released from the outer coating of the egg, a process called chemotaxis. In this paper the cellular pathway and the motor response during chemotaxis of sperm from sea urchin and starfish are briefly outlined.**

KEYWORDS: **calcium; cyclic nucleotides; ion channels; signaling; motor response**

INTRODUCTION

The study of chemotaxis of sperm is most advanced in sea urchin and starfish because the chemical nature of the chemoattractants and the identity of their respective receptors have been firmly established. Chemotaxis of mammalian sperm is established as well; however, the cellular pathways and the behavioral responses are only rudimentarily known. The interested reader is referred to a comprehensive up-to-date review by Eisenbach and Giojalas.[1]

CELLULAR SIGNALING PATHWAY

In marine invertebrates, notably sea urchin and starfish, small peptides have been identified that are chemotactically active,[2,3] for example, asterosap, a 34-amino-acid peptide from the starfish *Asteria amurensis,*[4,5] and resact, a 14-amino-acid peptide from the sea urchin *Arbacia punctulata.*[5–9] The peptides bind to receptor-type guanylyl cyclases (GC) on the cell membrane of sperm and thereby stimulate the synthesis of cGMP from GTP.[10,11] The ensuing rapid rise of cGMP represents the primary signaling event.[9]

The rise of cGMP has been proposed to hyperpolarize the cell due to efflux of K^+ ions via a K^+-selective ion channel in the membrane of the flagellum.[12–14] This channel has not yet been identified. An attractive hypothesis is that it belongs to the family of cyclic nucleotide-gated (CNG) channels[15] that are directly gated by cAMP and cGMP.[16] In vertebrates, CNG channels are nonselective cation channels that let

Address for correspondence: Prof. U.B. Kaupp, FZJ-IBI-1, 52425 Jülich, Germany. Voice: 492461614041; fax: 492461614216.
a.eckert@fz-juelich.de

**Ann. N.Y. Acad. Sci. 1061: 221–225 (2005). © 2005 New York Academy of Sciences.
doi: 10.1196/annals.1336.024**

pass K^+ and Na^+ ions equally well. Thus, opening these CNG channels does not result in a hyperpolarization. However, in ciliary photoreceptors of the scallop *Pecten*, a K^+-selective cGMP-gated channel mediates a hyperpolarizing light response.[17] Furthermore, a K^+-selective CNG channel has been identified in prokaryotes.[18,19]

The stimulation with chemoattractants initiates a rapid influx of Ca^{2+} ions from the extracellular medium.[9,20–23] The molecular identity of the Ca^{2+} channel and the mechanism of its activation are ill-defined. The Ca^{2+} channel probably belongs to the family of voltage-dependent $Ca^{2+}{}_v$ channels.[24,25] Perhaps sperm are partially depolarized, that is, the membrane voltage (V_m) is significantly more positive than the K^+ equilibrium potential ($E_K \sim -90$ mV). At this depolarized "resting state," voltage-dependent $Ca^{2+}{}_v$ channels are inactivated (for review, see Refs. 24 and 25). Upon hyperpolarization, $Ca^{2+}{}_v$ channels recover from inactivation and open when the membrane again becomes depolarized.

Interestingly, sea urchin sperm harbor in the flagellum an ion channel that belongs to the family of hyperpolarization-activated and cyclic nucleotide-gated (HCN) ion channels.[26] HCN channels display a reversal potential of about –20 to –30 mV and, when activated by hyperpolarization, these channels carry a depolarizing inward current carried by Na^+ ions.[27] The open probability of the HCN channel from *Strongylocentrotus purpuratus* is substantial at a V_m of ca. –40 mV;[26] furthermore, a small fraction of HCN channels are constantly open, independent of voltage.[26,28] Therefore, the HCN channel, as in neurons, may codetermine the resting voltage. The Na^+ inward current may contribute to the recovery from the hyperpolarizing response. Finally, this channel is exquisitely sensitive to cAMP; therefore, cAMP may modulate its physiological function. Peptides evoke a host of additional cellular reactions, notably a rise of intracellular pH (pH_i) and cAMP (for review, see Refs. 29 to 32). The functional significance of these reactions is controversial.[33,34] Recent kinetic studies argue that changes in pH_i and cAMP do not participate directly in chemotaxis[9,35,36] but may subserve other physiological functions. It is beyond the scope of this short account to discuss these issues adequately. A more detailed discussion is given in Kaupp *et al.* (in preparation).

MOTOR RESPONSE AND BEHAVIOR IN A GRADIENT OF CHEMOATTRACTANT

The Ca^{2+} dynamics in the flagellum controls the flagellar beat and, thereby, the swimming trajectory of sperm. The study of swimming behavior during chemotaxis has been guided by a simple yet plausible concept. Sperm swim on a straight or curvilinear path towards the source of chemoattractant, and they turn when swimming away from the source.[2,21] This model implies that the swimming mode ("straight" versus "turn") is determined by the direction of swimming with respect to the gradient. Furthermore, the flagellar beat of a sperm whose membrane has been rendered leaky with mild detergents is more asymmetrical at high than at low $[Ca^{2+}]$.[37] Asymmetrical and symmetrical beating results in high and low curvature of the trajectory, respectively.[9] By inference, when sperm swim up-gradient on a straight trajectory, $[Ca^{2+}]_i$ should be low and should become high when sperm swim down-gradient.

It was not until the introduction of caged compounds that the chemotactic behavior could be systematically studied. "Caged compound" is the generic term for molecules that have been functionally disabled by covalent modification with a photolabile group ("phototrigger") (for review, see Ref. 38). On irradiation with a flash of UV light, the molecule is photolytically cleaved and its active form is released. Two different classes of caged compounds were developed for the study of sperm chemotaxis: caged cyclic nucleotides and caged chemoattractants.[5,9,23,38–43] Caged cGMP inside the cell permits bypass of the receptor GC and initiation of downstream signaling events by a step increase of cGMP. Caged resact allows the rapid establishment of a gradient of the chemoattractant with a well-defined concentration profile and also allows following of the trajectory of sperm in the gradient.[5,9]

In a shallow experimental chamber, sperm swim in circles of about 60 μm in diameter.[9] When stimulated by a step increase of cGMP, sperm leave their circular swimming path and make a turn that is followed by a period of straight or less-curved swimming ("turn-and-run").[5,9] This motor response unit is brought about by an increase of the asymmetry of the flagellar beat followed by a decrease. A similar response is elicited by resact released from a caged compound.[5] The motor response is repeated several times, until sperm resume their circular swimming pattern. Thus, the trajectory is strung together from turn-and-run units. The turn-and-run pattern is produced by cGMP-induced Ca^{2+} spikes. Each turn is preceded by a rise of $[Ca^{2+}]_i$,[5] and straightening of the trajectory is initiated by a decrease of $[Ca^{2+}]_i$.

Recently, it has become feasible to follow the swimming trajectory of sperm in a resact gradient produced by UV irradiation of a caged compound.[5] Sperm approach the source by moving in circular loops along a curved or straight path. This pattern consists of alternating periods of high and low curvature of the trajectory. The Ca^{2+} spikes appear synchronously with the stimulus function, that is, with the resact concentration along the trajectory.[5]

SUMMARY AND OUTLOOK

Ca^{2+} dynamics control the flagellar beat and thereby the swimming trajectory of sperm. At least three ion channels—a K^+-selective CNG channel, an HCN channel, and a voltage-dependent Ca^+ channel—determine the resting voltage and the Ca^{2+} entry. Activation of the chemoreceptor GC initiates at least two signaling events: a rapid rise of the cGMP concentration followed by an increase of $[Ca^{2+}]_i$. This scheme is still incomplete, because some of the channels have not yet been identified on a molecular level.

While a coherent picture of the excitatory signaling events is emerging, the mechanisms that terminate signaling and reset the sensitivity to the ligand are largely unknown. In order to escape saturation of both the receptor occupancy and the Ca^{2+} response, the cell must dispose of mechanisms that regulate its sensitivity. While swimming up-gradient, a cell probably continuously adapts to the ever-increasing concentration of the chemoattractant. A comprehensive model of chemotaxis must, therefore, incorporate the interplay between excitation, adaptation, and recovery.

REFERENCES

1. EISENBACH, M., J.W. LENGELER, M. VARON, *et al.* 2004. Chemotaxis. Imperial College Press. London.

2. MILLER, R.L. 1985. Sperm chemo-orientation in the metazoa. *In* Biology of Fertilization, vol. 2: Biology of the Sperm. C.B. Metz & A. Monroy, Eds.: 275–337. Academic Press. New York.
3. SUZUKI, N. 1995. Structure, function and biosynthesis of sperm-activating peptides and fucose sulfate glycoconjugate in the extracellular coat of sea urchin eggs. Zool. Sci. **12:** 13–27.
4. NISHIGAKI, T., K. CHIBA, W. MIKI & M. HOSHI. 1996. Structure and function of asterosaps, sperm-activating peptides from the jelly coat of starfish eggs. Zygote **4:** 237–245.
5. BÖHMER, M., Q. VAN, I. WEYAND, *et al.* 2005. Ca^{2+} spikes in the flagellum control chemotactic behavior of sperm. EMBO J. **24:** 2741–2752.
6. SUZUKI, N. & D.L. GARBERS. 1984. Stimulation of sperm respiration rates by speract and resact at alkaline extracellular pH. Biol. Reprod. **30:** 1167–1174.
7. SUZUKI, N., H. SHIMOMURA, E.W. RADANY, *et al.* 1984. A peptide associated with eggs causes a mobility shift in a major plasma membrane protein of spermatozoa. J. Biol. Chem. **259:** 14874–14879.
8. WARD, G.E., C.J. BROKAW, D.L. GARBERS & V.D. VACQUIER. 1985. Chemotaxis of *Arbacia punctulata* spermatozoa to resact, a peptide from the egg jelly layer. J. Cell Biol. **101:** 2324–2329.
9. KAUPP, U.B., J. SOLZIN, E. HILDEBRAND, *et al.* 2003. The signal flow and motor response controling chemotaxis of sea urchin sperm. Nat. Cell Biol. **5:** 109–117.
10. RAMARAO, C.S. & D.L. GARBERS. 1985. Receptor-mediated regulation of guanylate cyclase activity in spermatozoa. J. Biol. Chem. **260:** 8390–8396.
11. SHIMOMURA, H., L.J. DANGOTT & D.L. GARBERS. 1986. Covalent coupling of a resact analogue to guanylate cyclase. J. Biol. Chem. **261:** 15778–15782.
12. LEE, H.C. & D.L. GARBERS. 1986. Modulation of the voltage-sensitive Na^+/H^+ exchange in sea urchin spermatozoa through membrane potential changes induced by the egg peptide speract. J. Biol. Chem. **261:** 16026–16032.
13. BABCOCK, D.F., M.M. BOSMA, D.E. BATTAGLIA & A. DARSZON. 1992. Early persistent activation of sperm K^+ channels by the egg peptide speract. Proc. Natl. Acad. Sci. USA **89:** 6001–6005.
14. COOK, S.P. & D.F. BABCOCK. 1993. Selective modulation by cGMP of the K^+ channel activated by speract. J. Biol. Chem. **268:** 22402–22407.
15. GALINDO, B.E., C. BELTRÁN, E.J. CRAGOE, JR., & A. DARSZON. 2000. Participation of a K^+ channel modulated directly by cGMP in the speract-induced signaling cascade of *Strongylocentrotus purpuratus* sea urchin sperm. Dev. Biol. **221:** 285–294.
16. KAUPP, U.B. & R. SEIFERT. 2002. Cyclic nucleotide-gated ion channels. Physiol. Rev. **82:** 769–824.
17. DEL PILAR GOMEZ, M. & E. NASI. 1995. Activation of light-dependent K^+ channels in ciliary invertebrate photoreceptors involves cGMP but not the IP_3/Ca^{2+} cascade. Neuron **15:** 607–618.
18. NIMIGEAN, C.M., T. SHANE & C. MILLER. 2004. A cyclic nucleotide modulated prokaryotic K^+ channel. J. Gen. Physiol. **124:** 203–210.
19. CLAYTON, G.M., W.R. SILVERMAN, L. HEGINBOTHAM & J.H. MORAIS-CABRAL. 2004. Structural basis of ligand activation in a cyclic nucleotide regulated potassium channel. Cell **119:** 615–627.
20. SCHACKMANN, R.W. & P.B. CHOCK. 1986. Alteration of intracellular $[Ca^{2+}]$ in sea urchin sperm by the egg peptide speract. J. Biol. Chem. **261:** 8719–8728.
21. COOK, S.P., C.J. BROKAW, C.H. MULLER & D.F. BABCOCK. 1994. Sperm chemotaxis: egg peptides control cytosolic calcium to regulate flagellar responses. Dev. Biol. **165:** 10–19.
22. COOK, S.P. & D.F. BABCOCK. 1993. Activation of Ca^{2+} permeability by cAMP is coordinated through the pH_i increase induced by speract. J. Biol. Chem. **268:** 22408–22413.
23. NISHIGAKI, T., C.D. WOOD, Y. TATSU, *et al.* 2004. A sea urchin egg jelly peptide induces a cGMP-mediated decrease in sperm intracellular Ca^{2+} before its increase. Dev. Biol. **272:** 376–388.
24. CENS, T., M. ROUSSET, J.P. LEYRIS, *et al.* 2005. Voltage- and calcium-dependent inactivation in high voltage-gated Ca^{2+} channels. Prog. Biophys. Mol. Biol. In press.

25. PEREZ-REYES, E. 2003. Molecular physiology of low-voltage-activated T-type calcium channels. Physiol. Rev. **83:** 117–161.
26. GAUSS, R., R. SEIFERT & U.B. KAUPP. 1998. Molecular identification of a hyperpolarization-activated channel in sea urchin sperm. Nature **393:** 583–587.
27. PAPE, H.-C. 1996. Queer current and pacemaker: the hyperpolarization-activated cation current in neurons. Annu. Rev. Physiol. **58:** 299–327.
28. PROENZA, C., D. ANGOLI, E. AGRANOVICH, *et al.* 2002. Pacemaker channels produce an instantaneous current. J. Biol. Chem. **277:** 5101–5109.
29. GARBERS, D.L. 1989. Molecular basis of fertilization. Annu. Rev. Biochem. **58:** 719–742.
30. DARSZON, A., P. LABARCA, T. NISHIGAKI & F. ESPINOSA. 1999. Ion channels in sperm physiology. Physiol. Rev. **79:** 481–510.
31. DARSZON, A., C. BELTRÁN, R. FELIX, *et al.* 2001. Ion transport in sperm signaling. Dev. Biol. **240:** 1–14.
32. DARSZON, A., T. NISHIGAKI, C. WOOD *et al.* 2005. Calcium channels and Ca^{2+} fluctuations in sperm physiology. Int. Rev. Cytol. **243:** 79–172.
33. KIRKMAN-BROWN, J.C., K.A. SUTTON & H.M. FLORMAN. 2003. How to attract sperm. Nat. Cell Biol. **5:** 93–96.
34. EISENBACH, M. 2004. Towards understanding the molecular mechanism of sperm chemotaxis. J. Gen. Physiol. **124:** 105–108.
35. MATSUMOTO, M., J. SOLZIN, A. HELBIG, *et al.* 2003. A sperm-activating peptide controls a cGMP-signaling pathway in starfish sperm. Dev. Biol. **260:** 314–324.
36. SOLZIN, J., A. HELBIG, Q. VAN, *et al.* 2004. Revisiting the role of H^+ in chemotactic signaling of sperm. J. Gen. Physiol. **124:** 115–124.
37. BROKAW, C.J., R. JOSSLIN & L. BOBROW. 1974. Calcium ion regulation of flagellar beat symmetry in reactivated sea urchin spermatozoa. Biochem. Biophys. Res. Commun. **58:** 795–800.
38. HAGEN, V., K. BENNDORF & U.B. KAUPP. 2005. Control of cellular activity. *In* Dynamic Studies in Biology. M. Goeldner & R.S. Givens, Eds.: 155-178. Wiley-VCH. Weinheim.
39. HAGEN, V., J. BENDIG, S. FRINGS, *et al.* 2001. Highly efficient and ultrafast phototriggers for cAMP and cGMP by using long-wavelength UV/Vis-activation. Angew. Chem. Int. Ed. **40:** 1046–1048.
40. TATSU, Y., T. NISHIGAKI, A. DARSZON & N. YUMOTO. 2002. A caged sperm-activating peptide that has a photocleavable protecting group on the backbone amide. FEBS Lett. **525:** 20–24.
41. SHIBA, K., J. OHMURO, Y. MOGAMI, *et al.* 2005. Sperm-activating peptide induces asymmetric flagellar bending in sea urchin sperm. Zool. Sci. **22:** 293–299.
42. WOOD, C.D., T. NISHIGAKI, T. FURUTA, *et al.* 2005. Real-time analysis of the role of Ca^{2+} in flagellar movement and motility in single sea urchin sperm. J. Cell Biol. **169:** 725–731.
43. HAGEN, V., S. FRINGS, J. BENDIG, *et al.* 2002. Fluorescence spectroscopic quantification of the release of cyclic nucleotides from photocleavable [bis(carboxymethoxy)coumarin-4-yl]methyl esters inside cells. Angew. Chem. Int. Ed. **41:** 3625–3628.

Epigenetic Regulation of Testis-Specific Gene Expression

JOHN R. McCARREY,[a,b] CHRISTOPHER B. GEYER,[b] AND HIROTAKA YOSHIOKA[a]

[a]*Department of Biology, University of Texas at San Antonio, San Antonio, Texas, USA*

[b]*Department of Cellular and Structural Biology, University of Texas Health Science Center at San Antonio, San Antonio, Texas, USA*

ABSTRACT: Cellular differentiation is mediated by differential gene expression. The cells of the testis are no exception. Indeed, recent studies based on microarray and expressed sequence tag analyses have revealed dynamic changes in gene expression during spermatogenesis. The autosomal phosphoglycerate kinase gene *Pgk2* is an example of a gene that is tightly regulated during spermatogenesis, with transcription initiating in primary spermatocytes and ceasing in postmeiotic spermatids. Our studies show that this tissue-specific transcription is regulated at multiple levels, including binding of ubiquitous and testis-specific transcription factors, reconfiguration of nucleosomes, decondensation of chromatin, modifications of specific residues in histones, and demethylation of DNA. Our ability to isolate relatively pure populations of different spermatogenic cell types has allowed us to determine the order of these events that lead up to initiation of transcription of the *Pgk2* gene. By examining similar parameters in transgenic mice carrying various portions of *Pgk2* regulatory sequence, we have localized specific signals responsible for each of these regulatory events. Together these results demonstrate that activation of testis-specific transcription of the *Pgk2* gene involves a multifaceted, ordered cascade of epigenetic events that potentiates this locus in preparation for the initiation of transcription.

KEYWORDS: spermatogenesis; *Pgk2*; transcription

INTRODUCTION

The testis is a complex organ made up of a variety of different cell types, including various types of somatic cells (e.g., Sertoli cells, Leydig cells, and peritubular cells) as well as germ cells. In the testis, germ cells undergo the process of spermatogenesis to give rise to the male gametes, spermatozoa, and so are collectively known as spermatogenic cells. In addition, the testicular somatic cell types perform a variety of distinct functions, including direct paracrine support of spermatogenic cells provided by Sertoli cells, production of androgens provided by Leydig cells, and enclosure of seminiferous tubules performed by peritubular cells. Among the vari-

Address for correspondence: Dr. John R. McCarrey, Department of Biology, University of Texas at San Antonio, 6900 N. Loop 1604 West, San Antonio, TX 78249. Voice: 210-458-4507; fax: 210-458-5658.

jmccarrey@utsa.edu

Ann. N.Y. Acad. Sci. 1061: 226–242 (2005). © 2005 New York Academy of Sciences. doi: 10.1196/annals.1336.025

ous spermatogenic cell types, prospermatogonia develop in the fetus as the first sex-specific manifestation of the male germ-cell lineage. These give rise to premeiotic spermatogonia in the prepuberal animal, including the formation of self-renewing (type A) spermatogonial stem cells at the base of the seminiferous epithelium and differentiating (type B) spermatogonia that are committed to enter meiosis. Spermatocytes are formed upon entry of male germ cells into meiosis, including primary spermatocytes (undergoing the first meiotic division) and secondary spermatocytes (undergoing the second meiotic division). Following meiosis, haploid spermatids are formed. These are first manifest as round spermatids that then undergo dramatic differentiation during the process of spermiogenesis to give rise to condensing or elongated spermatids and, finally, to spermatozoa. Following spermiogenesis, the spermatozoa are shed from the seminiferous epithelium, move to the rete testis, and then exit the testis, traverse the epididymis, and collect in the vas deferens.

As with any other tissue or organ in the body, the characteristic differentiation of specific cell types in the mammalian testis (somatic and germ cells alike) is based upon differential gene expression. The various cell types in the testis carry the same complete genome as that found in all other cell types in the body. Therefore, it is the unique patterns of expression of these genes that give rise to the unique cell types in the testis. Among testicular cell types, gene expression patterns have been most extensively studied in spermatogenic cells.[1,2] Indeed, recent studies have revealed that spermatogenic cells express a surprisingly large proportion of known genes in the genome. Results from a microarray study of gene expression in the mouse testis indicate that up to 50% of the murine genome may be involved in testis development and/or spermatogenesis at some point between birth and adulthood and that approximately 30% of the murine genome is differentially expressed in a significant manner during testis development.[1] Microarray studies of genes expressed in purified populations of premeiotic spermatogonia, meiotic spermatocytes, and postmeiotic spermatids show that the majority of the differential gene expression observed in the testis occurs in spermatogenic cells and that dynamic changes in gene expression accompany transitions from premeiotic to meiotic and from meiotic to postmeiotic stages.[1] Similarly, an ongoing study of expressed sequence tags found in cDNA libraries produced from purified populations of spermatogenic cell types suggests that as much as 43% of the genes expressed in spermatogenic cells are germ cell-specific genes, and confirms the occurrence of dramatic transitions in gene expression patterns in spermatogonia, spermatocytes, and spermatids, respectively (E.M. Eddy, personal communication).

The evidence for dynamic differential gene expression in the testis and particularly during spermatogenesis implies the ongoing function of gene regulation mechanisms. Evidence exists for regulatory mechanisms functioning at various levels of gene expression in spermatogenic cells. For instance, there are well-documented examples of post-transcriptional regulation that results in a critical delay between production of certain mRNAs and their subsequent translation.[3,4] However, an initial point of control over differential gene expression is imposed at the level of transcription, including regulation of (1) whether or not a gene is transcribed at all, (2) at what level a gene is transcribed, and (3) exactly when (in which cell types) a gene is transcribed.

Among the many germ-cell–specific genes expressed during spermatogenesis is a set of genes encoding isozymes that function during glycolysis, primarily in postmeiotic spermatogenic cells and spermatozoa. These include the *LdhC4* (also known as *Ldh3*), *Gapds*, *Pgm2*, and *Pgk2* genes, which encode sperm-specific forms of lactate dehydrogenase, glyceraldehyde-3-phosphate dehydrogenase, phosphoglycerate mutase, and phosphoglycerate kinase, respectively.[5] The process of glycolysis appears to play a critical role in determining motility and, hence, fertility of spermatozoa, as evidenced by the phenotype of a knockout of the gene encoding one of these sperm-specific isozymes, *Gapds*.[6] Each of these genes initiates expression of the encoded mRNA in spermatocytes, although protein products from each gene accumulate primarily in spermatids.

We have previously shown that spermatogenic cell-specific transcription of the *Pgk2* gene is regulated by the promoter of this gene and that as little as 232 bp of sequence 5′ to the translational start site is sufficient to drive appropriate spermatogenic cell type- and developmental stage-specific transcription of a reporter gene in transgenic mice.[7,8] In addition, using *in vitro* electrophoretic mobility shift assays (EMSAs), we showed that specific sequences in the core promoter of the *Pgk2* gene can bind ubiquitous transcription factors and that other sequences in an upstream enhancer region of the *Pgk2* promoter can bind testis-specific factors (putative transcriptional activators) as well as one factor found in somatic cells (a putative transcriptional repressor).[9] Factor binding activity was mapped to the GC-box and the CAAT-box in the *Pgk2* core promoter, both of which are known to bind ubiquitous factors Sp1 and NF-Y or CTF-1, respectively. However, this core promoter region alone is insufficient to drive expression of a reporter gene *in vivo* in transgenic mice. For *in vivo* expression, an additional 42 bp of 5′ sequence is minimally required.[8] This includes the enhancer region that binds multiple anonymous testis-specific factors detected in testis nuclear extracts by EMSAs and a single anonymous somatic factor found in somatic cell extracts.[9]

The results from our previous transgenic and EMSA experiments are consistent with the idea that the *Pgk2* gene is maintained in a repressed state in somatic cells by binding of a transcriptional repressor, while this gene is activated in spermatogenic cells by binding of one or more testis-specific transcriptional activators. However, while interactions between both cell type-specific and ubiquitous factors with gene-specific promoter sequences are critical to the formation of the preinitiation complex required to initiate transcription, these represent only one of several mechanisms that regulate tissue-specific transcription. Thus, in order for transcription factors to bind to their cognate recognition sites in gene promoters, the regions of DNA to which these bind must be present in an accessible chromatin structure. Genes that are not expressed in a particular tissue are typically maintained in a highly condensed chromatin structure that contributes to transcriptional repression by inhibiting access of transcription factors to their binding sites in gene promoters. Therefore, activation of tissue-specific transcription requires an initial decondensation of chromatin in the promoter region of a gene in order to facilitate subsequent factor binding. As this process predisposes transcriptional activation, it has been termed "potentiation."[10]

Gene potentiation results in a change in chromatin structure of a gene that directly influences function of that gene by predisposing transcriptional activation. Because this change in gene function is not based on a change in the primary DNA sequence of the gene but does represent a heritable change in the functional state of the gene, it represents an epigenetic mechanism.[11] Several key questions exist regarding the manner in which gene potentiation is accomplished. These include (1) how changes in chromatin structure are achieved, (2) how these events are signaled in a gene-specific manner, and (3) how these effects are propagated in a heritable fashion through cellular division.

The mammalian spermatogenic cell lineage is particularly amenable to studies of the gene potentiation process and the epigenetic mechanisms that regulate this process. This lineage is very well-characterized with respect to the sequential appearance of specific spermatogenic cell types (FIG. 1), and many of these cell types can be recovered in relatively pure populations such that they can be analyzed at the molecular level for characteristics of gene potentiation. We have taken advantage of

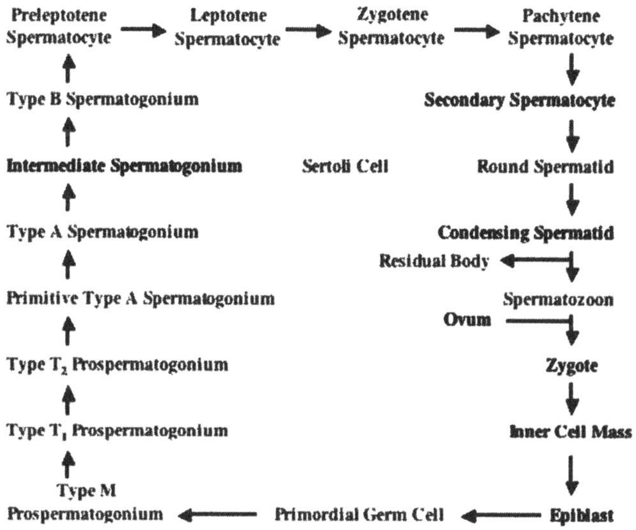

FIGURE 1. The mammalian spermatogenic cell lineage. The spermatogenic cell lineage begins with the setting aside of the primordial germ cells (PGCs), which are first detectable in mouse embryos at 7–7.25 dpc. The PGCs then give rise, in the testis, to type M prospermatogonia. These cells divide mitotically and then arrest as type T_1 prospermatogonia. Following birth in the mouse, prospermatogonia reinitiate mitotic activity as type T_2 prospermatogonia. A subset migrates to the periphery of the seminiferous epithelium, where they form primitive type A spermatogonia and then type A spermatogonia, which include the spermatogenic stem cells of the adult testis. These cells divide asymmetrically to continually populate the stem cell niche as well as to give rise to differentiating type B spermatogonia, which are committed to progress through spermatogenesis. Meiosis begins in primary spermatocytes and is complete with the appearance of haploid round spermatids. Spermatozoa are formed through the process of spermiogenesis, in which excess cytoplasm is removed to form residual bodies. The process of spermatogenesis in the seminiferous epithelium is supported by somatic Sertoli cells. Haploid spermatozoa are capable of fertilizing mature ova once they have exited the epididymis. As a result of fertilization of an ovum by a spermatozoon, a diploid zygote is formed, and this gives rise to an inner cell mass and then to the epiblast from which PGCs are derived, such that the cycle is reinitiated. It is possible, using various dissection and separation techniques, to isolate relativelypure populations of all of the cell types indicated in *red* [in online version].

this system to determine the order of epigenetic events that take place to achieve potentiation of the *Pgk2* gene and to study how these events are, themselves, regulated or signaled in a gene- and cell type-specific manner. The results of these studies have revealed that a cascade of epigenetic events occurs in an ordered and highly regulated manner, leading to potentiation of the *Pgk2* gene, prior to initiation of transcription of this gene in primary spermatocytes. We believe that this cascade of events represents an epigenetic regulatory hierarchy that is critical to the proper control of tissue-specific expression of the *Pgk2* gene during spermatogenesis.

MATERIAL AND METHODS

Isolation of Enriched Populations of Specific Spermatogenic Cell Types from the Mouse

Enriched populations of pre- and postnatal prospermatogonia (type T_1 and type T_2), premeiotic spermatogonia (primitive type A, type A, and type B), meiotic spermatocytes (preleptotene, leptotene plus zygotene, juvenile pachytene, and adult pachytene), and postmeiotic spermatids (round spermatids) and residual cytoplasmic bodies were recovered by sedimentation of dissociated seminiferous tubule cells through a 2–4% continuous StaPut gradient of bovine serum albumin at unit gravity as described elsewhere.[12,13] Spermatozoa were recovered from the testis, epididymides, and/or vas deferens by physically extruding cells from these tissues and subjecting these to sonication to lyse all but the sperm heads, which remained intact and were selectively recovered by centrifugation as described previously.[13] Purities of recovered cell types were assessed on the basis of morphological characteristics when viewed under phase optics and were ≥85% for prospermatogonia, spermatogonia, and juvenile spermatocytes (preleptotene, leptotene plus zygotene, and juvenile pachytene) and >95% for adult pachytene spermatocytes, round spermatids, residual cytoplasmic bodies, and testicular, epididymal, or vas deferens spermatozoa.

Analysis of Gene Expression at the RNA Level

Northern blot analysis was originally used to assess expression of the *Pgk2* gene in RNA samples prepared from purified populations of spermatogenic cells.[13] Subsequent experiments using reverse transcriptase PCR (RT-PCR) as described elsewhere[8,13] confirmed the Northern blotting results.

Production of PGK2/CAT Transgenic Mice

Transgene constructs consisting of varying portions of the human *PGK2* promoter sequence ligated to the chloramphenicol acetyltransferase (CAT) reporter gene were produced and introduced into mouse embryos by pronuclear injection to produce transgenic mice as described elsewhere.[7,8] Importantly, expression of *PGK2*/CAT transgenes bearing sufficient portions of the human *PGK2* promoter precisely mimicked the cell type- and developmental stage-specific pattern of expression of the endogenous mouse *Pgk2* gene. Thus, we concluded that regulation of transcription of the *PGK2*/CAT transgenes provides an accurate representation of the endogenous mouse *Pgk2* gene and, therefore, represents a valid model for studies of the mechanisms that regulate transcription directed by the *Pgk2* promoter.

Analysis of Protein–DNA Interactions

Protein–DNA interactions at the *Pgk2* promoter were initially assessed using an EMSA to determine the capacity of specific regions of the *Pgk2* promoter to bind factors in nuclear extracts from either testicular or somatic cell extracts *in vitro*.[9] Ongoing experiments to examine extant protein–DNA interactions at the *Pgk2* promoter as they exist in specific spermatogenic and somatic cell types *in vivo* are being performed by use of *in vivo* footprinting with the modifying agent dimethyl sulfate, followed by the ligation-mediated polymerase chain reaction and genomic sequencing to specifically amplify and examine the *Pgk2* promoter essentially as described previously.[14]

Analysis of Chromatin Structure

An assessment of general chromatin structure at the *Pgk2* locus in spermatogenic and somatic cell types was accomplished by digestion of nuclei with DNase I followed by specific amplification of the *Pgk2* gene by PCR as described elsewhere.[10] Relatively rapid digestion provided an indication of chromatin in a decondensed (10-nm fiber) state, whereas chromatin that was more resistant to nuclease was assumed to be present in a more condensed (30-nm fiber) structure. To detect reconfiguration of nucleosomes in the promoter region of the *Pgk2* gene, we treated nuclei from purified populations of specific spermatogenic or somatic cell types to detect a DNase I-hypersensitive site as described previously.[15]

Analysis of Histone Modifications

Ongoing studies of specific modifications of histones in chromatin at the *Pgk2* locus in spermatogenic or somatic cell types are being conducted using the chromatin immunoprecipitation assay essentially as described elsewhere.[16,17]

Analysis of DNA Methylation

Methylation of cytosines in CpG dinucleotides within recognition sites for methyl-sensitive restriction endonucleases at the *Pgk2* locus was originally assessed by preparing DNA from populations of purified cell types, digesting with an appropriate methyl-sensitive restriction endonuclease, and examining the digestion products by either Southern blot analysis or PCR.[18–20] Recently, a more thorough analysis of methylation at 23 CpG dinucleotides throughout the *Pgk2* locus was performed by bisulfite genomic sequencing as described elsewhere.[21] Briefly, genomic DNA from purified populations of spermatogenic or somatic cell types was treated with sodium bisulfite, which converts unmethylated cytosines to thymines but leaves methylated cytosines intact. By comparing the sequence of the *Pgk2* gene from a treated DNA sample with that from an untreated sample, the initial methylation status can then be directly deduced.

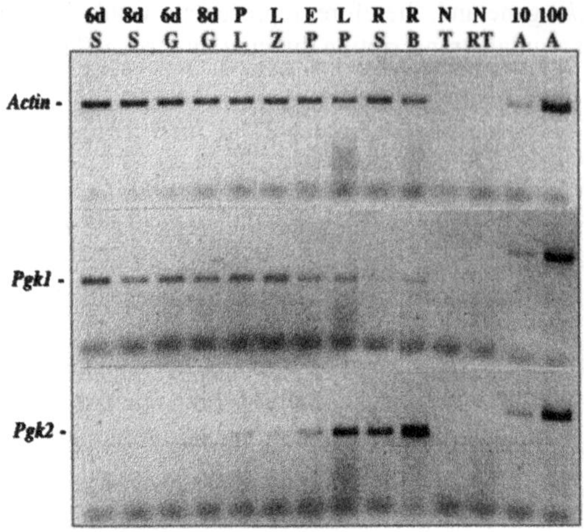

FIGURE 2. Expression of *Pgk* genes during spermatogenesis in the mouse. Semiquantitative RT-PCR was used to selectively detect RNA from the *Pgk1* and *Pgk2* genes in purified spermatogenic cell types. Detection of the RNA from the ubiquitously expressed β-actin gene (*Actin*) was used as a positive control and gel-loading control. *Pgk1* mRNA was detected in somatic Sertoli cells and premeiotic spermatogonia and, at declining levels, in meiotic spermatocytes. *Pgk2* mRNA was first detected at low levels in early meiotic spermatocytes and then at high levels in adult pachytene spermatocytes and postmeiotic spermatids. *6dS,* Sertoli cells from testes of 6-day-old mice; *8dS,* Sertoli cells from testes of 8-day-old mice; *6dG,* spermatogonia from testes of 6-day-old mice; *8dG,* spermatogonia from testes of 8-day-old mice; *PL, LZ,* and *EP,* preleptotene, leptotene plus zygotene, and early pachytene spermatocytes, respectively, from testes of 18-day-old mice; *LP, RS,* and *RB,* late pachytene speramatocytes, round spermatids, and residual bodies, respectively, from testes of adult (60- to 70-day-old) mice. PCR controls included *NT* (no template) and *NRT* (no reverse transcriptase), and loading controls included products derived from 10 ng (*10 A*) and 100 ng (*100 A*) of actin RNA.

RESULTS

Expression of the Pgk2 *Gene during Spermatogenesis in the Mouse*

mRNA from the *Pgk2* gene first becomes detectable during spermatogenesis in primary spermatocytes. The earliest spermatogenic cell type in which *Pgk2* mRNA is detectable is preleptotene spermatocytes (FIG. 2).[13] This observation correlates with results from whole testis RNA recovered from mice at different ages postpartum. In that case, *Pgk2* mRNA was first detected in testes from 12- to 14-day-old mice.[13] This corresponds to the age at which primary spermatocytes first appear in the maturing testis[12] and the time when active transcription of the *Pgk1* gene ceases.[13,15] *Pgk2* mRNA persists in postmeiotic round spermatids and is also detectable in residual cytoplasmic bodies, although evidence of some degradation of the mRNA was apparent on Northern blots of RNA from this latter cell type.[13] *Pgk2* mRNA was not detected in premeiotic spermatogonia or in any somatic cell type tested, including Sertoli cells.[13] Interestingly, as shown in FIGURE 2, transcription of the *Pgk2* gene initiates essentially coincident with cessation of transcription of the X-linked *Pgk1* gene due to meiotic sex chromosome inactivation, thus providing the PGK enzyme critically required in spermatozoa.[13,22] These data show that transcription of the *Pgk2* gene is testis specific and, within the testis, is limited to spermatocytes and spermatids. These expression data were further confirmed by nuclear runoff experiments, which demonstrated ongoing transcriptional activity of the *Pgk2* gene in both spermatocytes and round spermatids in the mouse.[15] These studies established that initiation of transcription of the *Pgk2* gene occurs specifically in primary spermatocytes and continues throughout meiosis and into postmeiotic round spermatids.

Analysis of Transgenic Mice to Identify Promoter Elements Required for Spermatogenic-Specific Transcription of the Pgk2 Gene

The sequence of the promoter has been determined for both the human and mouse *Pgk2* genes.[23,24] These promoters show significant similarity.[25] In both species the *Pgk2* promoter consists of a core promoter plus an upstream enhancer region, as summarized diagrammatically in FIGURE 3.

Recognizable elements within the "TATA-less" *Pgk2* core promoter include a GC-box and a CAAT-box. An enhancer region ("E1-E4") was found immediately upstream of the core promoter, and this region has been shown to bind up to three putative activators from nuclear extracts of mouse testicular cells at sites E1/E2 and E4 to direct appropriate testis-specific expression in transgenic mice. This region

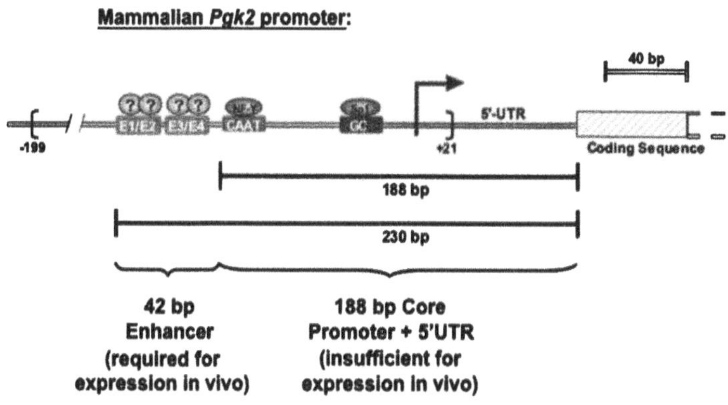

FIGURE 3. The mammalian *Pgk2* gene promoter. A schematic representation of the mammalian (mouse or human) *Pgk2* promoter is shown. The transcription start site is represented as a bent arrow, followed by the 5′-untranslated region (5′-UTR) and the beginning of the coding sequence. The core promoter as defined in transgenic mice includes 188 bp of sequence 5′ to the translational start site. (Note: it is not known whether the *Pgk2* 5′-UTR is required to direct appropriate tissue-specific transcription of this gene.) The core promoter includes a GC-box that binds the Sp1 factor and a CAAT-box that can bind NF-Y and/or CTF-1. Immediately upstream from the core promoter is an additional 42-bp enhancer region that has been shown to bind up to four anonymous factors, including three putative testis-specific activators of *Pgk2* transcription (E1, E2, and E4) and one putative somatic repressor of *Pgk2* transcription (E3). A DNase I-hypersensitive site has been detected over the *Pgk2* promoter in expressing cell types and has been mapped from –199 to +21 (brackets).

also binds a factor at site E3 in nonexpressing somatic cells that may represent a repressor.[8,9] The structure of the human and mouse *Pgk2* promoters is consistent with the suggestion that the core promoter is derived from a retroposed copy of a transcript from the X-linked *Pgkl* gene, along with the coding sequence of this gene,[23,26] and that the enhancer region subsequently evolved as expression of this retroposon became testis specific.[25,27] In both humans and mice, the *Pgk2* promoter lacks a CpG island and a TATA-box.[25] Although the core promoter of the human *Pgk2* gene is able to direct transcription of a downstream reporter gene in transient-transfection assays,[24] this portion of the *Pgk2* promoter alone is insufficient to direct expression of a reporter gene *in vivo* in transgenic mice.[7,8] Indeed, an additional 42 bp of enhancer sequence immediately upstream of the core promoter, along with the core promoter itself, is required to direct testis-specific expression *in vivo*.[8]

Analysis of Protein–DNA Interactions at the Pgk2 Promoter in Expressing Spermatogenic Cells and Nonexpressing Somatic Cells

EMSAs indicate that the GC-box and CAAT-box in the *Pgk2* core promoter have the capacity to bind the ubiquitous factors Sp1 and CTF-1 and/or NF-Y, respectively, while the *Pgk2* enhancer region can bind up to three or four anonymous, testis-specific activators (FIG. 3).[9] The occurrence of these protein–DNA interactions *in vivo* has been largely confirmed by ongoing *in vivo* footprinting experiments. *In vivo* footprints have been detected at the GC-box in the core promoter and at the E3/E4 enhancer region in spermatogenic cell types actively expressing the *Pgk2* gene. Although our *in vivo* footprinting experiments have not detected protein-DNA interactions at the E1/E2 enhancer region or at the CAAT box in expressing spermatogenic cell types, it may be that factors are actually bound to this region as well but escape detection by the *in vivo* methods used.

Analysis of Chromatin Structure and Histone Modifications at the Pgk2 Locus in Expressing Spermatogenic Cells and Nonexpressing Somatic Cells

A general increase in nuclease sensitivity, indicative of a transition from a condensed (repressed) chromatin structure to a decondensed (potentiated) chromatin structure, was observed in late spermatogonia just prior to the initiation of transcription of the *Pgk2* gene in spermatocytes.[10] The comparison of nuclease sensitivity within the *Pgk2* locus in premeiotic spermatogonia and postmeiotic spermatids shown in FIGURE 4 exemplifies this transition. Coincident with this decondensation of chromatin in the *Pgk2* gene, ongoing analyses of histone modifications indicate simultaneous changes, including an increase in acetylation of histones H3 and H4, as well as an increase in methylation at lysine 4 in histone H3.

A DNase I-hypersensitive site forms over the *Pgk2* promoter (FIG. 3) at about the same time or immediately following the transition in general nuclease sensitivity.[15] This hypersensitivity is believed to reflect a reconfiguration of nucleosomes (chromatin remodeling) over the *Pgk2* promoter that is presumably required to facilitate subsequent binding of transcription factors associated with initiation of transcription. Indeed, ongoing *in vivo* footprinting experiments suggest that protein-DNA interactions associated with active transcription are not established in the *Pgk2* promoter until *after* these changes in chromatin structure have initiated. However, these

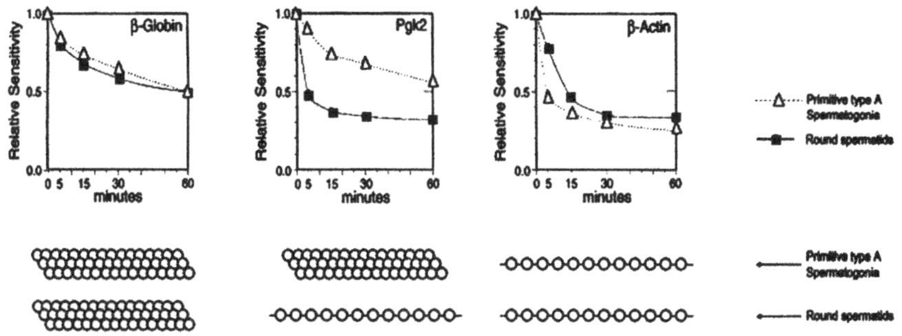

FIGURE 4. Cell type-specific changes in the chromatin structure of the *Pgk2* gene. A schematic representation of nuclease digestion experiments is shown based on data from Kramer *et al.*[10] The status of the chromatin structures of three different genes, β-globin (repressed in both spermatogonia and spermatids), *Pgk2* (repressed in spermatogonia, expressed in spermatids), and β-actin (expressed in both spermatogonia and spermatids), was examined in two spermatogenic cell types, premeiotic primitive type A spermatogonia and postmeiotic round spermatids. Relatively resistant digestion profiles for the repressed β-globin gene indicated that this gene is in a condensed chromatin configuration in both cell types (schematically represented below the digestion profiles). Conversely, relatively sensitive digestion profiles for the expressed β-actin gene indicated that this gene is in a decondensed chromatin configuration in both cell types. By comparison, the *Pgk2* gene appears to be in a condensed chromatin configuration in primitive type A spermatogonia, where it is still transcriptionally repressed, but is then in a decondensed chromatin configuration in round spermatids, where it is actively transcribed.

same studies reveal other protein–DNA interactions that are distinct from those associated with active transcription of this gene and that occur coincident with changes in chromatin structure prior to initiation of transcription. These may be indicative of factors involved in chromatin remodeling.

Analysis of Changes in DNA Methylation That Precede Activation of the Pgk2 *Gene in Spermatogenic Cells*

Interestingly, decondensation of chromatin, changes in histone modifications, reconfiguration of nucleosomes, and binding of transcription factors associated with the initiation of transcription are all preceded by a change in the status of DNA methylation over the 5′-half of the intronless *Pgk2* gene. Thus, we have shown that a process of apparently active demethylation of DNA is initiated over the promoter of the *Pgk2* gene in type T_1 prospermatogonia at approximately 18.5 days postcoitum (dpc) in the mouse and that this wave of demethylation then extends bidirectionally, such that by the time the cells have developed into primitive type A spermatogonia at 6 days postpartum (dpp), a domain of demethylation covering approximately 1.8 kb of sequence is found to be centered over the *Pgk2* promoter (FIG. 5).

Studies of transgenic mice carrying various portions of the *Pgk2* promoter ligated to a CAT reporter gene have demonstrated that the *Pgk2* core promoter is the site of a signal that directs this cell type- and developmental stage-specific demethyla-

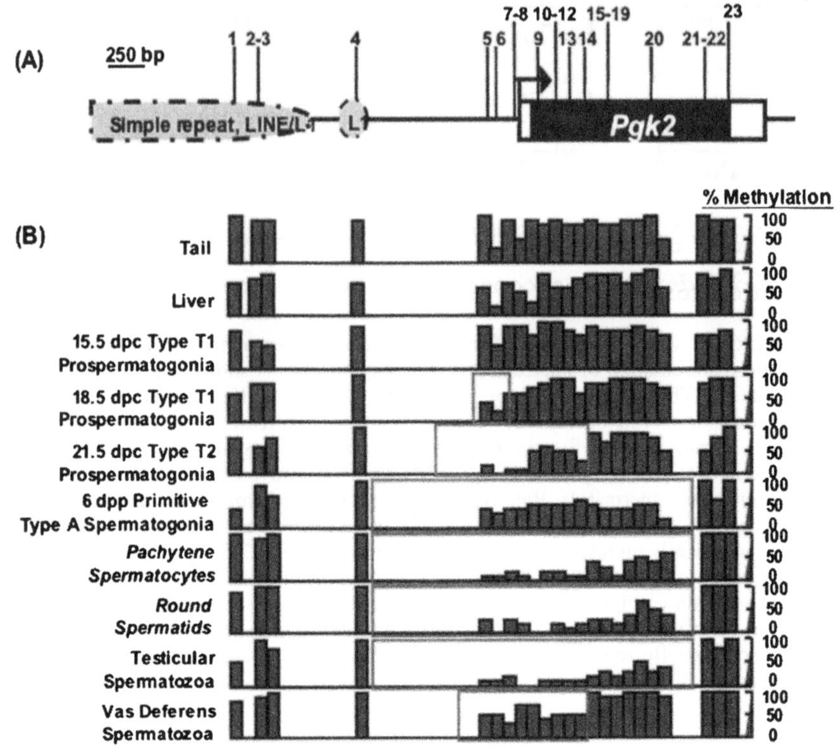

FIGURE 5. Status of DNA methylation across the mouse *Pgk2* gene locus in spermatogenic cells of the mouse. **(A)** Genomic structure of the mouse *Pgk2* gene locus. The locations of 23 CpGs are shown throughout the intronless *Pgk2* gene locus. These were analyzed for DNA methylation in different spermatogenic cell types by bisulfite genomic sequencing. **(B)** Spermatogenic cells were separated by unit gravity sedimentation (StaPut gradient). Bar graphs indicate the percentage of methylated CpGs among 10 separate subclones sequenced for each site. The domain of demethylation is indicated by red rectangles. Note the gradual development of a demethylation domain (*red boxes* in online version) prior to the activation of transcription in expressing cell types (spermatocytes and spermatids, *listed in italics*) and the subsequent remethylation of this domain in vas deferens spermatocytes after the cessation of transcription.

tion.[21,28] Specifically, this signal appears to reside in a region of approximately 32 bp in which the CAAT-box is the only recognizable and well-conserved sequence element. As noted above, the *Pgk2* core promoter alone (without the upstream enhancer sequence) is insufficient to direct expression of the *Pgk2* gene *in vivo*. Thus, it appears that the gene-specific signals required for (1) demethylation of the 5′-portion of the gene and (2) initiation of transcription of this same gene can be functionally uncoupled.

DISCUSSION

Because proper cellular differentiation and function are based largely upon properly regulated differential gene expression, it is as important to maintain well-regulated *repression* of genes that are not supposed to be expressed in a specific cell type as it is to maintain regulated *activity* of genes that are supposed to be expressed in the same cell type. Multiple mechanisms regulate differential gene expression. At the transcriptional level these include cell type-specific availability of key transcription factors, especially tissue-specific activators that typically bind to enhancer regions in gene promoters. However, many genes also contain core promoters that contain binding sites for ubiquitously expressed transcription factors which have the potential, even in the absence of tissue-specific activators binding to upstream enhancer sequences, to direct ectopic transcription at a basal level.[29] Thus, it is perhaps not surprising that additional mechanisms have evolved to fully repress transcriptional activity of genes in cell types in which they are not supposed to be expressed. Primary among these is local chromatin structure, which, when highly condensed, strongly inhibits access of transcription factors to their cognate binding sites. It is by this mechanism of chromatin condensation that a tissue-specific gene can remain fully repressed even in the presence of factors that might otherwise bind to the core promoter of that gene. However, to the extent that this mechanism is used to maintain transcriptional repression, it must be reversed to facilitate transcriptional activation of a tissue-specific gene.

Our studies of the status of the *Pgk2* gene during spermatogenesis indicate that an ordered cascade of molecular events is required to derepress or potentiate activation of transcription of this gene in primary spermatocytes. These events are summarized in FIGURE 6 and suggest that it is necessary to decondense chromatin at this locus prior to binding of transcription factors required for the initiation of transcription. Potentiation involves decondensation of chromatin and reconfiguration of nucleosomes, both of which are presumably required to ultimately expose DNA sequence-specific binding sites for key components of the preinitiation complex, the function of which is to locate and orient RNA polymerase for the initiation of transcription.

While it clearly appears to be the case that the molecular events involved in derepressing the *Pgk2* locus are critical regulators of transcriptional activation of this gene, it is also apparent that these events are, themselves, individually regulated. This is reflected in the fact that there is a specific order to the series of events that lead first to potentiation and then to initiation of transcription of this gene, as well as by the fact that these events occur in a cell type-, developmental stage-, and gene-specific manner. In this regard it is interesting that the earliest event we have observed is demethylation of DNA. The active development of a domain of demethylation over the 5′-half of the intronless *Pgk2* gene begins in type T_1 prospermatogonia at approximately 18.5 days of gestation in the mouse, at least 10 to 12 days prior to the initiation of transcription of this gene in primary spermatocytes.[21] This domain appears to be fully formed by the time stem spermatogonia are formed at about 6 dpp, which is still at least 6 days before initiation of transcription of this gene. An advantage of establishing this domain of demethylation at this early stage is that the active demethylation event need only occur once, *prior* to the spermatogonial stem cell stage, and then it can simply be propagated, rather than actively recreated, dur-

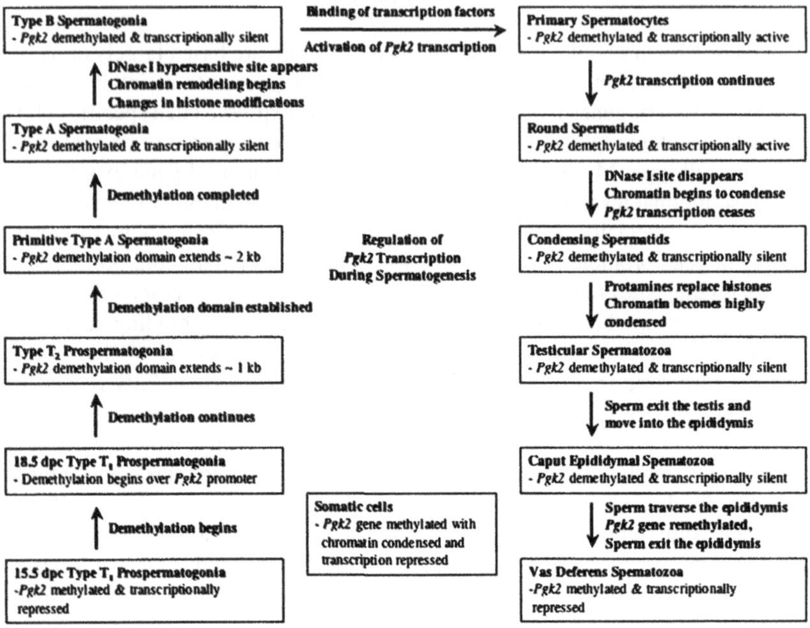

FIGURE 6. Regulation of *Pgk2* gene transcription during spermatogenesis in the mouse. A schematic representation is shown depicting tissues that give rise to the germ-cell lineage and the specific cell types in the murine spermatogenic lineage, along with key molecular events that occur in each cell type to potentiate and then initiate transcription of the testis-specific *Pgk2* gene in spermatocytes and round spermatids. Also indicated are molecular events that follow cessation of active transcription of the *Pgk2* gene in condensing spermatids, as well as the status of the *Pgk2* gene in nonexpressing somatic cells.

ing each successive wave of spermatogenesis in the mature testis. In addition, however, we believe this early demethylation event functions to signal the beginning of testis-specific derepression and activation of the *Pgk2* locus in spermatogenic cells and that this sets in motion the entire subsequent cascade of molecular changes required to fully potentiate and activate transcription of the *Pgk2* gene.

DNA methylation is a well-established epigenetic mechanism that differentially marks genes in a heritable fashion without changing the DNA sequence.[30] However, this modification of DNA also functions to signal changes in histone modifications and, subsequently, chromatin structure. Thus, it has been shown that methylated DNA binds "methyl-DNA binding proteins" (e.g., MeCP2) that, in turn, attract histone-modifying enzymes such as histone deacetylases and/or histone methylases.[31,32] The histone-modifying enzymes that associate with methyl-DNA binding proteins impose modifications (e.g., deacetylation) that stabilize a condensed (repressed) chromatin structure. It is therefore not surprising that many repressed genes are characterized by hypermethylation of DNA, especially in the promoter region.[30] This is also the case for the *Pgk2* gene.[21] Although not a "CpG island gene," the intronless *Pgk2* locus in the mouse includes 23 CpG dinucleotides spanning the region from upstream of the enhancer plus core promoter to downstream of the transcriptional stop signal. In somatic cells where this gene is terminally repressed, all of these CpG dinucleotides are predominantly methylated. The same is true of the *Pgk2* gene in type T_1 prospermatogonia at 15.5 dpc. However, beginning just 3 days later, in type T_1 prospermatogonia at 18.5 dpc, a domain of demethylation begins to form over the promoter of the *Pgk2* gene, and this domain then spreads bidirectionally such that by 6 dpp, a fully developed domain of demethylation is found covering a region of approximately 1.8 kb centered over the *Pgk2* promoter.[21]

It is only after the domain of demethylation has formed over the *Pgk2* promoter that we observe other changes associated with derepression of the locus. Thus, although demethylation is complete in primitive type A spermatogonia at 6 dpp, at this same stage the *Pgk2* locus continues to display a condensed chromatin structure and histone modifications that are commonly associated with a repressed gene. However, by approximately 2 days later in type B spermatogonia, we begin to observe changes in histone modifications and associated changes in chromatin structure indicating that the locus is undergoing decondensation of chromatin to transition to a potentiated state. At about the same time, we observe the formation of a DNase I-hypersensitive site, indicating reconfiguration of nucleosomes to expose sequence-specific binding sites for key transcription factors involved in the establishment of the preinitiation complex. That the preinitiation complex does indeed form only following the events of demethylation of DNA, modifications of histones favoring decondensation of chromatin, decondensation of chromatin, and reconfiguration of nucleosomes to expose transcription factor binding sites is confirmed by results of our ongoing *in vivo* footprinting experiments indicating that footprints of factors bound to the GC-box and *Pgk2* enhancer region observed in actively expressing spermatocytes and spermatids are not seen in premeiotic spermatogonia but do form in meiotic spermatocytes and then persist in postmeiotic round spermatids.

Interestingly, these same *in vivo* footprinting experiments have revealed distinct protein–DNA interactions in type A and type B spermatogonia that clearly differ from those associated with the actively transcribed *Pgk2* gene in spermatocytes and spermatids. It is known that specific factors are typically involved in chromatin remodeling prior to initiation of transcription.[33] Because the footprints we observe in the *Pgk2* promoter in premeiotic spermatogonia occur just prior to and during the time this gene undergoes decondensation of chromatin and reconfiguration of nucleosomes, these footprints may represent the presence of such chromatin remodeling factors.

Although many tissue-specific genes have been shown to be hypomethylated in cell types in which the gene is expressed and hypermethylated in cell types in which the gene is repressed,[34] very little is known about the mechanism by which these genes become demethylated in association with (or prior to) activation of transcription. The demethylation we have observed in the 5′-portion of the *Pgk2* gene clearly results from a mechanism of *active* demethylation, since it occurs predominantly in type T_1 prospermatogonia that are in a state of mitotic arrest. We previously showed that a signal that directs this gene-, developmental stage-, and cell type-specific demethylation event resides in the core promoter of the *Pgk2* gene.[21] Ongoing experiments with transgenic mice will allow us to further delineate the portion of the *Pgk2* promoter from which this signal emanates, and ongoing *in vivo* footprinting experiments will allow us to examine potential protein–DNA interactions that occur in this same region at the time when demethylation begins, to determine if key factors may signal and/or regulate this process.

CONCLUSIONS

In summary, by examining molecular parameters associated with the testis-specific *Pgk2* gene in purified populations of spermatogenic cell types beginning in the fetal period and extending throughout spermatogenesis in the adult, we have demonstrated a series of molecular events that lead up to activation of transcription of this gene in primary spermatocytes (FIG. 6). These events represent an ordered cascade that initially facilitates derepression of this gene and subsequently leads to initiation of transcription. Our results are consistent with the notion that each of these events is itself signaled by specific sequences in either the core promoter or the enhancer region of the 5′-regulatory region of the *Pgk2* gene. Moreover, we suggest that key events in this cascade may be necessary prerequisites for subsequent events. Thus, we hypothesize that demethylation of DNA over the 5′-portion of the *Pgk2* gene is required to facilitate subsequent changes in chromatin modifications and chromatin structure in this same region and that only after these events have occurred is it possible for reconfiguration of nucleosomes to expose binding sites for factors required to form the preinitiation complex that ultimately facilitates initiation of transcription. Our ongoing studies are designed to reveal the specific mechanisms by which these events are signaled in a gene- and cell type-specific manner and the extent to which each event is indeed required to facilitate the following event in this cascade.

ACKNOWLEDGMENTS

This work has been supported by National Institutes of Health Grants HD 42417 and HD 046637 to J.R.M.

REFERENCES

1. SHIMA, J.E., D.J. MCLEAN, J.R. MCCARREY & M.D. GRISWOLD. 2004. The murine testicular transcriptome: characterizing gene expression in the testis during the progression of spermatogenesis. Biol. Reprod. **71:** 319–330.
2. SCHULTZ, N., F.K. HAMRA & D.L. GARBERS. 2003. A multitude of genes expressed solely in meiotic or postmeiotic spermatogenic cells offers a myriad of contraceptive targets. Proc. Natl. Acad. Sci. USA **100:** 12201–12206.
3. YANG, J., S. MEDVEDEV, P.P. REDID, *et al.* 2005. The DNA/RNA-binding protein MSY2 marks specific transcripts for cytoplasmic storage in mouse male germ cells. Proc. Natl. Acad. Sci. USA **102:** 1513–1518.
4. BRAUN, R.E. 2000. Temporal control of protein synthesis during spermatogenesis. Int. J. Androl. **23**(Suppl. 2)**:** 92–94.
5. EDDY, E.M. 2002. Male germ cell gene expression. Recent Progress Hormone Res. **57:** 103–128.
6. MIKI, K., W. OU, E.H. GOULDING, *et al.* 2004. Glyceraldehyde 3-phosphate dehydrogenase-S, a sperm-specific glycolytic enzyme, is required for sperm motility and male fertility. Proc. Natl. Acad. Sci. USA **101:** 16501–16506.

7. ROBINSON, M.O., J.R. MCCARREY & M.I. SIMON. 1989. Transcriptional regulatory regions of testis-specific *PGK2* defined in transgenic mice. Proc. Natl. Acad. Sci. USA **86:** 8437–8441.
8. ZHANG, L.P., J. STROUD, C.A. EDDY, *et al.* 1999. Multiple elements influence transcriptional regulation from the human testis-specific *PGK2* promoter in transgenic mice. Biol. Reprod. **60:** 1329–1337.
9. GEBARA, M.M. & J.R. MCCARREY. 1992. Protein-DNA interactions associated with the onset of testis-specific expression of the mammalian *Pgk-2* gene. Mol. Cell. Biol. **12:** 1422–1431.
10. KRAMER, J.A., J.R. MCCARREY, D. DJAKIEW & S.A. KRAWETZ. 1998. Differentiation: the selective potentiation of chromatin domains. Development **125:** 4749–4755.
11. MCCARREY, J.R. 2003. Epigenetic mechanisms regulating gene expression. *In* Introduction to Bioinformatics. S.A. Krawetz & D.D. Womble, Eds.: 123–150. Humana Press Inc. Totowa, NJ.
12. BELLVÉ, A.R. 1993. Purification culture, and fractionation of spermatogenic cells. Methods Enzymol. **225:** 84–113.
13. MCCARREY, J.R., W.M. BERG, S.J. PARAGIOUDAKIS, *et al.* 1992. Differential transcription of *Pgk* genes during spermatogenesis in the mouse. Dev. Biol. **154:** 160–168.
14. HORNSTRA, I. K & T.P. YANG. 1993. *In vivo* footprinting and genomic sequencing by ligation-mediated PCR. Anal. Biochem. **213:** 179–193.
15. KUMARI, M., J.C. STROUD, A. ANJI & J.R. MCCARREY. 1996. Differential appearance of DNase I-hypersensitive sites correlates with differential transcription of *Pgk* genes during spermatogenesis in the mouse. J. Biol. Chem. **271:** 14390–14397.
16. FISCHLE, W., Y. WANG & C.D. ALLIS. 2003. Binary switches and modification cassettes in histone biology and beyond. Nature **425:** 475–479.
17. PEREZ-BURGOS, L., A.H. PETERS, S. OPRAVIL, *et al.* 2004. Generation and characterization of methyl-lysine histone antibodies. Methods Enzymol. **376:** 234–254.
18. ARIEL, M., J. MCCARREY & H. CEDAR. 1991. Methylation patterns of testis-specific genes. Proc. Natl. Acad. Sci. USA **88:** 2317–2321.
19. ARIEL, M., H. CEDAR & J. MCCARREY. 1994. Developmental changes in methylation of spermatogenesis-specific genes include reprogramming in the epididymis. Nat. Genet. **7:** 59–63.
20. KAFRI, T., M. ARIEL, M. BRANDEIS, *et al.* 1992. Developmental pattern of gene-specific DNA methylation in the mouse embryo and germ line. Genes Dev. **6:** 705–714.
21. GEYER, C.B., C.M. KIEFER, T.P. YANG & J.R. MCCARREY. 2004. Ontogeny of a demethylation domain and its relationship to activation of tissue-specific transcription. Biol. Reprod. **71:** 837–844.
22. MCCARREY, J.R., M. KUMARI, M.J. AIVALIOTIS *et al.* 1996. Analysis of the cDNA and encoded protein of the human testis-specific *PGK-2* gene. Dev. Genet. **19:** 321–332.
23. MCCARREY, J.R. 1987. Nucleotide sequence of the promoter region of a tissue-specific human retroposon: comparison with its housekeeping progenitor. Gene **61:** 291–298.
24. BOER, P.H., C.N. ADRA, Y.F LAU & M.W. MCBURNEY. 1987. The testis-specific phosphoglycerate kinase gene *pgk-2* is a recruited retroposon. Mol. Cell. Biol. **7:** 3107–3112.
25. MCCARREY, J.R. 1990. Molecular evolution of the human *Pgk-2* retroposon. Nucleic Acids Res. **18:** 949–955.
26. MCCARREY, J.R. & K. THOMAS. 1987. Human testis-specific *PGK* gene lacks introns and possesses characteristics of a processed gene. Nature **326:** 501–505.
27. MCCARREY, J.R. 1994. Evolution of tissue-specific gene expression in mammals: a new gene is formed and refined. Bioscience **44:** 20–27.
28. ZHANG, L.P., J.C. STROUD, C.A. WALTER, *et al.* 1998. A gene-specific promoter in transgenic mice directs testis-specific demethylation prior to transcriptional activation *in vivo*. Biol. Reprod. **59:** 284–292.
29. SIEGFRIED, Z., S. EDEN, M. MENDELSOHN, *et al.* 1999. DNA methylation represses transcription *in vivo*. Nat. Genet. **22:** 203–206.
30. HASHIMSHONY, T., J. ZHANG, I. KESHET, *et al.* 2003. The role of DNA methylation in setting up chromatin structure during development. Nat. Genet. **34:** 187–192.
31. FUKS, F., P.J. HURD, D. WOLF, *et al.* 2003. The methyl-CpG-binding protein MeCP2 links DNA methylation to histone methylation. J. Biol. Chem. **278:** 4035–4040.

32. JONES, P.L., P.A. WADE & A.P. WOLFFE. 2002. Purification of MeCP2-containing deacetylase from *Xenopus laevis*. Methods Mol. Biol. **200:** 131–141.
33. MELLOR, J. 2005. The dynamics of chromatin remodeling at promoters. Mol. Cell **19:** 147–157.
34. RAZIN, A. 1998. CpG methylation, chromatin structure and gene silencing—a three-way connection. EMBO J. **17:** 4905–4908.

Index of Contributors

Addlagatta, A., 135–148
Agrawal, V., 100–108
Allen, J.A., 120–134
Angelova, K., 41–54
Anway, M.D., 18–32

Baker, P.J., 109–119
Baughman, J.M., 33–40
Behringer, R.R., 9–17
Braydich-Stolle, L., 94–99
Brown, C.W., 173–182
Buck, S., 120–134

Chaboissier, M.-C., 9–17
Chang, H., 9–17
Collins, F., 65–76

Dass, B., 163–172
DeMars, G., 41–54
Diemer, T., 120–134
Duax, W.L., 135–148
Dym, M., 94–99

Efthymiadis, A., 173–182

Fanelli, F., 41–54
Flück, C.E., 100–108

Geijsen, N., 33–40
Geyer, C.B., 226–242
Griswold, M., ix
Gupta, A., 77–93

Habegger, L., 135–148
Hales, D.B., 120–134
Hales, K.H., 120–134
Hardy, M., ix
Heckert, L.L., 55–64
Hildebrand, E., 221–225
Hofmann, M.-C., 94–99
Hogarth, C., 173–182
Huang, N., 100–108
Huber, Z., 163–172
Huether, R., 135–148

Itman, C., 173–182

Jans, D.A., 173–182
Janus, P., 120–134
Jarow, J.P., 208–220
Johnston, D.S., 190–202
Johnston, H., 109–119

Karpova, T., 55–64
Kaupp, U.B., 221–225
Kobayashi, A., 9–17
Kopf, G.S., 190–202

Li, Y., 41–54
Loveland, K.L., 173–182
Ly, J., 173–182

MacDonald, C.C., 163–172
Maran, R.R.M., 55–64
Marshall, G.R., 149–162
Matusik, R.J., 77–93
McCarrey, J.R., 226–242
Meachem, S., 173–182
Meehan, T.P., 41–54
Mendis, S., 173–182
Millar, M.R., 65–76
Miller, W.L., 100–108
Mirosevich, J., 77–93
Monarez, R.R., 163–172

Narayan, P., 41–54
Nef, S., 1–8
Nolan, C., 94–99

O'Shaughnessy, P.J., 109–119

Orgebin-Crist, M.-C., 77–93

Pandey, A.V., 100–108
Plant, T.M., 149–162
Pletnev, V., 135–148
Presley, J., 55–64
Puett, D., 41–54

Qiu, Y., 190–202

Ramaswamy, S., 149–162
Robert, N.M., 183–189
Roberts, K.P., 190–202

Saunders, P.T.K., 65–76
Schedl, A., 9–17
Scherrer, S.P., 55–64
Shankara, T., 120–134
Sierens, J.E., 65–76
Simorangkir, D., 149–162
Skinner, M.K., 18–32
Sneddon, S.F., 65–76
Suzuki, K., 77–93

Tejada, L., 55–64
Thomas, J., 135–148
Tremblay, J.J., 183–189

Verma-Kurvari, S., 1–8

Wang, Y., 77–93
Weeks, C.M., 135–148
Wooters, J., 190–202

Yanagimachi, R., 203–207
Yoshioka, H., 226–242
Yu, X., 77–93

Zirkin, B.R., 208–220